COMPUTER NUMERICAL CONTROL

MACHINING AND TURNING CENTERS

ROBERT QUESADA

Milwaukee Area Technical College

PEARSON

Prentice Hall

Upper Saddle River, New Jersey
Columbus, Ohio

Library of Congress Cataloging-in-Publication Data
Quesada, Robert.

  Computer numerical control : machining and turning centers / Robert Quesada.
    p. cm.
  Includes index.
  ISBN 0-13-048867-4
  1. Machine-tools—Numerical control.   I. Title.

TJ1189.Q47 2005
621.9′023—dc22

2004008273

Vice President and Publisher: Stephen Helba
Executive Editor: Ed Francis
Production Editor: Christine M. Buckendahl
Production Coordination and Text Design: *The GTS Companies*/York, PA Campus
Design Coordinator: Diane Ernsberger
Cover Designer: Bryan Huber
Production Manager: Matt Ottenweller
Marketing Manager: Mark Marsden

This book was set in Stone Serif by *The GTS Companies*/York, PA Campus. It was printed and bound by Courier Kendallville, Inc. The cover was printed by The Lehigh Press, Inc.

Pearson Education Ltd.
Pearson Education Singapore Pte. Ltd.
Pearson Education Canada, Ltd.
Pearson Education—Japan

Pearson Education Australia Pty. Limited
Pearson Education North Asia Ltd.
Pearson Educación de Mexico, S.A. de C.V.
Pearson Education Malaysia Pte. Ltd.

10 9 8 7 6 5 4 3 2 1
ISBN 0-13-048867-4

PREFACE

This textbook was developed to introduce the student to the basics of CNC technology, and to prepare him or her for future CNC machining and programming courses that may follow. Included is related tooling data, machining data, shop mathematics, and CNC machine and control information. *Computer Numerical Control* clearly distinguishes between CNC machining centers and CNC lathes and will prepare students exceptionally well for CNC machining in each respective area.

First and foremost, this textbook is intended for students with diverse backgrounds. This includes students just out of high school, or students who may have had some training in industry or in the military. These students may just be beginning alternative or second careers. Other students may already be regular employees or trainees in industry. They may need to learn CNC machining in order to improve their skills as a programmer–operator. They all need a timely but meaningful introduction to CNC. Although there is no prerequisite to this book, it is always helpful to have some minimal exposure to CNC machine tools in the workplace.

This book was written because existing books either assume that the reader has extensive knowledge of CNC or are too specialized for a complete understanding of CNC technology. Other books tend to use examples that put too much emphasis on a specific machine, or may concentrate on one part of CNC, such as machine control or programming. These books may be too theoretical or verbose, may include material that is not work oriented, or may not fully provide students a hands-on experience with the knowledge and skills they need to get started on the job.

The pedagogical approach in this book is to stress hands-on application over theoretical material. It applies a step-by-step method that is straightforward, usable, and easy to understand. Because numerous machine tool brands exist, it presents generic examples and illustrations of their common characteristics, including CNC setup, operation, tooling, program structure, and machining. After introducing each of these areas, the book shows how they are intertwined in typical CNC machining centers and CNC lathes. These generic examples give students to-the-point coverage of what they need to know (without material that is just nice to know) to get up and running on the job—efficiently, productively, and safely.

Several chapters include CNC program samples that the student will learn to read block by block. The program examples progress in a spiral fashion. This gives students a good understanding of the unit just completed, making the next unit easier to undertake. Some students may have had courses in mathematics (including trigonometry), blueprint reading, and metrology, as well as other machining-related courses, while others may need a review of these materials. Therefore, these topics coupled with data from the appendixes will be brought into the coverage as needed.

This textbook, then, provides a learning tool that can be used in both educational and in-plant training environments. In education, it will certainly help students develop a better understanding of all the components that make up CNC machining technology. Whenever possible, it is recommended that a CNC laboratory component be included with this course. This will give the student a greater understanding of the material covered in the textbook.

This text material and the methodology have been class tested at Milwaukee Area Technical College (MATC) in the CNC machine-related programs. Feedback was obtained from students, teachers, and employers who subsequently employ these students. After reading this book students, regardless of their background and institutional affiliation, will have a fundamental understanding of CNC technology, CNC terminology, and CNC machine and control operations, and will be able to decipher CNC programs.

■ ■ Distinguishing Features

- An initial and overall emphasis is placed on safety.

- Each unit begins with objectives that clearly identify the unit expectations.

- Useful and effective illustrations visually support the explanations and program samples.

- Students are not directed to create CNC programs, but rather to understand the format and concepts of CNC programs.

- A focus is placed on G codes, M codes, and the various other letter address codes. This also will prepare students who will study CAM system courses because CNC downloads are typically in those formats.

- A focus is placed on operational-type codes such as tool changing, offsets, spindle speed, coolant, and program stops.

- Summary learning tables and charts are included for easy reference.

- CNC machine types and components are clearly described and illustrated.

- CNC toolholders and cutting tools are clearly described and illustrated.

- CNC workholding devices and setup methods are clearly described and illustrated.

- CNC control features and operations are clearly described and illustrated.

- Related technical data such as shop mathematics, feedrate, spindle speed, surface finish, material types, cutter inserts, and geometric dimensioning and tolerancing are linked to the overall CNC process.

- The appendixes and glossary contain valuable and pertinent CNC and machining-related information.

ACKNOWLEDGMENTS

To my family—for their patience, support, and understanding throughout the development of this project.

I would like to acknowledge the many individuals whom I have worked with in industry, where I gained my diverse knowledge of CNC machining.

I would like to acknowledge Milwaukee Area Technical College, where I developed the pedagogical methodology that was instrumental in directing me to develop this project.

I would also like to acknowledge the helpful comments and suggestions from the reviewers of this text: Jeffrey B. Hellwig, Alfred State College; Lloyd Pulsifer, Central Maine Technical College; and Michael Reynolds, Corning Community College.

CONTENTS

chapter 1

INTRODUCTION TO CNC MACHINING CENTERS

Chapter Objectives

After studying and completing this chapter, the student should have knowledge of the following:

- **The history and evolution of CNC**

- **The advantages of CNC**

- **The safety rules for CNC**

- **The CNC process**

- **The CNC documentation**

INTRODUCTION TO CNC

The History of Manufacturing

American manufacturing technology has been evolving ever since the early 1900s when the steel mills began to spring up in Pittsburgh. Ultimately, Henry Ford introduced the "mass production" philosophy and began producing automobiles that almost all Americans could afford. His philosophy is still a major manufacturing method that is used in present-day industries throughout the world. Today, however, this philosophy has evolved to include high-technology computer numerical control (*CNC*) machines, CNC controls, computers, and some innovative procedures.

In the early 1930s W. Edwards Deming helped introduce the statistical revolution to American manufacturing industries. This was the start of research for data collection, which is commonly known today as statistical process control (SPC).

The 1940s introduced the United States to World War II and shifted American industrial companies into high gear. Many high-production techniques, machines, and equipment still used today were developed during this period. After World War II product demand shifted back to consumer goods such as automobiles, appliances, radios, and televisions. Around this time, Dr. Deming was invited to Japan so that he could help with the reconstruction effort.

NC Technology

The 1950s through the 1960s introduced the philosophy of Total Quality Control (TQC). Also, the first *numerical control* (NC) machine became available to industry. A few years earlier, John Parsons had been contracted by the U.S. Air Force to design and build a three-axis machine that would enable us to machine complex contours. This was the dawn of our present-day machining technology. This NC technology was not welcomed with open arms at first. Most manufacturers and workers resisted this new technology in machining because it was revolutionary and raised various unfounded fears. Today, the list of manufacturing equipment that has evolved to include NC are metal machining, wood routes, water jet cutting, laser cutting, welding, flame cutting, and so on.

With the influx of computers in the 1960s and 1970s, industrial companies began to apply them to any area that needed enhancement, accuracy, and/or speed. The NC machines were ideally suited for this, and rapidly evolved to what is now CNC; computer-aided design (*CAD*) and computer-aided manufacturing (*CAM*) computer-integrated manufacturing (CIM), and robotics.

During the 1970s American manufacturers were riding the crest of the consumer-glut wave. They basically had no competition from foreign manufacturers. Therefore, most American manufacturers were not initially motivated to invest in any costly quality programs or new technologies. The 1980s served as a wake-up call for American manufacturing companies like Ford, General Motors, Chrysler Corporation, and U.S. machine toolbuilders. Suddenly, foreign products, mainly Japanese, were becoming attractive to American consumers and U.S. sales started to take a plunge. Consequently, many industries either "downsized" or went out of business.

This prompted the surviving American manufacturers to adopt the same technologies that put Japanese manufacturers in the lead. These technologies included Just-In-Time (JIT) to reduce inventories and Work In Process (WIP), SPC, Total Quality Management (TQM), Quality Assurance (QA), quality circles, and cellular manufacturing in the form of Flexible Manufacturing Systems (FMS). They also started investing more in CNC machines, *CAD/CAM* systems, computer-aided programming, robots, and CIM systems, which incorporate and manage various machines and/or robots.

Advantages of CNC

- CNC machines run automatically without operator intervention.
- CNC machines give a manufacturing shop greater machine uptime.
- CNC machines increase overall productivity.
- CNC machines output maximum part accuracy.
- CNC machines reduce scrap and rework.
- Less inspection time is required.
- Tooling costs can be minimized.
- WIP and inventory are minimal.
- Complex parts can be quickly and efficiently produced.

CNC Technology

Today, CNC technology has evolved to include "conversational programming," which is software that allows the CNC operator to create CNC programs at the machine. However, the most common method of programming is still computer-aided software. *APT* is one software system that is used for creating CNC programs. There are many other such software systems that can perform the same CNC processes. The postprocessor is the part of the software that converts computer language to CNC code to match a specific CNC machine. The APT software is therefore a big advantage to companies with various CNC machines of different makes, models, and options. This is usually the case in most manufacturing plants.

The types of machines and equipment that utilize CNC are numerous. For example, the CNC router is a machine originally designed for wood product manufacturing industries. This machine is used to route designs on wooden furniture components. It is also utilized for machining, Polyvinyl Chloride (PVC) plastic parts, which have similar machining characteristics. This is an example of how versatile the CNC machine tool industry has become. The CNC router utilizes a vacuum system to hold the work material during the machining cycle. The table is made of metal with a top plate that is made of balsa wood. The balsa wood is used because it is porous and allows air to suck the work material (PVC) down.

The CNC water jet machine is a unique type of machine. It is used to machine parts with a grit-filled water stream that is forced through a small diameter nozzle at a very high pressure. The types of materials that are ideal for this machine include soft flexible materials such as aluminum. It has also proven to be an ideal method of cutting sponge forms. The water jet traverses at a rate in inches per minute to cut the work material. The type and thickness of the material being cut determine the traverse rate.

The CNC laser machine is used for cutting small sheet metal parts. The laser is more accurate than CNC punch presses because there is virtually no distortion from cutting with the laser. The CNC laser machine is more efficient than the CNC punch press machines and produces parts at a higher level of quality. However, the CNC laser machine has limitations that CNC punch presses do not have. For example, the laser is not suitable for cutting reflective materials that bounce the laser beam back, causing damage to the laser. Brass is one type of reflective material that the laser cannot cut. Therefore, the laser machine cannot completely replace CNC punch presses. Similar to the CNC water jet machine, the laser traverses at a rate determined by type and thickness of the material.

Some manufacturing departments utilize a process that is called "nesting." This process minimizes material waste by using CAD/CAM to arrange various parts on a standard sheet of material. The nesting process is used for the CNC water jet, CNC laser, and CNC punch presses.

CNC Turning Centers

The second part of this textbook focuses on turning machines. The machine toolbuilders of CNC turning machines have redesigned CNC lathes to include two or more spindles and live tooling. The dual spindle allows turning of two parts at the same time. The live tooling allows secondary machining such as milling slots and drilling off-center holes to be completed in the same operation. The advantages of this type of lathe are reduced set-up time, reduced handling (which improves quality), reduced cycle time, reduced WIP, and facilitates JIT. However, there are also some limitations and disadvantages. For example, the live tooling usually is powered by a low-power motor of some type. This limits the size of secondary milling and drilling that can be performed. Another consideration is that usually only a few parts are suitable for machining on these higher cost machines. Therefore, these machines will sometimes sit idle because of a lack of production requirements.

CNC Machining Centers

The first part of this textbook focuses on *machining centers,* which perform the same cutting functions that manufacturing industries have performed on manual mills for many decades. The major difference and the principal advantage of CNC equipment are the increased control and repeatability of the cutting tool. CNC machining allows for the manufacturing of parts that would otherwise be difficult or impossible to produce with conventional machines and methods.

CNC has proven to be an economical and efficient method of controlling and operating machine tools. Therefore, most machine shops typically utilize *horizontal machining centers* (HMC), *vertical machining centers* (VMC), and CNC lathes.

CNC-coded programs provide the information used by the machine control unit (MCU) to control and position the cutting tools. The MCU is the brain of the CNC machine. It reads, interprets, and converts programmed input into appropriate movement. It also controls various accessories such as spindle motion, coolant, automatic tool changes, and graphics. The MCU (also called the controller) converts the CNC-coded part program information into voltage or current pulses of varying frequency or magnitude. The converted pulses are used to position and control the operation of the CNC machine.

Most CNC machines store the CNC program into the control's memory when the program is first read into it. Most MCUs are equipped with extended memory that can store many CNC part programs. CNC machines have two primary types of memory: *RAM* (random-access memory) and *ROM* (read-only memory).

The CNC part program is input into the MCU and usually stored in RAM. The MCU may be equipped with ROM that contains information concerning diagnostic, start-up parameters, operating instructions, online help, and similar repeated functions. The advantage of storing these functions in ROM is that they only have to be programmed once and they are not deleted when the power is turned off or disconnected.

CNC technology is the same as NC technology except for the addition of the computer to aid program processing. CNC uses the added capabilities of the computer to read, store, edit, and process programmed information. It also provides graphical capabilities, diagnostic procedures, alarm messages, and system troubleshooting.

There are two basic types of machining centers—the vertical type and the horizontal type (see Figure 1-1). Their names are derived from their respective spindle designs, which are either in a vertical position or horizontal position. There are other designs and variations of both the vertical and the horizontal machines that are designed for other application needs.

The application or production part requirements are the factors that determine which machine design is appropriate. It may be determined that the horizontal, vertical, or both are required to meet the production needs of the company. Some of the determining factors that must be considered include the size of the part and the part material such as a casting or a plate. Most CNC machine shops usually use a variety of CNC machine types as well as some conventional machines. The two CNC machine types are described in the following text.

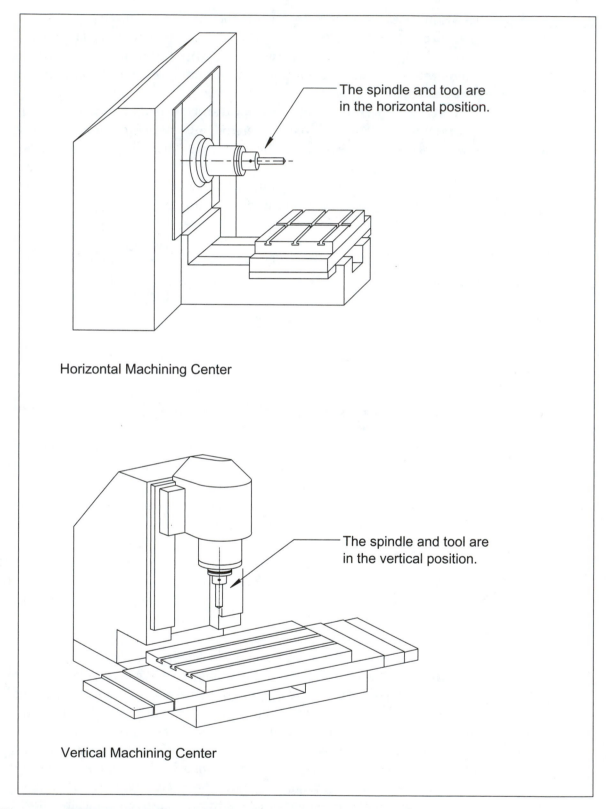

The spindle and tool are in the horizontal position.

Horizontal Machining Center

The spindle and tool are in the vertical position.

Vertical Machining Center

*Figure 1-1 **Horizontal and vertical machining centers***

- The VMC *spindle* holds the cutting tool in a vertical position. VMCs are generally used to perform operations on flat parts that require cutting on the top surface of the part. The VMC is programmed to position and cut in the *X-, Y-,* and *Z-axes* (three-axes). Other options can be added that will increase the flexibility and productivity of the VMC. Some of the options that are typically added to VMCs include an indexer or a shuttle table. The indexer, which is mounted on the machine table, can rotate the part along the X-axis for 360° machining (fourth *axis*). The shuttle table option is used for shuttling the parts, which allows the operator to load and unload parts without interrupting production.

- The HMC spindle holds the cutting tool in a horizontal position. The HMC machine table can be programmed to rotate 360° in a circular motion (*B-axis*). HMCs can machine parts on more than one side in one clamping, and find wide use in flexible manufacturing systems. They are generally used to perform operations on parts that require machining on various sides such as castings. The HMC is programmed to position and cut in the X-, Y-, Z-, and B-axes (four-axes).

CNC MACHINING CENTER SAFETY RULES

CNC machines are extremely fast and powerful machine tools; therefore, it is imperative that all safety rules and operating instructions are adhered to in every situation. Figures 1-2 through 1-5 are a categorized list of recommended safety rules that the reader is strongly advised to fully read and understand before attempting to operate CNC machines or enter the shop area. Additionally, each CNC machine has an operator's manual, which includes both operating instructions and safety rules. Most CNC machines also have warning labels to alert the operator of dangers that may occur. Due to the higher spindle speeds that are typical of CNC machines, most machine tool manufacturers have designed standard safety features into the CNC machine. For example, a standard CNC machine enclosure prevents metal chips from being ejected and causing injuries to the operator or other personnel. The enclosure usually includes doors and viewing panels. The doors are usually hardwired with switches, which prevent the CNC machine spindle from operating whenever the doors are open.

These safety rules are generally applicable in most CNC machine operating situations. However, due to the many variables that exist in manufacturing, there may be other safety concerns not listed. For example, due to a situation that is specific to a company product or process, other safety factors may arise, such as from a situation specific to the CNC machine, the CNC control, the tooling, or the workholding method.

In any case, each CNC machining situation must be evaluated individually to determine every safety concern before operating the CNC machine.

Wear American National Standards Institute (*ANSI*)-approved safety glasses with side shields at all times in designated areas.
Wear safety shoes when working with heavy tools and equipment.
Wear hearing protection for noise levels that exceed Occupational Safety and Health Administration (*OSHA*) specifications.
Wear an approved face mask for dust levels that exceed OSHA specifications.
Keep long hair covered when operating or standing near machines.
Do not wear jewelry or loose clothing while operating or standing near machines.
Lift with the legs, not the back.
Avoid skin contact with any cutting fluids or oils.
Do not operate any machines or equipment while under the influence of drugs (prescribed or otherwise).
Always report any injury and apply first-aid treatment.

Figure 1-2 ***Personal safeguards***

Store tools in their appropriate tool trays and racks.
Keep tools sharp and in good working condition.
Ensure that safety guards and devices are in place and working before operating any machine.
Keep all electrical and mechanical panels secured in place.
Do not handle loose wires or electrical components.
Check that all oil levels are maintained.
Check that all compressed air equipment is in good working condition.
Do not use compressed air to clean machine slides.
Keep tools, parts, and any other items off the machine and part.
Use gloves when handling tools by their cutting edges.
Never use gloves when operating a machine.
Check that the machine is electrically sound and use lock-out tag-out practices.
Check that all lights in the machine work area are in good working condition.
Keep clear from obstructions and sharp tools when leaning into the machine work area.

Figure 1-3 ***Machine and tool handling safeguards***

In case of any emergency while operating a machine, immediately press the Emergency Stop button.
Keep hands away from the spindle while it is rotating.
Do not open the electrical panel or control doors.
Keep hands clear of all moving machine components.
Check each tool for possible collisions with the part or machine components before starting any operation.
Do not operate controls unless you have been properly instructed about the Emergency Stop button, Feed Hold button, Spindle Stop button, and the various other operation functions.
Use caution to avoid inadvertently bumping any CNC control buttons.
Proveout new programs in Dry Run mode before actual cutting.
Setup the workholding device and cutting tools as rigidly as possible.
Do not remove chips or debris by hand or while the spindle is on or in operation.
Check that all feeds and speeds do not exceed recommended values.
Maintain a continuous flow of coolant to the cutting tool when it is required.
Always consult with an authorized person if you are uncertain or unfamiliar with any operation.

Figure 1-4 *Safe machining practices*

Remove chips from the floor.
Clean all liquid, oil, and grease spills immediately.
Keep floors and walk aisles clean.
Report any fumes or odors.

Figure 1-5 *Shop environment safeguards*

An unsafe practice or situation can affect any person in the shop area directly due to an injury. Other persons associated with the company can be affected indirectly due to loss of wages, insurance rate increases, and other factors.

In summary, there are various safety factors that the CNC operator must know, understand, and follow whenever he or she is in the shop area or operating CNC machines. When an accident occurs because of an unsafe condition or practice, it affects everyone directly or indirectly. Therefore, every person directly or indirectly associated with CNC shops or machines must be concerned with safety.

CNC MACHINING CENTER PROCESSING

Planning and documenting the CNC process is required to efficiently complete a job for CNC machining. Basically, the process begins when the CNC programmer reviews the engineering drawing to determine which machining operations are required. Generally, the machining determination follows an order similar to that described in Table 1-1.

Table 1-1 basically indicates that rough milling, drilling, and boring operations are performed first, and then finish milling, drilling, tapping, boring and reaming operations follow. Whenever possible, in an effort to increase productivity, the machining process should be reduced to the minimum number of operations. If roughing operations are not required, they can be eliminated.

Besides the obvious reason that rough machining is performed before finish machining, there are valid reasons that necessitate roughing operations. Some common reasons that necessitate rough machining before finish machining include the following:

- When the material removal amount exceeds the designed safe cutting limit of the tool or machine.

- To assure that a minimum amount of material is machined in order to maintain a required *tolerance* throughout a production run.

- To assure that a minimum amount of material is machined in order to maintain a required surface finish throughout a production run.

- When the material removal amount exceeds the horsepower of the machine.

■___Table 1-1 *CNC machining determination*

SEQUENCE	OPERATION DESCRIPTION
1	Rough face milling operations
2	Rough end milling operations
3	Rough drilling operations
4	Rough boring operations
5	Finish face milling operations
6	Finish end milling operations
7	Finish drilling operations
8	Finish tapping operations
9	Finish boring operations
10	Finish reaming operations

- If the part structure is flimsy or weak, the material removal amount will be limited to whatever the part structure can withstand.

- If the cutting tool structure is flimsy or weak, the material removal amount will be limited to whatever the tool structure can withstand.

- When the material removal amount exceeds the capability of the work-holding device.

- In situations where the hardness or toughness of the part material limits the material removal amount to whatever the cutting tool can effectively machine.

There are many machining applications throughout industry. Therefore, there are certainly other reasons that necessitate rough machining. The common rough machining reasons listed should be considered whenever processing a part to be machined.

The CNC programmer also determines and specifies how the part will be held based on the part drawing *datums* and requirements. If a special fixture is required, it should be immediately designed and ordered to be built. This is because special fixtures require longer lead times.

CNC MACHINING CENTER DOCUMENTATION

After the machining process has been determined, the CNC programmer can begin selecting the tools required to machine the part. *Tool drawings* are typically used to specify standard and nonstandard tool dimensions and features (see Figure 1-6). The person who assembles the tools—the CNC operator, CNC *setup* person, or the *tool setter*—uses these drawings as a reference.

Most CNC machining is usually processed using standard tools and workholding devices. It is important to note that in addition to longer lead time, special fixtures or tools are very costly to design and build. Therefore, the first choice with regard to tooling or workholding should always be to use standard items.

At this point the CNC programmer can start creating the CNC program. CNC programs can be created either manually or with the assistance of computer CNC programming software. There are numerous types of computer software systems for creating CNC programs. CAD/CAM, which is one method of programming, has become very popular with most manufacturing plants. This allows the CNC programmer or other manufacturing personnel to access the CAD part design file, duplicate it, and then use the geometry section to create the CNC program.

There are various forms of documentation for the setup, such as the set-up plan, the tool list, and the CNC program *manuscript*. Most are documented on paper, while others may use the computer and cathode-ray tube (*CRT*) to document these items. The examples in Figures 1-7 and

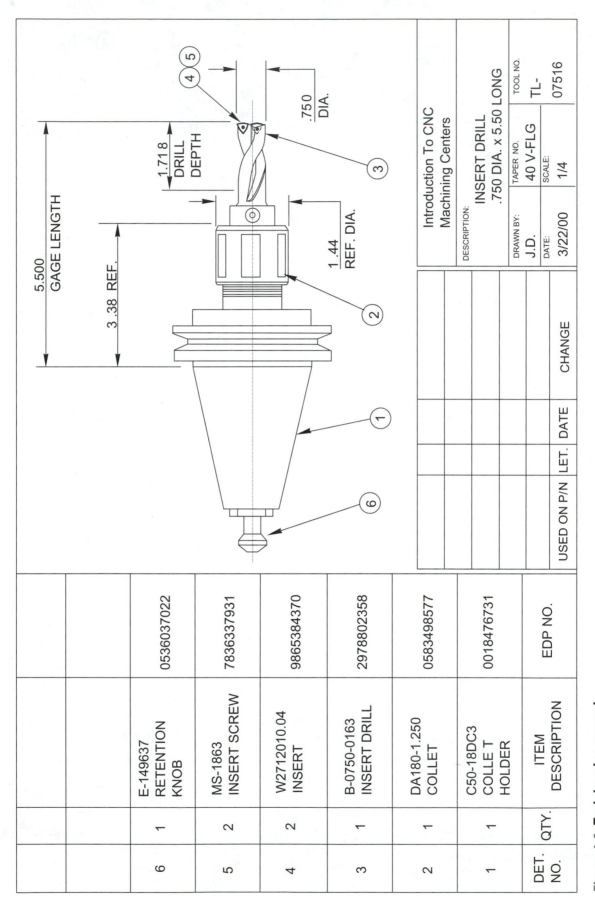

Figure 1-6 *Tool drawing example*

Introduction To CNC
Machining Centers

DESCRIPTION:			
INSERT DRILL			
.750 DIA. x 5.50 LONG			
DRAWN BY:		TAPER NO.	TOOL NO.
J.D.		40 V-FLG	TL-
DATE:		SCALE:	07516
3/22/00		1/4	

			CHANGE		
USED ON P/N	LET.	DATE			

.750 DIA.

1.718 DRILL DEPTH

5.500 GAGE LENGTH

3.38 REF.

1.44 REF. DIA.

DET. NO.	QTY.	ITEM DESCRIPTION	EDP NO.
6	1	E-149637 RETENTION KNOB	0536037022
5	2	MS-1863 INSERT SCREW	7836337931
4	2	W2712010.04 INSERT	9865384370
3	1	B-0750-0163 INSERT DRILL	2978802358
2	1	DA180-1.250 COLLET	0583498577
1	1	C50-18DC3 COLLE T HOLDER	0018476731

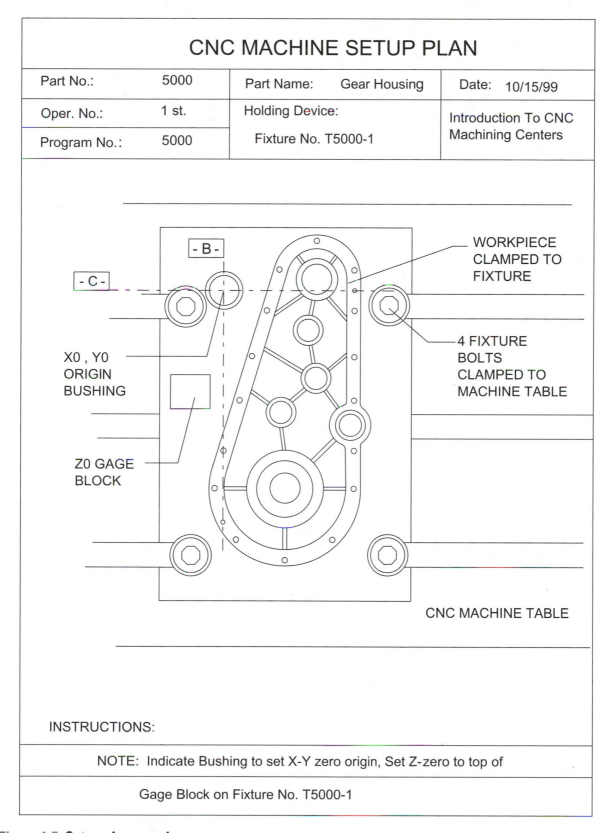

Figure 1-7 Setup plan sample

1-8 are methods that can be utilized in either paper form or on a computer CRT screen.

The CNC programmer must document the machining process, the set-up plan, the tool list, any special fixture or tool drawings, and the CNC program. This information must be documented and relayed to respective personnel such as the CNC setup person or the CNC operator. This will ensure that the part will be set up and machined as efficiently as possible the next time it is scheduled for a production run. This will also make the process consistent throughout future runs and thereby eliminate most chronic quality problems. At this point, the part process is complete and ready to be scheduled for machining.

Once the process and documentation are completed, the setup is initiated. The setup person is responsible for assuring that the following setup items are completed:

- Workholding device is clamped to the machine as specified.

- Tools are loaded into the specified magazine pockets.

- The CNC program is loaded into the control.

- The X-, Y-, and Z-zero origins are set as specified.

- The tool length offset values are stored under the specified numbers.

- The tool diameter offset values are stored under the specified numbers.

- The part sizes are adjusted to drawing specifications.

- The CNC program is *debugged* and edited if necessary.

- The parts are produced to the drawing specifications and tolerances.

The proper process determination and the documentation are essential to the efficiency of the CNC production process. Therefore, it is important to properly utilize the CNC information available. Table 1-2 describes the typical information that can be obtained from each CNC document.

The documentation types previously described are directly related to CNC machining applications. However, there are other documents used in machining manufacturing that deal with overall production efficiency and product quality. Some of these common types of production and quality documentation typically include

- Inspection instruction forms

- Inspection report forms

- SPC charts

- Process instruction forms

- Process drawings

- Routing forms

The documentation forms listed above are used for controlling and/or recording the entire manufacturing process of a production run. They are

CNC TOOL LIST SHEET

Part No.:	08000		Prog. No.:	O1000	Date: 10/15/99
Part Name:		End Plate	Oper. No.:	10	Page 1 of 1

Pocket No.	Length Offset	Radius Offset	Tool No.	Tool Dia.	Tool Description and Operation
T01	H01	D	1001	3/4	90 Degree Spot Drill spot (8) holes
T02	H02	D	1008	5/16	HSS Drill drill (8) holes
T03	H03	D	1019	1/2	HSS Drill drill (2) holes
T04	H04	D	2015	3/8−16	HSS Tap Spiral flute tap (8) holes

*Figure 1-8 **Tool list sample***

■___*Table 1-2 **CNC documentation information***

DOCUMENT NAME	DESCRIPTION OF INFORMATION
Engineering drawing	The engineering drawing specifies what *features* are to be machined. It specifies the feature dimensions, tolerances, and finish requirements. Also included are the part number, revision level, and part material specifications.
Set-up plan	The set-up plan specifies how the part is orientated and held on the machine table. It also specifies where the part zero origins are located. It may specify use of a common holding device such as a milling machine vise, a special or modular fixture, a 360° axial positioning indexer, or a 3-jaw chuck for holding round parts.
Tool list	The tool list specifies which tools have been selected to perform the machining of the part. Other information may include tool diameters, tool lengths, insert identification number, and insert grade. Also specified are tool materials such as *high-speed steel* (HSS), coated or uncoated carbide grade, cobalt, diamond, and ceramic.
CNC program manuscript	The CNC program manuscript is the document where all the CNC codes needed to machine the part are specified. It includes all the specific codes to control the miscellaneous machine functions, the coordinates that control machine axis movement, and other auxiliary command functions.
Tool drawing	The tool drawing specifies the tool number, style, components, insert number, insert grade, and any required dimensions.
Tool library	The tool library is a company's catalog of all the CNC tools available. This is used for ready access and to reduce redundant tool inventory (see appendix A).

used throughout the entire process, which may include CNC machines, conventional machines, and other miscellaneous equipment.

The design and application of these document forms varies from company to company depending on the specific needs of the company. Each individual company basically decides what document forms are required for their production and quality needs. These document forms can be generated on either paper or displayed on a computer CRT.

Companies usually require their production personnel to read, update, record, and maintain these forms. They will typically train production personnel to ensure that the documentation is utilized accurately and efficiently.

It is important to correctly understand and properly communicate the information on these document forms. CNC personnel are also required to incorporate these document forms into their CNC manufacturing process. It is equally important that CNC personnel also identify the other various types of document forms and use them accordingly.

■ ■

CHAPTER SUMMARY

The key concepts presented in this chapter include the following:

- The history and development of CNC machining technology.
- The advantages of utilizing CNC technology.

- The various types of CNC applications in manufacturing.
- The common types of CNC machining centers.
- The CNC vertical machining center features.
- The CNC horizontal machining center features.
- The safe CNC machining practices.
- The CNC machining process determination.
- The CNC machining documentation types.
- The information obtained from CNC machining documentation.
- The other types of manufacturing documentation forms.

■ ■ REVIEW QUESTIONS

1. Why was NC developed and applied to production machining?

2. Describe the production milling machine types that utilize CNC.

3. List the advantages of utilizing CNC for production.

4. List the personal safety equipment that is required in the shop.

5. List the personal safety practices that are required in the shop.

6. Why is it important for everyone to follow all shop safety practices?

7. List the general order of operations that can be performed on CNC machining centers.

8. List the documentation that is typically required for CNC machining centers.

9. List the typical items that are documented on the CNC tool list.

10. List the basic items that are documented on the CNC setup sheet.

11. List the basic items that are documented on the CNC tool drawing.

12. List the basic items that are documented on the CNC tool library.

13. List the basic items that are documented on the engineering drawing.

14. List the basic items that are documented on the CNC program manuscript.

15. List the setup items that are important and should be checked for completion.

16. List other types of quality documents that are also used for CNC machining.

chapter 2

CNC MACHINE FUNDAMENTALS

Chapter Objectives

After studying and completing this chapter, the student should have knowledge of the following:

- *CNC machining center components*
- *Cartesian coordinate system*
- *Absolute and incremental coordinates*
- *X-Y coordinate calculation*
- *CNC program format*
- *CNC command codes*

This chapter identifies and describes the basic CNC machining center components and fundamentals. The items covered include the CNC machine components, the *Cartesian coordinate system,* CNC program format, and CNC command codes.

CNC MACHINING CENTER COMPONENTS

CNC machining centers have the same basic components that conventional milling machines have, such as the main motor, the spindle, the table, the frame, and the way system. The CNC machining center, in addition, has a computerized control and servomotors to operate the machine. Thus, the vertical and horizontal machining centers are numerically controlled milling machines that change tools, position the spindle, and cut material automatically (see Figures 2-1 and 2-2). When milling-type machines are coupled with CNC, they become the "high-tech production CNC centers" of the machine shop. Complex contours such as tapers and radii, as well as, repetitive operations. Drilling, tapping, reaming, milling, and boring are perfect machining applications for CNC machining centers.

The main components of CNC machining centers are the frame, column, spindle, table, tool magazine, operator control panel, servomotors, ball screws, hydraulic and lubrication systems, and the MCU.

There are three primary axes on a machining center, the X, Y, and Z. The X- and Y-axes, which control the table, position the spindle to the CNC program coordinates. The Z-axis, which controls the spindle, moves the spindle to the program coordinates (see Figures 2-1 and 2-2).

CNC Control

The modern CNC machine tool is software driven. The computer controls are programmed instead of being hardwired. The control panel is where the machine operation buttons and knobs are located. Some typical features may include the following:

- Power On button
- Cycle Start button
- Feed Hold button
- Axis Select knob
- Feed Override knob
- Keyboard pad
- Emergency Stop button
- Mode Select knob
- Miscellaneous Function buttons
- Spindle override knob

CRT Display

The CRT is also located on the control panel. The CRT display allows the operator easy visual access to program and machine information. On the screen of the CNC control, the operator can see the program, tools and cutter size

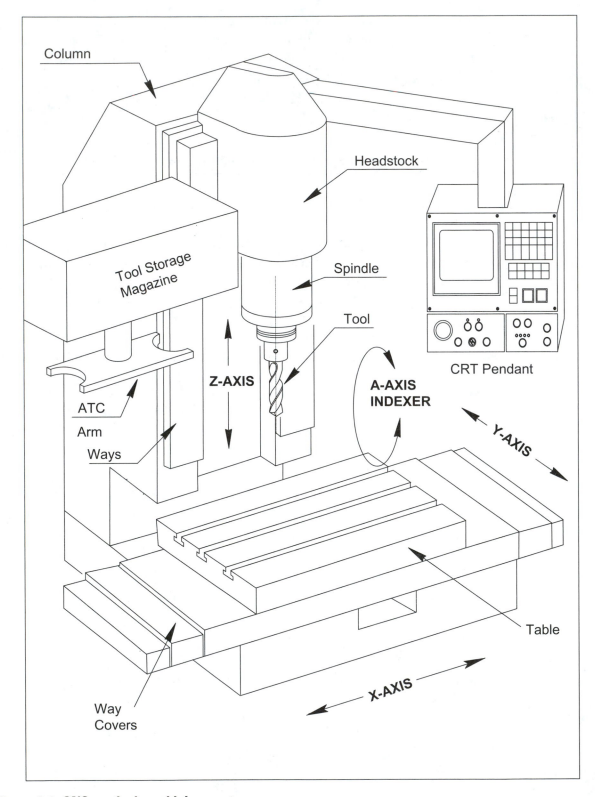

Figure 2-1 **CNC vertical machining center**

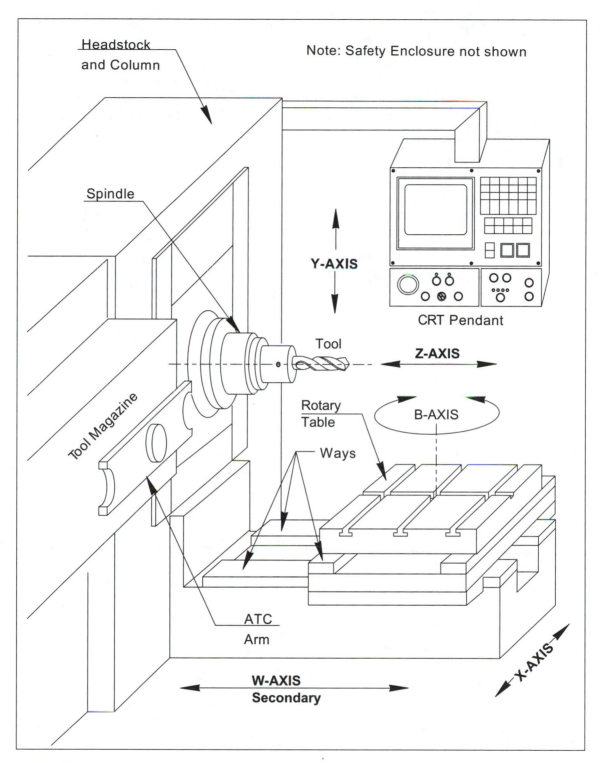

Figure 2-2 **CNC horizontal machining center**

offsets, machine positions, variables, alarms, error messages, spindle revolutions per minute *(RPM)*, and horsepower. Some also offer a graphic tool path simulation. The tool path simulation can be used to check the program before the program is run on an actual part. This simulation can minimize or eliminate programming errors that could injure the operator or damage the tools, parts, and/or the machine. Some simulation options can also calculate and display the machining cycle time of each individual cutting tool in the CNC program.

Frame

The frame supports and aligns the axis and cutting tool components of the machine. The frame will absorb the shock and vibration associated with metal cutting conditions. Frames are designed in one of two ways: either cast iron or fabricated. Most machining centers have a fabricated frame design.

Headstock

The headstock contains the spindle and transmission gearing, which rotate the cutting tool. The headstock spindle is driven by a variable-speed motor, which is programmable in RPM. Headstocks are equipped with a variety of motor sizes ranging from 5 to 30 horsepower, and spindle speeds from 32 to 10,000 RPM.

Spindle/Tool Taper

The spindle is designed with a taper bore that accurately aligns each tool that is clamped into it (see Figure 2-3). The tool is clamped using a drawbar mechanism and retention knob. The spindle is designed with two keys that align and drive the tools. The tool change arm holds the tools on the V-flange when loading and unloading.

Table/Pallet

The table is designed to hold the workholding device, which holds the workpiece while it is being machined (see Figures 2-1 and 2-2). The workholding device, such as a milling machine vise or a fixture, is clamped to the table using bolts and nuts. The table can be programmed or manually operated to move in the X- and the Y-axes. Additionally, the table for HMCs can be rotated about the B-axis. A pallet system allows the operator to perform work such as loading and unloading the workpiece while the machine is machining another part.

Tool Magazine

The basic function of the tool magazine is to hold and quickly index the CNC cutting tools. Tool magazines are designed in various styles and sizes. CNC machining centers are usually equipped with a tool magazine

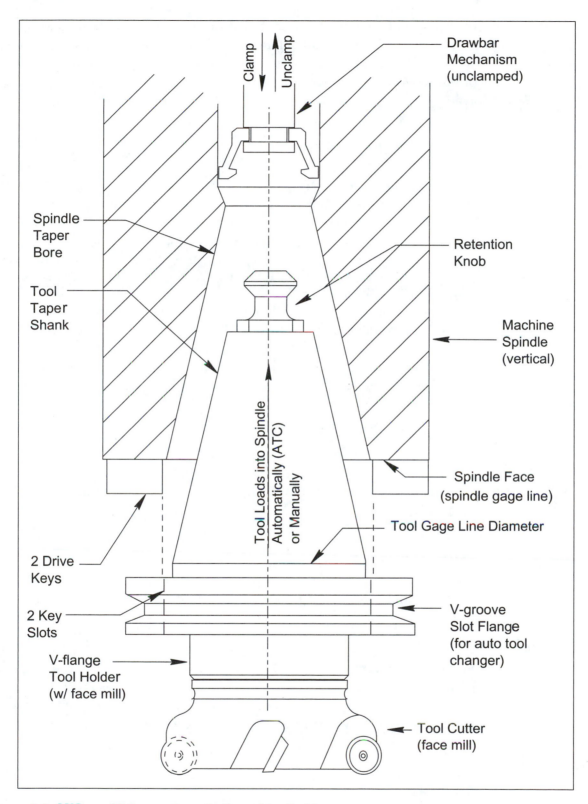

Figure 2-3 CNC machining center spindle and toolholder

that can hold a range of twenty to over one hundred tools. The tool magazine can automatically change the cutting tools with the tool change arm. In order to increase productivity, most tool magazines are capable of bidirectional movement to select the tools using the quickest path available.

Automatic Tool Change Arm

The *automatic tool change* (ATC) arm is designed to grasp and then remove the CNC tool from the tool magazine and insert it into the taper bore of the spindle. The ATC arm must accurately align each tool with the bore and the drive keys. The ATC arm can complete a tool change in either manual or automatic mode in approximately five to ten seconds.

Ways and Way Covers

The ways are precision-hardened rails that allow the table and spindle to travel in rapid traverse and feed motion (X-, Y-, and Z-axes). The ways are leveled and aligned for accuracy. The way covers are a means of protecting the way rail surfaces from damage due to scraping from chips, or due to dents from metal items such as tools or parts.

Way Lube System

The way lube system is designed to keep the ways lubricated, which will reduce wear from friction and eliminate early machine failure. The operator and maintenance personnel must check the way lube system reservoir level on a daily basis.

Electrical Control Panel

The electrical control panel is where all the electric components, including fuses, are enclosed for safety, and also where the main power switch is located.

Servodrive Motors

CNC machines use electric servomotors that turn ball screws, which in turn drive the different axes of the machine tool. Each axis has a separate servomotor and ball screw to control the X-, Y-, and Z-axes independent of each other.

Ball Screw

The rotary motion generated by the drive motors is converted to linear motion by recirculating ball screws. The ball lead screw uses rolling motion rather than the sliding motion of a normal lead screw. Sliding motion is used on conventional acme lead screws. Unlike the ball screw, the motion principle of an acme lead screw is based on friction and backlash. Here are some advantages of the ball screw versus the acme lead screw:

- Less wear
- Precise position and repeatability
- High-speed capability
- Longer life

CNC machine tool manufacturers have designed and incorporated an electronic pitch error that compensates for backlash error. This backlash compensator is used on most CNC machines.

Open-Loop Systems

The stepping motor is an electric motor that rotates a calculated amount every time the motor receives an electronic pulse from the MCU. The stepping motor's rotary motion is converted into the linear motion of the machine axes through the use of lead screws. There is no sensing or feedback system to verify and monitor the machine's actual positioning. The *open-loop system* is simpler than the closed-loop system.

Closed-Loop Systems

Servomotors permit automatic operation of the machine and the closed-loop control system verifies that the machine accurately positions the CNC program commands (see Figure 2-4). The computer makes it possible to continuously monitor the machine's position and velocity while it is operating. The advantages of the servomotor are increased accuracy and repeatability. CNC machines can generally position to an accuracy of ± 0.0002 inch and have a repeatability of ± 0.00006 inch.

Point-to-Point and Continuous Path

Most CNC machining centers rapid-position (G00 rapid traverse) in point-to-point mode, which means that the tool does not follow an exact path to a coordinate point. However, when the tool feeds or cuts (G01 linear *interpolation*), it then operates in a *continuous path* mode.

Input Media

CNC programs created by methods external to the CNC machine, such as with CAD/CAM, must be *input* into the machine control memory. There are various ways that have been developed to perform this input. Earlier input media methods included punched cards and then punched tape. Punched tape, which was designed as one-inch-wide tape made of plain paper, paper-Mylar plastic, or aluminum-Mylar laminates required a tape reader. The tape reader processed the punched pattern on the tape in a sequential manner and converted it into a corresponding electrical code.

The development of other more efficient and flexible methods have followed. *Magnetic tape,* which usually comes in the form of a tape

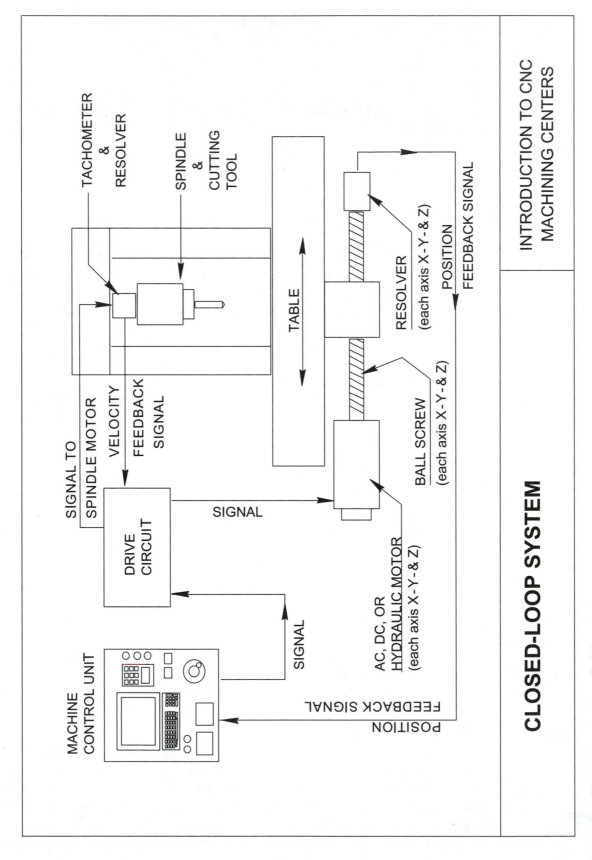

CLOSED-LOOP SYSTEM

Figure 2-4 Closed-loop system diagram

cassette, is another form of input media. This type of input media stores the CNC program in the form of a magnetic pattern on a cassette tape. Diskette devices, which are circular in shape, also store programs in the form of a magnetic pattern. They are designed to spin when operating as they are read by recording heads in the disk drive unit. Diskettes, or floppy disks, are a popular method of storing and inputing media. They are loaded through the CNC machine control panel or with the use of micro-computers and workstations. The storage capacity of a disk is much greater than the storage capacity of both tapes and cassettes. The floppy disk, which is a random-access medium, can find and retrieve files almost instantaneously. Another method of loading CNC programs is to use a remote device that connects to the CNC control and accesses the CNC files from a diskette.

CNC Machine Options

CNC machines can be designed with various options. Some options may be included to reduce setup and cycle time. One option is probe capability, which allows the CNC machine to set the part origin, align stock to be machined, and inspect machined features. Another method of reducing cycle time is to arrange a number of CNC machines in such a way that one or two operators can maintain more than one machine. This method is called "cellular manufacturing," and may include robots, special machines, and/or other machines that will perform machining simultaneously with the CNC machine cycle. High-production machining centers can be designed with multiple pallets (HMC) or table shuttle (VMC) that reduces handling time.

Spindle Speed

The spindle speed is an important factor to consider when using CNC machines. The maximum spindle speed of a CNC machine tool can vary depending on the need or application. The factors that determine the spindle speed range for a CNC machine tool include part materials, part sizes, and annual production quantities. Another factor that influences spindle speed range is the type of tooling to be used, such as HSS, coated carbide, diamond, or ceramic.

Tool Setter

The tool setter is a sensing device on the machine table that automatically references each tool in a setup. The operator manually moves each tool to the tool setter and touches-off. The control then automatically records the values in the offset storage memory. This device minimizes downtime and improves the quality of the parts produced.

■ ■ ■ ■

CARTESIAN COORDINATE SYSTEM

The Cartesian coordinate system is the basis for all machine axis movement. This coordinate system is described in Figures 2-5 and 2-6. The figures illustrate the labeling of each axis (X, Y, and Z). It is important to note the positive and negative signs, which specify the direction of the coordinate. The point where all three axis lines intersect is called the origin or X-zero, Y-zero, and Z-zero point.

The other motion applied to CNC machine tools is rotary motion. There are three rotary axes (see Figure 2-7). They are labeled *A-axis*, B-axis, and C-axis. The clockwise or counterclockwise direction of rotation is specified by either a positive or negative sign. The Cartesian coordinate system is applied to CNC machine tools in various ways. The most common are the X-axis, Y-axis, and Z-axis.

On a CNC horizontal machining center, the X-axis travels left to right, the Y-axis travels up and down and the Z-axis travels towards and away from you. These axis movements are illustrated in Figure 2-6.

On a CNC vertical machining center, the X-axis also travels from left to right, the Y-axis travels toward and away from you, and the Z-axis travels up and down. This is illustrated in Figure 2-5.

There are some similarities between the vertical and the horizontal machining centers. For example, the spindle travel is always the Z-axis on both types of machines. The left-to-right table movement is always the X-axis. Another item common to both machining centers is the cutting toolholder or V-flange holder. Figure 2-4 describes some features associated with the toolholder, such as the V-flange, the taper shank, the retention knob, and the gage length.

All CNC machining centers have an X-, Y-, and Z-reference point home position. This machine home position is fixed. The part origin is separate from the machine home position. The part origin position will vary depending on where the part is located on the CNC machine table.

Absolute and Incremental Programs

Absolute and incremental program coordinates specify the motion or tool path with respect to the program zero or the tool moving distance from its current position. Programs can be written in the absolute format, the incremental format, or both formats in the same program.

Absolute programs specify a coordinate position or path from the workpiece coordinate zero, or origin. The positive and negative signs are derived from the part origin (see Figure 2-8).

Incremental programs specify the same positioning or path from the point where the machine spindle is currently located. A move to the right or up is always a positive (+); a move to the left or down is always a negative move (−). This method is best for programming parts with patterns that repeat numerous times (see Figure 2-9).

VMC Cartesian Coordinate System

Figure 2-5 illustrates how the Cartesian coordinate system is applied to VMCs. The table utilizes the X- and Y-axes, while the spindle utilizes the Z-axis in the vertical position.

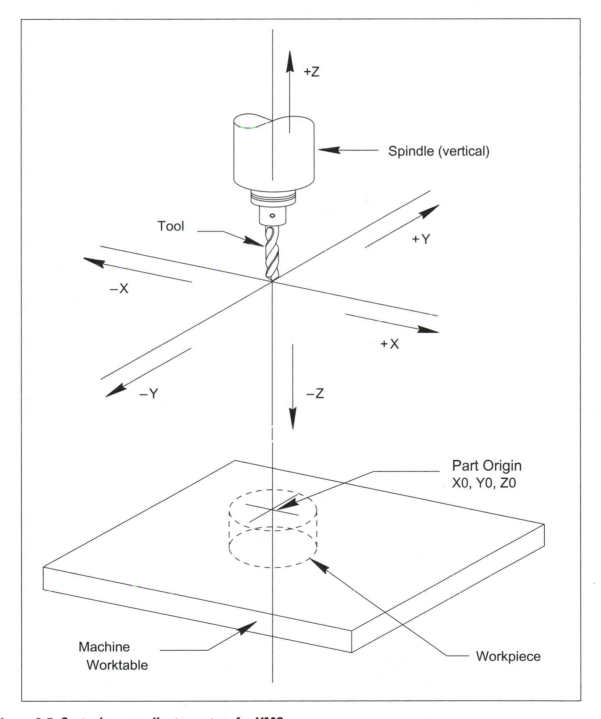

Figure 2-5 Cartesian coordinate system for VMCs

HMC Cartesian Coordinate System

Figure 2-6 illustrates how the Cartesian coordinate system is applied to HMCs. The table utilizes the X- and Y-axes, while the spindle utilizes the Z-axis in the horizontal position.

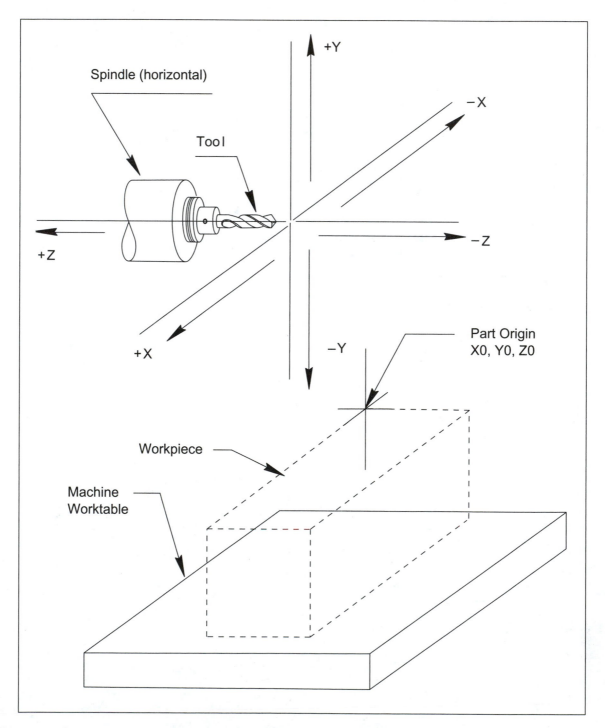

Figure 2-6 Cartesian coordinate system for HMCs

Rotary Cartesian Coordinate System

Figure 2-7 illustrates how the rotary Cartesian coordinate system is applied to CNC machining centers. The X-, Y-, and Z-axes utilize the A-, B-, and *C-axes*, respectively.

The rotational axes A, B, and C are used to control circular motion about each of these axis. Some common applications include rotary indexers, which are commonly used on CNC vertical machining centers. This option allows the CNC machine to machine the part on all sides. Another application is the rotary table on CNC horizontal machining centers. The rotary B-axis table of the CNC horizontal machining centers is a standard operational feature.

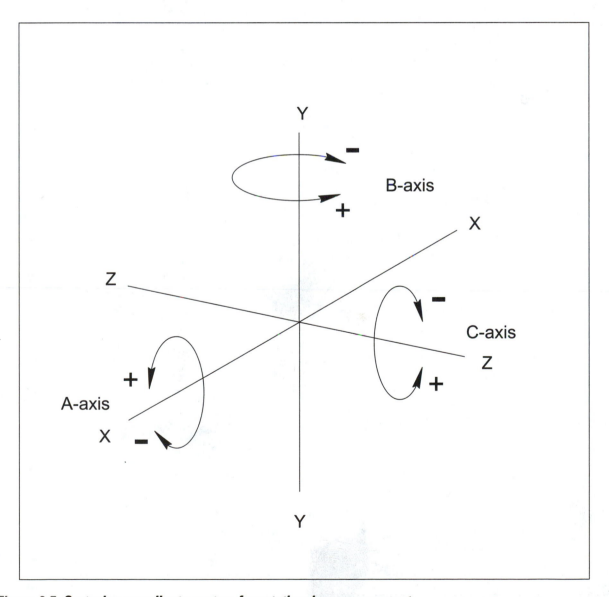

Figure 2-7 Cartesian coordinate system for rotational axes movement

Cartesian Quadrants

Figure 2-8 illustrates the sign (plus or minus) that the X-Y coordinate points are assigned for quadrants 1–4 in absolute mode. Note that all the X-Y values in *quadrant* 1 are plus. In quadrant 2 all X values are minus and all Y values are plus. In quadrant 3 all the X-Y values are minus. In quadrant 4 all X values are plus and all Y values are minus.

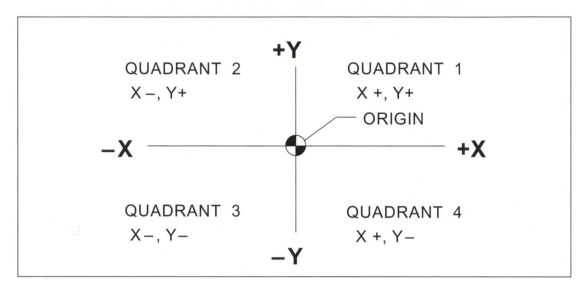

Figure 2-8 Cartesian quadrants in absolute values

Figure 2-9 illustrates the sign (plus or minus) that the X-Y coordinate points are assigned for quadrants 1–4 in incremental mode. Note that in all four quadrants direction determines the plus or minus sign. For the X-axis, left is minus and right is plus. For the Y-axis down is minus, and up is plus.

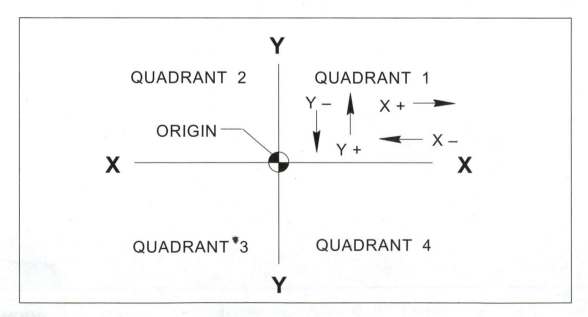

Figure 2-9 Cartesian quadrants in incremental values

■ ■ ■ ■
ABSOLUTE VERSUS INCREMENTAL WORKSHEET

Calculate the X-Y coordinate values for each chart (Figures 2-10 and 2-11) by plotting the points (A–E) on the following graph.

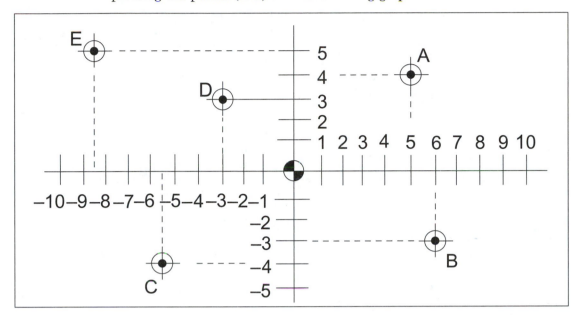

Calculate the X-Y coordinate values for each point in absolute mode; include the plus or minus sign. Start at the *origin* and plot points A–E.

Point	X-coordinate	Y-coordinate
A		
B		
C		
D		
E		

Figure 2-10 *Coordinate values chart in absolute mode*

Calculate the X-Y coordinate values for each point in incremental mode; include the plus or minus sign. Start at the origin and plot points A–E.

Point	X-coordinate	Y-coordinate
A		
B		
C		
D		
E		

Figure 2-11 *Coordinate values chart in incremental mode*

CNC MACHINING CENTER PROGRAM FORMAT

The sample CNC program (Figure 2-12) describes the format and the main components that are typically included in a CNC program. Note that each line of the CNC program contains various codes and coordinates. These items are explained in more detail in the sections that follow. For Figure 2-12, the items on the left are the CNC program codes and the explanations are on the right side.

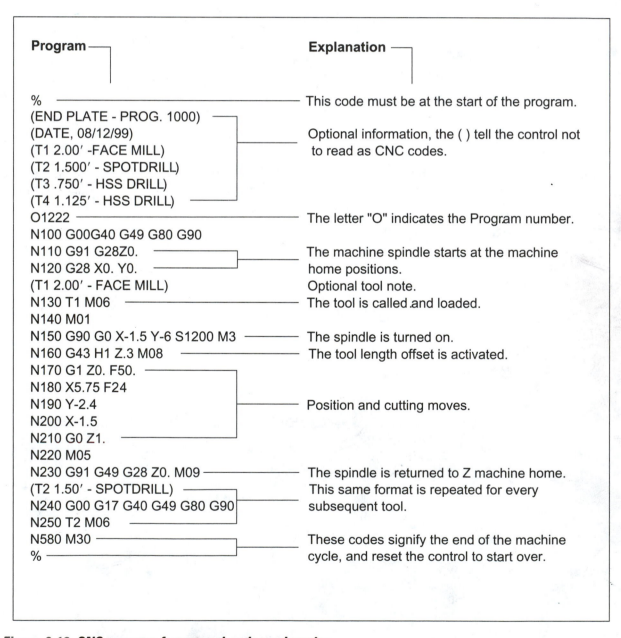

Program

Explanation

```
%                                      This code must be at the start of the program.
(END PLATE - PROG. 1000)
(DATE, 08/12/99)                       Optional information, the ( ) tell the control not
(T1 2.00' -FACE MILL)                  to read as CNC codes.
(T2 1.500' - SPOTDRILL)
(T3 .750' - HSS DRILL)
(T4 1.125' - HSS DRILL)
O1222                                  The letter "O" indicates the Program number.
N100 G00G40 G49 G80 G90
N110 G91 G28Z0.                        The machine spindle starts at the machine
N120 G28 X0. Y0.                       home positions.
(T1 2.00' - FACE MILL)                 Optional tool note.
N130 T1 M06                            The tool is called and loaded.
N140 M01
N150 G90 G0 X-1.5 Y-6 S1200 M3         The spindle is turned on.
N160 G43 H1 Z.3 M08                    The tool length offset is activated.
N170 G1 Z0. F50.
N180 X5.75 F24
N190 Y-2.4                             Position and cutting moves.
N200 X-1.5
N210 G0 Z1.
N220 M05
N230 G91 G49 G28 Z0. M09              The spindle is returned to Z machine home.
(T2 1.50' - SPOTDRILL)                 This same format is repeated for every
N240 G00 G17 G40 G49 G80 G90           subsequent tool.
N250 T2 M06
N580 M30                               These codes signify the end of the machine
%                                      cycle, and reset the control to start over.
```

Figure 2-12 CNC program format and code explanation

■ ■ ■ ■
CNC MACHINING CENTER COMMAND CODES

The CNC machine receives a series of commands via the CNC program as previously described in Figure 2-12 through the machine control. These commands are structured using letters and numbers such as G00, S2300, and M03. Each letter and number combination gives the control a specific command. There are some code differences between the various machine tool and control manufacturers. Therefore, each machine and control manufacturer provides a specific manual listing all the commands. Most CNC codes are identical, while some may vary slightly or entirely.

Letter Address Commands

Most letter *address* codes fall into two categories: *modal* or *nonmodal*. Nonmodal codes are command codes that are only active in the block in which they are specified and executed. Modal codes are command codes that once executed will remain active throughout the program until another code in the same group overrides or cancels it. Tables 2-1, 2-2, and 2-3 list and describe some common address letters that are used in CNC programs to control the CNC machine tool.

The *M codes* perform miscellaneous machine functions such as automatic tool changes, coolant control, and spindle operations. For example, a typical tool change requires the letter T and the tool number (T02), next the automatic tool change sequence is activated with the miscellaneous code (M06). The line would be programmed as follows: **N005 T02 M06.**

The *G codes* set preparatory machine functions such as rapid traverse mode, *feed* mode, and activate tool length offsets. For example, to activate a tool length offset, the line is typically programmed as follows: **N005 G43 Z1.0 H01.**

Spindle speeds are controlled with the S code followed by up to four digits to specify the RPM. For example, an RPM of 1575 would be programmed as **N005 S1575 M03.**

The axes of the CNC machine are controlled by the letters X, Y, and Z as previously illustrated in Figures 2-5 and 2-6. The X, Y, and Z letters require either a (plus) or (minus) sign to specify direction and a coordinate value to specify location. When a sign is positive (plus), it typically is not specified because the control defaults to a (plus) sign. The semicolon (;) is used to signify the *end-of-block* (EOB) or line of information. The example below describes how a typical CNC *block* of information is programmed and read.

N005 G00 X1.75 Y-0.75 M08;

N005	is the sequence block number
G00	is the rapid mode preparatory code
X1.75 Y-0.75	is the X-Y coordinate location to rapid to
M08	is the coolant on miscellaneous code
;	is the EOB code

■____*Table 2-1 Word address code descriptions*

LETTER	DESCRIPTION	MODE
O	Identifies the CNC program number	
N	Identifies a line number of a CNC program	
X	Positions the table coordinate axis	
Y	Positions the table coordinate axis	
Z	Positions the spindle coordinate axis	
W	Positions the secondary spindle coordinate axis	
A	Rotary position coordinate X-axis	
B	Rotary position coordinate Y-axis	
C	Rotary position coordinate Z-axis	
F	Specifies the *feedrate* of an axis [*inches per minute* (IPM) or *inches per revolution* (IPR)]	Modal
S	Specifies the speed of the spindle [revolutions per minute (RPM)]	Modal
T	Specifies the tool number or pocket selected for tool change	
H	Specifies the tool length offset number	Modal
D	Specifies the tool radius or diameter offset number	Modal
I	Specifies X-axis *circular interpolation* coordinate	
J	Specifies Y-axis circular interpolation coordinate	
K	Specifies Z-axis circular interpolation coordinate	
R	Specifies circular interpolation (G02 or G03) radius value, and *canned cycle* (G81–G89) retract point	
L	Specifies *subprogram* line numbers	
P	Specifies subprogram number, or canned cycle parameters	
Q	Specifies canned cycle parameters	
U	Specifies a *dwell* time with (G04)	
G	Specifies a *preparatory function* command (see Table 2-2)	
M	Specifies a miscellaneous function command (see Table 2-3)	

The order that the codes and coordinates are programmed follow for each block does not have to be in an exact order, but the N sequence should be first and the semicolon must be last. The codes and coordinates can be placed in any order, but in the interest of simplification and organized structure, the order is typically as follows:

The letter **O** program number must be at the start of the program.

The letter **N** and a number **005,** although not required, is the first code of a block.

The letter **T** code specifies the tool change.

■___Table 2-2 Preparatory function "G" code descriptions

LETTER	DESCRIPTION	MODE
G00	Rapid traverse positioning	Modal
G01	Linear positioning at a feedrate	Modal
G02	Circular interpolation clockwise at a feedrate	Modal
G03	Circular interpolation counterclockwise at a feedrate	Modal
G04	Dwell	Nonmodal
G20	Inch programming format	Modal
G21	Metric programming format	Modal
G28	Zero or machine home return	Nonmodal
G40	Tool diameter compensation cancel	Modal
G41	Tool diameter compensation left	Modal
G42	Tool diameter compensation right	Modal
G43	Tool height offset activate	Modal
G49	Tool height offset cancel	Modal
G54–G59	Work coordinate presets	Nonmodal
G80	Canned cycle cancel	Modal
G81	Drilling canned cycle	Modal
G82	Counterbore canned cycle	Modal
G83	Peck drilling canned cycle	Modal
G84	Tapping canned cycle	Modal
G85–G89	Boring canned cycles	Modal
G90	Absolute coordinate positioning mode	Modal
G91	Incremental coordinate positioning mode	Modal
G92	Axis coordinate preset	Nonmodal
G94	IPM feedrate mode	Modal
G95	IPR feedrate mode	Modal

The letter **M** miscellaneous code is usually last; only one M code per block is allowed.

The letter **S** code with a number value is usually next.

The letter **G** preparatory code is usually next, multiple G codes per block are allowed.

The coordinate letters **X, Y,** or **Z** are next in this same order and they also require a number value with a decimal point after the letter.

The letter **F** code with a number value for feeding usually follows.

The last code for each block is the semicolon (;) EOB code.

■___*Table 2-3 Miscellaneous function "M" code descriptions*

LETTER	DESCRIPTION	MODE
M00	Program stop	Nonmodal
M01	*Optional program stop*	Nonmodal
M02	Rewind or end of program	Nonmodal
M03	Spindle start clockwise	Nonmodal
M04	Spindle start counterclockwise	Nonmodal
M05	Spindle stop	Nonmodal
M06	Automatic tool change	Nonmodal
M07	Mist coolant on	Modal
M08	Flood coolant on	Modal
M09	Coolant off	Modal
M30	*End-of-program* and reset to the program start	Nonmodal
M40	Spindle low range	Modal
M41	Spindle high range	Modal
M98	Subprogram call	Modal
M99	End subprogram and return to main program	Modal

Some CNC programs do not include block sequence numbers for every line. However, they typically include a block sequence number for each line that starts a new tool.

■ ■
CHAPTER SUMMARY

The key concepts presented in this chapter include the following:

- The main components of CNC machining centers.
- The Cartesian coordinate system applied to VMCs.
- The Cartesian coordinate system applied to HMCs.
- The Cartesian coordinate system applied to rotary axes.
- Incremental coordinate calculation.
- Absolute coordinate calculation.
- The workpiece origin locations.
- The Cartesian quadrants and their corresponding (plus and minus) signs.
- The CNC program structure and format.

- G code preparatory function types.

- M code miscellaneous function types.

■ ■ REVIEW QUESTIONS

1. List the main components of the CNC machining center.

2. What is the purpose of the tool magazine?

3. What is the purpose of the ATC arm?

4. List the main items that are displayed on the CRT.

5. How is the V-flange holder accurately aligned in the machine spindle?

6. What is the purpose of the tool retention knob?

7. What is the purpose of the lube system?

8. What is the difference between absolute G90 and incremental G91?

9. Which axis rotates the table on an HMC?

10. In which quadrant are both X and Y absolute coordinates positive?

11. What is the difference between modal codes and nonmodal codes?

12. Which letter code identifies the CNC program number?

13. Which letter code specifies the axis feedrate?

14. Which letter code specifies the spindle speed?

15. Which letter code specifies the tool length offset number?

16. Explain the function of the M06 code and list the other codes that are required with this command for it to perform a function.

chapter 3

CNC MACHINING CENTER CUTTING FUNDAMENTALS

Chapter Objectives

After studying and completing this chapter, the student should have knowledge of the following:

- *CNC machining center operation types*

- *CNC machining center hole descriptions*

- *CNC machining center toolholders*

- *CNC machining center cutting tools*

- *Carbide insert data*

- *CNC machining center applications*

CNC MACHINING CENTER OPERATION TYPES

This chapter identifies the basics of VMC and HMC machining fundamentals. The areas of machining include operation types, hole types, toolholder types, cutting tool types, and carbide inserts. The more common machining operations that are performed on VMCs and HMCs are described below:

Drilling Operations

Drilling, which is the most common machining operation performed includes various applications (see Figures 3-1 to 3-8). For this reason, there are various types of drill designs (see Figures 3-9 and 3-11). Drilling does not typically produce holes of high accuracy. Other tools must be used to produce holes to higher precision characteristics.

Chamfer

The *chamfer* operation is designed to cut a 45° lead angle, usually at the top of the drilled surface (see Figure 3-1). The chamfer can be cut before or after the drilling operation. The CNC spot drill (see Figure 3-10) is the tool that is used to align the drill and to machine the chamfer, which eliminates the sharp edge typically produced when a hole is drilled.

Countersink

The countersink operation is designed to cut a lead angle, usually at the top of the drilled surface (see Figure 3-3). The countersink, which can be cut before or after the drilling operation, is designed to cut an angle of 60°, 82°, 90°, or 100°. The countersink tool is similar in design to the CNC spot drill (see Figure 3-10). The main purpose for countersink operations is to allow the head of a countersunk head screw to sink below the top of the mounting surface.

Spotface

The spotface operation is designed to cut a shallow enlargement of a previously drilled hole, usually at the top of the drilled surface (see Figure 3-4). The *spotface* operation is sometimes also required at the bottom of a drilled hole (back spotface). This operation is performed after the drilling operation. The tool that is used to machine the spotface is designed with a pilot to stabilize it as it cuts. The main purpose for spotface operations is to allow the head of a screw to mount square on the mounting surface.

Counterbore

The counterbore operation is similar to spotfacing (see Figure 3-5). The difference is in the depth of the enlargement. This operation is also performed after drilling. The same type of tool with a pilot is used. The main purpose for counterbore operations is to allow the head of a screw to mount below the top of the mounting surface.

Spot Drilling

As previously mentioned, twist drills are not capable of locating programmed hole positions with sufficient accuracy due to several factors. To better locate a hole position location, a short drill called a CNC spot drill (see Figure 3-10) is used before drilling. The spot drill is also used to machine a chamfer, which eliminates the sharp edge produced when a hole is drilled. Spotting and chamfering are usually combined.

Spade Drilling

Spade drilling is an operation that removes larger amounts of material, usually hole sizes from 1- to 4-inches (see Figure 3-11). The spade drill design offers several advantages over twist drills for drilling holes 1 *inch* in diameter and up. Mainly, spade drill designs are stronger, create less cutting pressure, and produce better chip control. Their tooling costs are lower as well.

Additional hole operations that sometimes follow drilling include boring, reaming, and tapping. These operations are described below.

Boring

Boring is used for two main purposes: enlargement of an existing hole and accurate machining of the center axis location of the enlarged hole. Better hole straightness and *surface finish* can also be achieved by boring. Some typical boring tools are illustrated in Figure 3-12.

Reaming

Reaming is an operation similar to boring that produces holes to a precise *diameter* size and surface finish (see Figure 3-13). Reaming produces accurate holes at a much faster rate than boring. It should be noted that the reamer is guided by the existing hole. It will not correct errors in hole location or its straightness. If these problems exist, it is advisable to first bore, then ream.

Tapping

Tapping is the process of cutting threads on the inside of an existing hole with a *tap* (see Figures 3-14 and 3-15). Tapping is usually a delicate and sometimes troublesome process depending upon the material type and various other factors such as thread depth, tap diameter, percentage of thread, *cutting speed*, cutting lubricant, and effectively clearing chips from the hole.

Face Milling

Face milling is a type of operation where a large surface is milled flat at a right angle to the cutter axis. This operation is designed to remove material as quickly as possible from larger surface areas. There are many different face milling applications, and therfore there are various types of face mill designs (see Figures 3-16 and 3-17).

End Milling

End milling is a type of operation where the tool removes material on the periphery of the part surface. This type of cutting is also known as contour milling, and cuts parallel or perpendicular to the cutter axis. End milling is used to machine slots, keyways, pockets, and contours. There are many different end milling applications, and therefore there are various types of end mill designs (see Figure 3-18).

Slot Milling

Slot milling is a type of operation where the tool removes material to create a slot the size of the cutter width or larger. This type of cutting is also known as saw milling and cuts perpendicular to the cutter axis (see Figure 3-19).

■ ■ ■ ■
CNC MACHINING CENTER HOLE DESCRIPTIONS

Figures 3-1 through 3-8 describe the common hole types machined on CNC machining centers. They are specified on the engineering drawing as required by the part design.

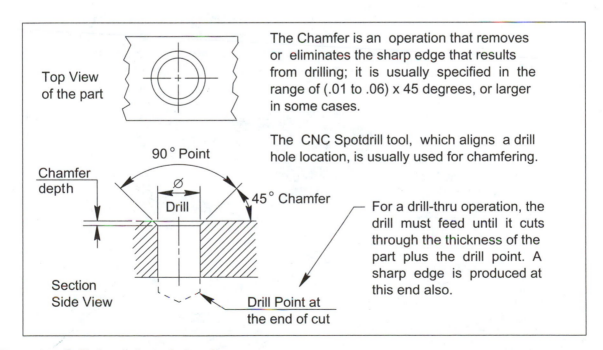

Top View of the part

The Chamfer is an operation that removes or eliminates the sharp edge that results from drilling; it is usually specified in the range of (.01 to .06) x 45 degrees, or larger in some cases.

The CNC Spotdrill tool, which aligns a drill hole location, is usually used for chamfering.

90° Point

Chamfer depth

Ø

Drill

45° Chamfer

For a drill-thru operation, the drill must feed until it cuts through the thickness of the part plus the drill point. A sharp edge is produced at this end also.

Section Side View

Drill Point at the end of cut

Figure 3-1 Drill-thru hole and chamfer

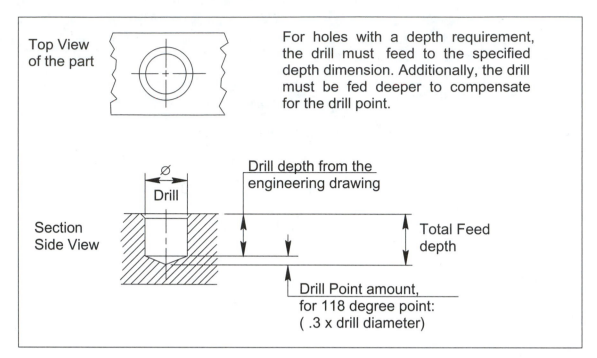

Top View of the part

Section Side View

∅
Drill

Drill depth from the engineering drawing

Total Feed depth

Drill Point amount, for 118 degree point: (.3 x drill diameter)

For holes with a depth requirement, the drill must feed to the specified depth dimension. Additionally, the drill must be fed deeper to compensate for the drill point.

Figure 3-2 **Drill hole to depth and chamfer**

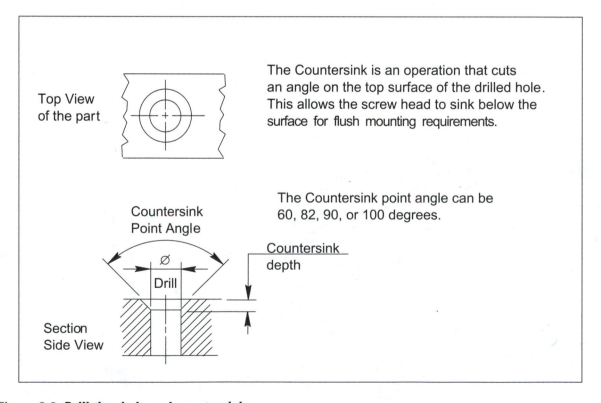

Top View of the part

Section Side View

Countersink Point Angle

∅
Drill

Countersink depth

The Countersink is an operation that cuts an angle on the top surface of the drilled hole. This allows the screw head to sink below the surface for flush mounting requirements.

The Countersink point angle can be 60, 82, 90, or 100 degrees.

Figure 3-3 **Drill-thru hole and countersink**

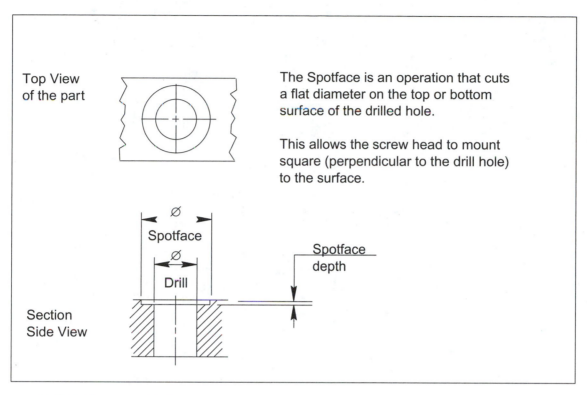

Top View
of the part

The Spotface is an operation that cuts a flat diameter on the top or bottom surface of the drilled hole.

This allows the screw head to mount square (perpendicular to the drill hole) to the surface.

Ø
Spotface
Ø
Drill

Spotface
depth

Section
Side View

Figure 3-4 *Drill-thru hole and spotface*

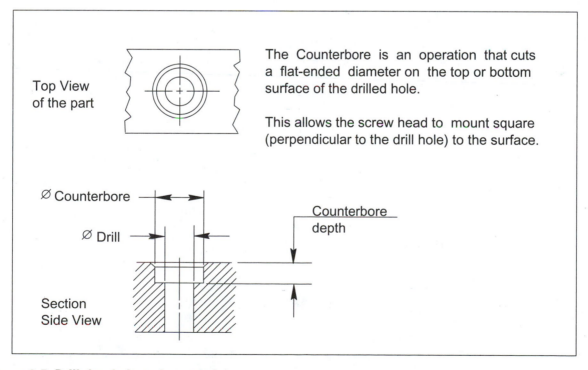

Top View
of the part

The Counterbore is an operation that cuts a flat-ended diameter on the top or bottom surface of the drilled hole.

This allows the screw head to mount square (perpendicular to the drill hole) to the surface.

Ø Counterbore

Ø Drill

Counterbore
depth

Section
Side View

Figure 3-5 *Drill-thru hole and counterbore*

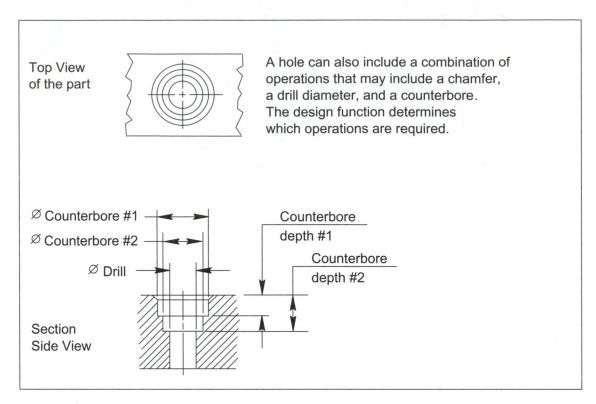

Top View of the part

A hole can also include a combination of operations that may include a chamfer, a drill diameter, and a counterbore. The design function determines which operations are required.

Ø Counterbore #1

Ø Counterbore #2

Ø Drill

Counterbore depth #1

Counterbore depth #2

Section Side View

Figure 3-6 **Drill-thru hole and combination counterbore**

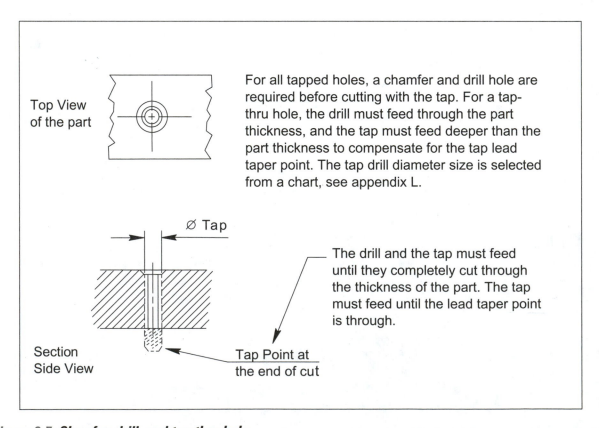

Top View of the part

For all tapped holes, a chamfer and drill hole are required before cutting with the tap. For a tap-thru hole, the drill must feed through the part thickness, and the tap must feed deeper than the part thickness to compensate for the tap lead taper point. The tap drill diameter size is selected from a chart, see appendix L.

Ø Tap

The drill and the tap must feed until they completely cut through the thickness of the part. The tap must feed until the lead taper point is through.

Section Side View

Tap Point at the end of cut

Figure 3-7 **Chamfer, drill, and tap-thru hole**

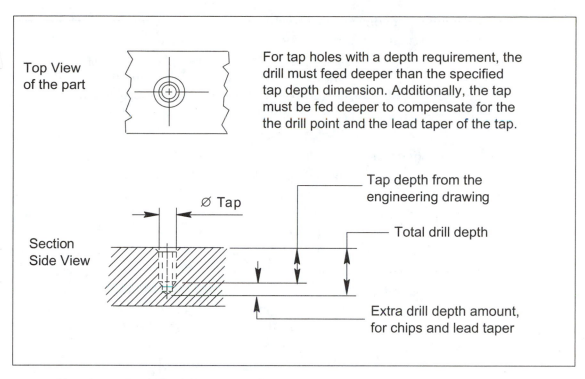

Figure 3-8 *Chamfer, drill, and tap tole to depth*

CNC MACHINING CENTER CUTTING TOOLS

The CNC operator must be thoroughly knowledgeable in machining operations that pertain to the type of CNC machine to be operated. The CNC tooling is directly linked to the machining operations planning process. This section describes and illustrates the most commonly used machining center tools. It presents a view of the various types of holders and tool types used in CNC machining operations. Drilling, which accounts for a majority of hole operations on CNC machining centers, is discussed and illustrated. Other hole operations to consider are tapping, boring, reaming, counterboring, and countersinking. Another type of cutting operation discussed is milling. The reader is introduced to the face milling and contour milling type of operations. The most common types of toolholders are discussed. Each toolholder type is followed by a discussion of the types of tools that are mounted to them.

CNC Toolholders

Toolholders are designed to be asembled offline. Once asembled, the tools are loaded into the machining center's tool storage mechanism (tool magazine) in preparation for running a CNC machining program. Modular tooling systems are a key factor in boosting CNC productivity uptime to approximately

30–70% over conventional tooling. The two most common V-flange tool-holder designs for gripping and centering tools are the end mill holder and the collet holder. Figure 3-9 describes and illustrates the major features and components of the V-flange holder. The more widely used CNC toolholders described and illustrated in Figures 3-10 to 3-19 include *collet* holders, floating holders, tap holders, end mill holders, and shell mill holders.

Clamping the Tools into the Machine Spindle

There are various designs used for loading and unloading the tools. In one system the tool interchange arm grips the toolholder and aligns it with the spindle. For another system, the spindle may descend on the toolholder or the arm may insert the toolholder into the spindle. The spindle and toolholder mate, a drawbar machanism inside the spindle locks onto a retention knob on the top of the toolholder and, then draws the toolholder up and into the spindle, creating positive clamping and centering action, as previously illustrated in Figure 2-4. V-grooved flange toolholders are used with side-gripping interchange arms. For these configurations, the holder is gripped from one side and placed into the spindle. The interchange arm disengages the toolholder by moving off to the side.

Collet Holders

This type of toolholder involves the use of a collet for gripping and centering the tool in the toolholder (see Figure 3-10). Collet-type toolholders are more expensive than other holders, such as end mill holders, but they generally provide better tool-holding capability and more accurate tool centering than the end mill type holders. Collets are partially slit all around so that they can flex diametrically when clamping or unclamping. They are available in two basic designs: double angle and single angle (slow taper). Double-angle collets can be used to hold many different tool shank shapes including tapers. Single-angle collets are especially useful for applying very high clamping forces on cutting tools. When selecting a collet, care should be taken to ensure that the diameter range is appropriate for the shank diameter of the tool.

CNC V-Flange Holder Description

The CNC toolholders that hold the cutting tools are designed in various styles. The shank size varies depending on the size of the CNC machine spindle and the cutting tools. Figure 3-9 illustrates the major features and components typical for each toolholder style. They include the retention knob, the taper shank, the V-flange, the key slots, and the gage line diameter. Additionally, most toolholders are designed with a through hole that directs the coolant to the cutting edge. The V-flange is designed so that an automatic tool change arm can quickly load and unload the toolholders. The second section of the toolholder design is determined by the type of machining operation or application. The common types, which are illustrated in Figures 3-10 to 3-19, include collet holder, end mill holder, reamer holder, tap holders, shell mill holder, and stub arbor holder.

Drilling Tools

Drilling, in most cases, is the first machining operation performed in the total production of a hole. The drill can vary in design, size, and material. The most common types of drilling tools are listed and described on the pages that follow.

Twist Drills

The most commonly used tool for drilling is the twist drill (see Figure 3-10). This end-cutting tool has two helical grooves or flutes that are cut around a

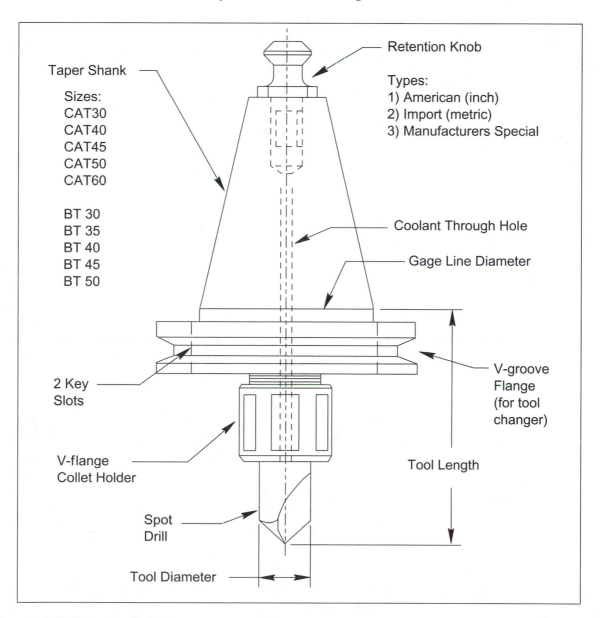

*Figure 3-9 **V-flange toolholder***

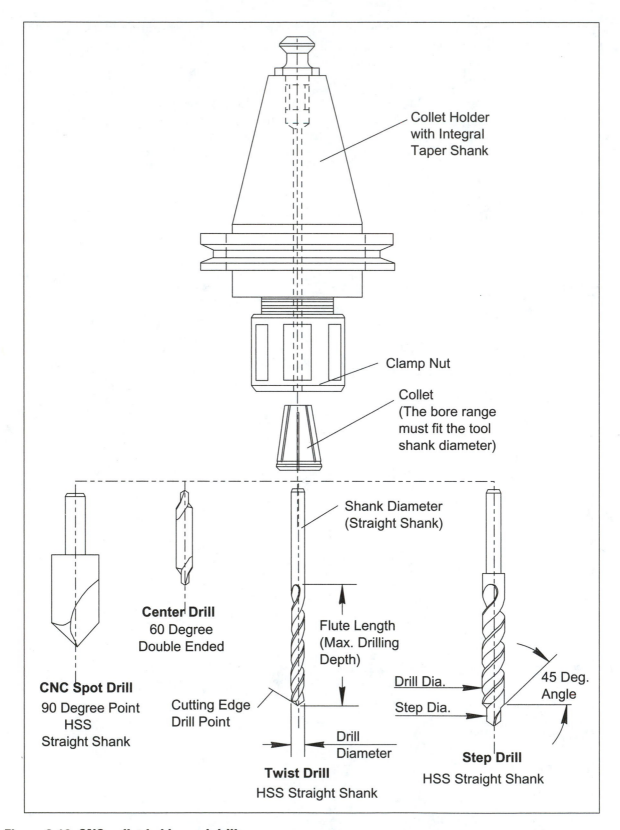

Figure 3-10 **CNC collet holder and drills**

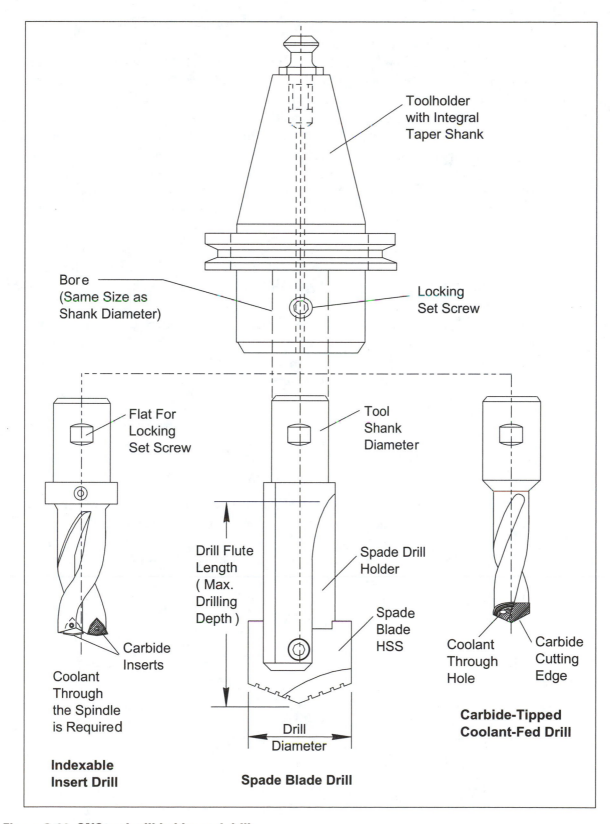

Toolholder
with Integral
Taper Shank

Bore
(Same Size as
Shank Diameter)

Locking
Set Screw

Flat For
Locking
Set Screw

Tool
Shank
Diameter

Drill Flute
Length
(Max.
Drilling
Depth)

Spade Drill
Holder

Spade
Blade
HSS

Carbide
Inserts

Coolant
Through
the Spindle
is Required

Coolant
Through
Hole

Carbide
Cutting
Edge

**Carbide-Tipped
Coolant-Fed Drill**

Drill
Diameter

**Indexable
Insert Drill**

Spade Blade Drill

Figure 3-11 **CNC end mill holder and drills**

center called a "web." The flutes act as cutting edges for feeding the tool into the material and as channels for admitting lubricant and carrying away the cut chips. The web gives the drill strength in resisting deflections. They are designed with either a straight or a tapered shank. Metal-cutting twist drills are made from a wide range of materials, which include HSS, cobalt, coated, carbide tipped, and solid carbide. Straight shanks are common for drills up to 0.500 inch. Larger drills can have straight or tapered shanks. The tang of the tapered shank prevents slipping of the drill while cutting larger holes. Drilling accuracy tends to decrease when either drill length or drill diameter is increased. Longer drills exhibit less stiffness and more torsional deflection. For these two reasons, drills with longer lengths or larger diameters are usually designed with a taper shank. A good machining practice is to always select the shortest drill possible for any hole operation. When drilling through material, a good practice is to allow one-third of the diameter of the drill ($0.3 \times$ drill diameter) plus 0.100 thousandths (0.100 inch) to extend beyond the part material.

Twist drills are standardized by diameter into three categories that include: Number size, Letter size, or Fractional size. A complete conversion of the number and letter sizes is listed in appendix K and the fractional sizes are listed in appendix J. The diameter ranges for each category are as follows (English system):

- Number sizes—from no. 1 (0.228 inch) to no. 80 (0.0135 inch)

- Letter sizes—from letter A (0.234 inch) to letter Z (0.413 inch)

- Fractional sizes—from 1/64 to 63/64 inch

Twist drills are also standardized by overall length (OAL) into three categories that include screw machine length, jobbers length, or taper length. Most drill manufacturers also produce extra long sizes. A complete listing with the exact lengths in each category is usually found in each manufacturer's catalog. The OAL ranges for each category are as follows (English system):

- Screw machine length—from 1-3/8 inches OAL to 8-1/2 inches OAL

- Jobbers length—from 3/4 inch OAL to 7-5/8 inches OAL

- Taper length—from 4-5/8 inches OAL to 29.0 inches OAL

Twist drills are also available in standard metric diameters, which range from 0.15 mm (0.0059 inch) diameter to 33.0 mm (1.2992 inches) diameter. The lengths are also available in metric sizes.

Spot Drills and Center Drills

As stated previously, twist drills are usually not capable of machining holes accurately on-center. Several factors including flute length, drill diameter, drill flexibility, cutting edge preparation, and material *hardness* will influence hole drilling.

To accurately locate a hole center, a center drill (Figure 3-10) must first drill a start point. A good practice is to produce a center drill point such that the countersunk diameter is approximately 0.010 to 0.030 inch larger than the corresponding twist drill diameter. This centers the drill as it cuts and creates a chamfer that eliminates the edge burr.

Coolant-Fed Drills

Coolant-fed drills have one or two holes passing from the shank to the cutting point (see Figure 3-11). Compressed air, oil, or cutting fluid is passed through the drill as it operates. This design enables the cutting point and work to be cooled as chips are flushed out. These drills are especially useful for drilling deep holes.

Spade Drills

A spade drill consists of two parts, a spade drill holder and a spade blade that is bolted to the holder (see Figure 3-11). Generally, spade drills offer several advantages over twist drills for drilling holes 1 inch in diameter and up. The larger web of the spade drill ensures that during penetration less flexing occurs and thus a more accurate hole is produced. Tooling costs are lower with spade drills because standard blade holders will accommodate a variety of blade diameters, normally ranging from 0.500 to 6.000 inches. Worn blades can be either resharpened or simply replaced with new ones. Job setup for CNC hole operations is also reduced. Spade drills are designed to machine a hole from the solid in one pass, eliminating the need for center drilling or multiple-pass drilling to gradually enlarge the hole size.

In order to appropriately use a spade drill, 50% or greater machine torque is needed beyond that which is normally used for drilling with a standard twist drill. The machine and the setup rigidity must also be increased.

Most spade drills operate with coolant flowing through the tool for heat dissipation and to flush out the chips. Thus, a high-pressure coolant system is usually required. The drilling depth capability is also limited with spade drills because spiral flutes that help eject chips from the cut do not exist. The cutting edges of the blade incorporate chip splitting and breaking action to reduce chip size and to facilitate chip removal.

Carbide-Indexable Insert Drills

Carbide-indexable insert drills (see Figure 3-11) represent the latest state-of-the-art advancements in CNC hole drilling. The indexable insert drill is sometimes used in place of a HSS twist drill. The indexable insert can drill holes at much higher rates than HSS twist drills. Insert drills are suited for hole diameters ranging from 0.625 to 3.000 inches. They offer all the advantages of spade drills including replaceable (indexable) inserts. They are capable of drilling from a solid at penetration rates of 5–10 times faster than twist drills or spade drills. The carbide inserts also allow the tool to be driven into harder materials. Insert drills require higher machining horsepower and a thru-the-tool high-pressure coolant system.

Today many vendors manufacture carbide inserts and each has its own grading system. Vendors also supply supplementary grades within a particular ANSI class. Vendor tool catalogs usually give a complete listing of their grade systems. The reader is advised to consult these catalogs for more detailed information.

Boring Bars

Boring bars can be generally classified into two groups: roughing or finishing (see Figure 3-12). Better hole straightness and surface finish can also be achieved by boring. As a general rule, the shortest boring bar should be selected for any operation. As with drills, the greater the length-to-diameter ratio the more flexible and error prone the boring bar will be. The finish of the surface inside the hole will also be affected because long bars tend to chatter.

Floating Reamer Holder

The floating reamer holder is designed to float approximately 0.030 inches in all directions (see Figure 3-13). The floating mechanism will compensate for any misalignment between the drill holder and the reamer holder.

Reamer

As stated previously, a twist will not consistently cut holes to an exact size or a smooth surface finish. When a higher degree of *accuracy* or finish is required, an additional operation called "reaming" must be included. A reamer is a cylindrical tool with straight or helical cutting edges. The majority of reamers are typically made of HSS. Reamers cut a small amount of material (0.005 to 0.030 inch) on the sides as well as the end. Hole type, size, and the number of holes to be reamed in a *production run* will determine the type of reamer used (see Figure 3-13). As stated previously, the reamer is guided by the existing hole. Therefore, it will not correct errors in hole location or straightness. If these problems exist, it is advisable to first bore, then ream.

Tap Holders

There are two types of tap holders—the tension and comprehension type, which is more expensive, and the rigid type, which requires a CNC machine control with synchronized tapping capability (see Figures 3-14 and 3-15). The tap holder is designed to specifically meet tapping characteristics that are more demanding than drilling. Tapping requires a positive drive at the shank end because of the high torque that is generated. Additionally, spindle compensation is required for the tap engagement and spindle reversal for feed out.

Taps

Taps are made from a variety of materials such as carbon tool steel, HSS, and even carbide. They are categorized according to size and type. Tap sizes cover the range from 0 (0.060 inch) to 1.0 inch. Thread milling is suggested for threads larger than 1.0 inch. The lead taper at the front of the tap is used to

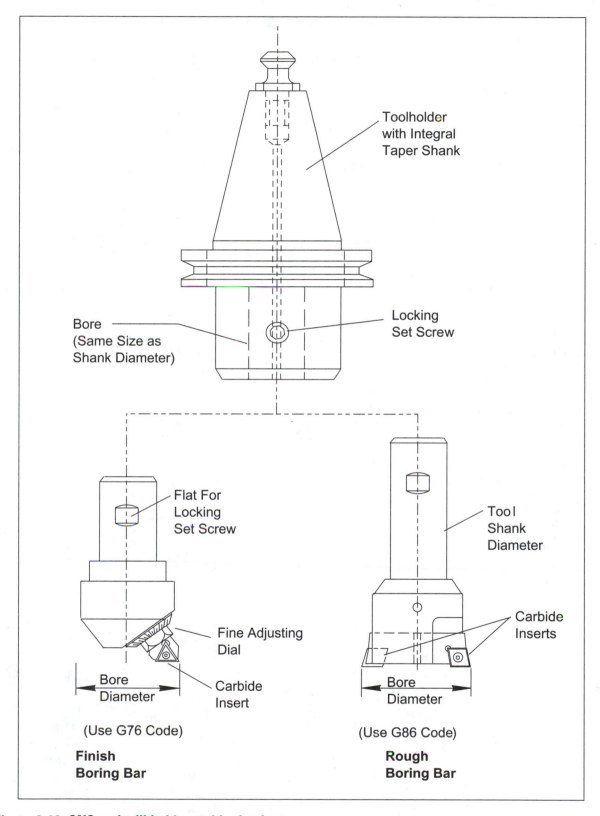

Figure 3-12 *CNC end mill holder and boring bars*

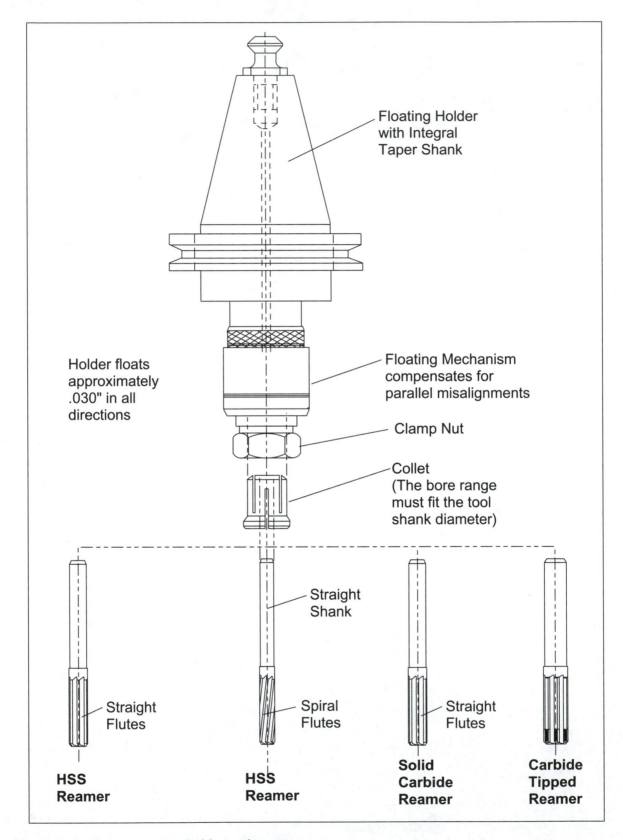

Figure 3-13 Floating reamer holder and reamers

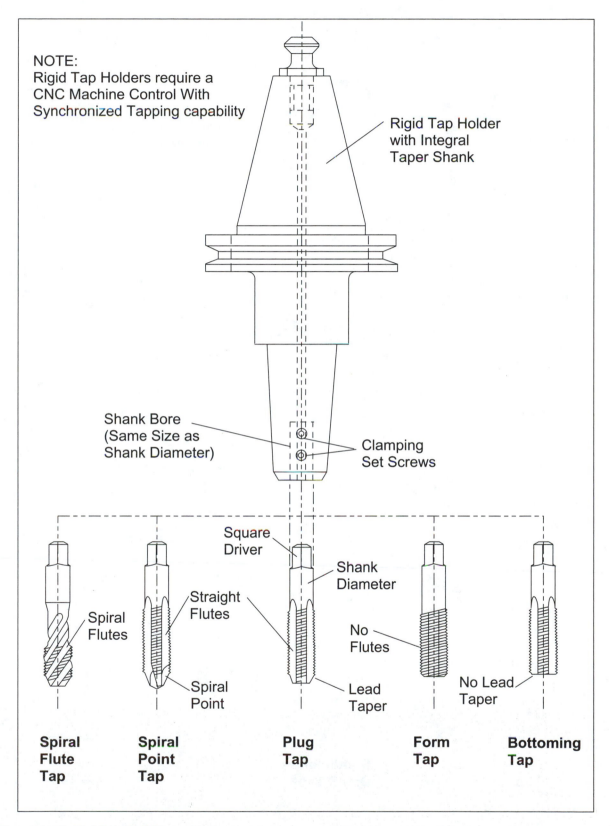

NOTE:
Rigid Tap Holders require a
CNC Machine Control With
Synchronized Tapping capability

Rigid Tap Holder
with Integral
Taper Shank

Shank Bore
(Same Size as
Shank Diameter)

Clamping
Set Screws

Square
Driver

Shank
Diameter

Straight
Flutes

Spiral
Flutes

No
Flutes

Spiral
Point

Lead
Taper

No Lead
Taper

**Spiral
Flute
Tap**

**Spiral
Point
Tap**

**Plug
Tap**

**Form
Tap**

**Bottoming
Tap**

Figure 3-14 Rigid tap holder and taps

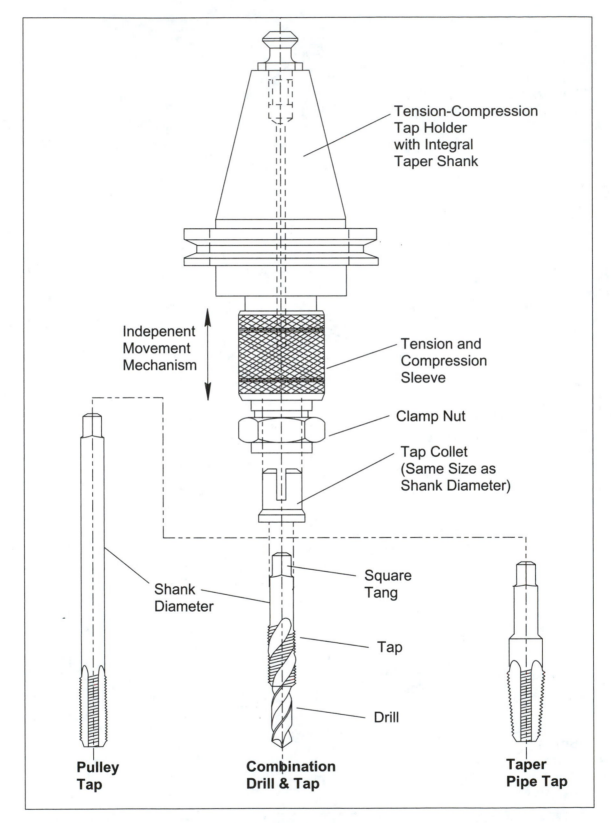

Figure 3-15 **Tension-compression tap holder and taps**

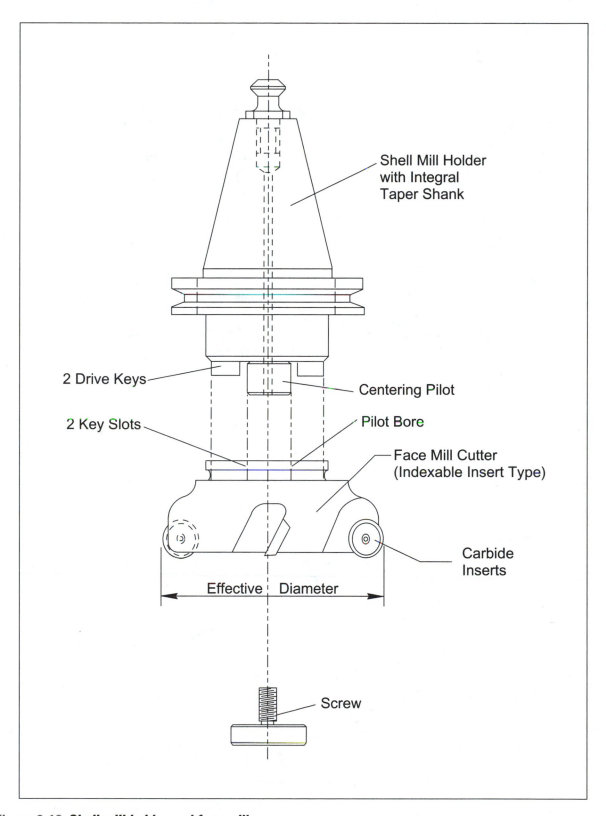

Shell Mill Holder
with Integral
Taper Shank

2 Drive Keys

Centering Pilot

2 Key Slots

Pilot Bore

Face Mill Cutter
(Indexable Insert Type)

Carbide
Inserts

Effective   Diameter

Screw

Figure 3-16 Shell mill holder and face mill

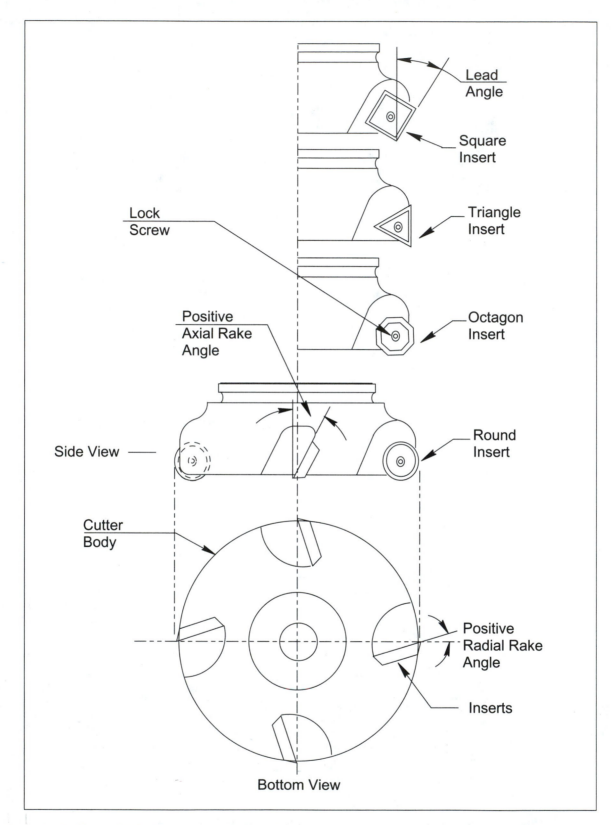

Figure 3-17 ***Face mill features and components***

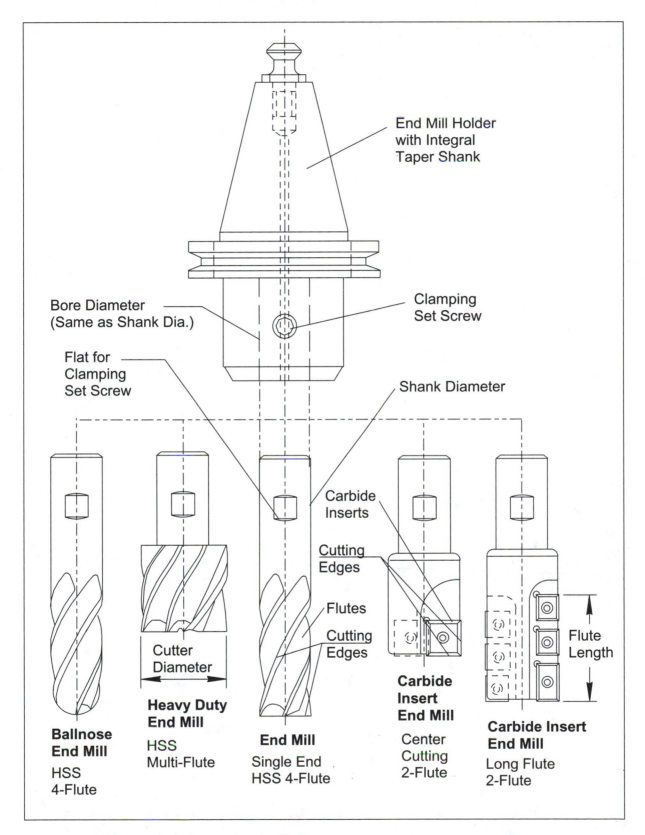

Figure 3-18 **CNC end mill holder and end mills**

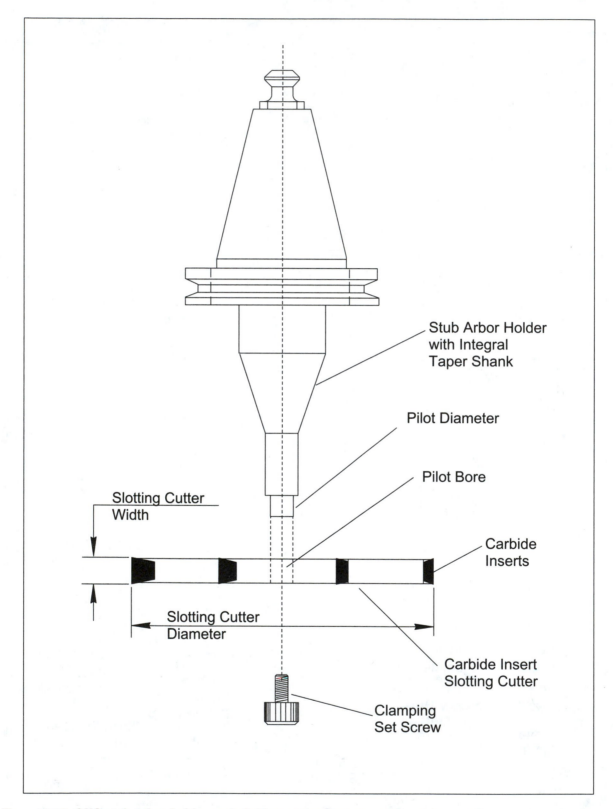

Figure 3-19 *CNC stub arbor holder and slotting cutter features and components*

start the thread. It ranges in design depending on the application. A plug tap has less taper and is used for through holes or blind holes that have minimal threaded depths. The bottoming tap is chamfered at the lead end but not tapered. It is used for applications where a full thread must be cut to the bottom of a blind hole. A spiral flute tap allows lubricant to flow into the cutting area and ejects the chips. A form tap, which is recommended for soft materials such as aluminum, shapes the material to the required thread rather than cutting chips. Other types, which include spiral point tap, pulley tap, pipe tap, and combination drill-tap, are illustrated in Figures 3-14 and 3-15. Likewise, most other tools such as inserts, drills, taps, and reamers are also coated to increase their efficiency and cutting performance.

Shell Mill Holders

Shell mill holders are specifically designed for holding and driving face mill cutters (see Figure 3-16). They position the holder on center with a centering plug and are equipped with two drive keys, which allow it to machine heavy cuts that generate higher torque and horsepower.

Face Mills

Face mill cutters are designed for various applications such as rough milling, finish milling, milling soft materials (aluminum, for example), or hard materials (*alloy steel*, for example). They are basically used to mill flat surfaces within specified tolerances of 0.001 to 0.010 inch and surface finish ranges from fine to rough. Face milling cutters are available in various cutter diameters sizes that include 2.0, 3.0, 4.0, 5.0, 6.0, 8.0, 10.0, and 12.0 inches. The face mill is designed with axial and radial rake angles that range from positive to negative (see Figure 3-17). The cutting is performed by the carbide inserts, which are available in a variety of geometry shapes. The geometry shapes include square, round, triangle, rectangle, and octagon.

End Mill Holders

End mill holders lock and center the cutting tool (end mill) by way of one or two *set screws*, which press on the flats of the tool shank. Because the screws are both located on one side, tool runout from the spindle axis is increased. These toolholders are relatively inexpensive in comparison to *collet-* or *chuck-* type holders. They are used to hold end mills, drills, boring bars, spade drills, and other tools with straight shank diameters of standard dimensions. They find wide applications in many milling and boring operations not requiring very close tolerances (see Figure 3-18).

End Mills

End mills are available in various styles and cutter materials (see Figure 3-18). End mills are used for various applications such as contour shapes, key slots, plunge cutting, and circular shapes. Some common styles include: 2-flute center cutting, 3-flute, 4-flute, single ended, double ended, and ballnose. The materials that they are made from include HSS, coated HSS, carbide tipped, solid carbide, carbide insert, and coated carbide. They generally range in diameter sizes from 0.061 to 3.000 inches.

Slot Mill Holders

Slot mill holders center and clamp the cutting tool by way of a pilot plug and one set screw (see Figure 3-19). These toolholders are relatively inexpensive in comparison to the other toolholders. They are used to hold slot mill cutters, and they find a wide range of slotting applications. They are available in various diameters and lengths.

Slot Mills

Slot mills are available in various styles and materials (see Figure 3-19). Slot mills are typically used for cutting slots with larger than average depths. Some common styles include narrow slot cutter, variable width cutter, half-side cutter, face cutter, and shell mount type. The materials that they are made from include HSS, coated HSS, carbide tipped, solid carbide, carbide insert, and coated carbide. They generally range in diameter sizes from 3.000 to 12.000 inches.

Milling Operation Description

The two illustrations in Figures 3-20 and 3-21 describe the "climb" mill cutting method and the "conventional" mill cutting method. The illustrations include the position of the milling cutter, the workpiece, the rotation, and the feed direction of the cutter.

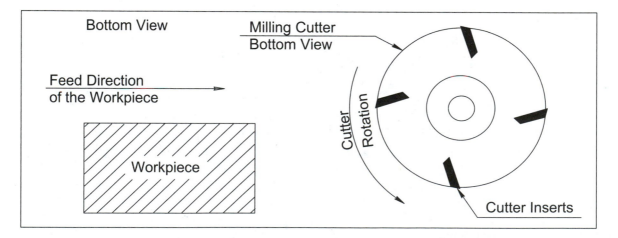

Figure 3-20 Climb milling illustration

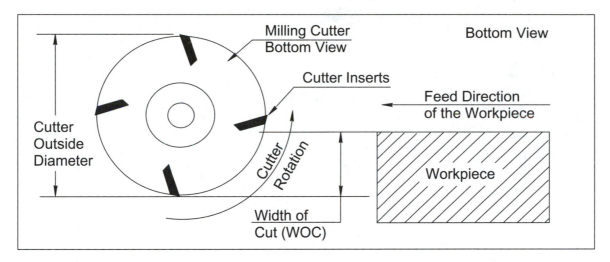

Figure 3-21 Conventional milling illustration

Climb Milling

Climb milling, which usually requires less *clamping force* on the workpiece, is usually recommended for machining parts that are thin, parts that are difficult to hold down, and when milling work-hardened materials. Also, machining marks or surface breakouts are minimized with the climb milling method.

Conventional Milling

Conventional milling, which usually requries maximum hold-down clamping force on the workpiece, is usually recommended for machining castings, forgings, and hot-rolled steel. Conventional milling is also recommended for finishing cuts and where long tool length may cause excessive chatter and poor surface finish.

CARBIDE INSERT DATA

Cutting tools are made from various materials such as HSS, carbide, ceramic, and diamond. Most manufacturing today is done with carbide-inserted tools. The indexable carbide insert has some distinct advantages over the other cutting tool materials.

The first advantage is that carbide tools can machine metal at a faster rate than HSS tools. The second advantage is that when the carbide insert tool wears, it can be quickly changed by simply indexing the insert or changing to a new insert. The carbide insert can also be coated to reduce wear and increase tool life. Carbide inserts are available in various sizes and shapes (geometry) (see Figures 3-22 through 3-24). All these characteristics make carbide-indexable inserts ideal for satisfying the wide variety of CNC machining needs and applications.

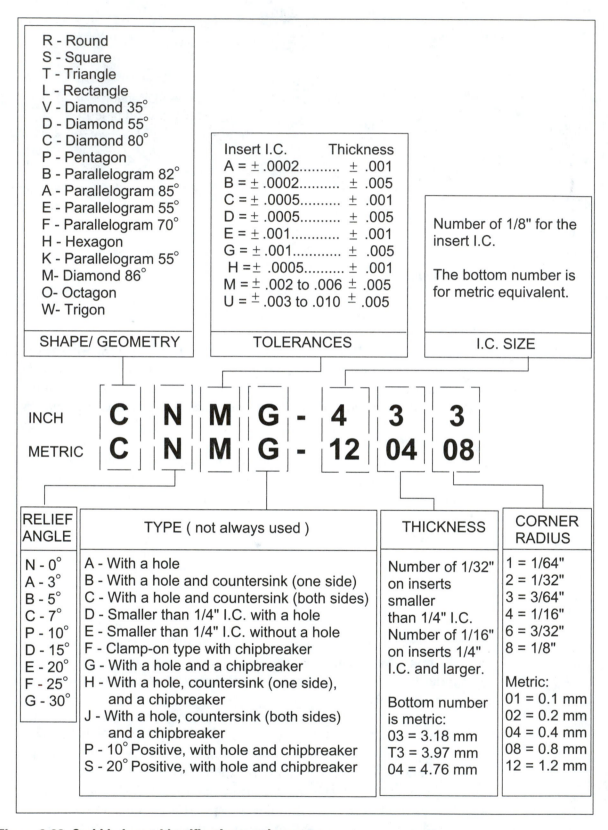

SHAPE/ GEOMETRY

R - Round
S - Square
T - Triangle
L - Rectangle
V - Diamond 35°
D - Diamond 55°
C - Diamond 80°
P - Pentagon
B - Parallelogram 82°
A - Parallelogram 85°
E - Parallelogram 55°
F - Parallelogram 70°
H - Hexagon
K - Parallelogram 55°
M- Diamond 86°
O- Octagon
W- Trigon

TOLERANCES

	Insert I.C.	Thickness
A =	± .0002	± .001
B =	± .0002	± .005
C =	± .0005	± .001
D =	± .0005	± .005
E =	± .001	± .001
G =	± .001	± .005
H =	± .0005	± .001
M =	± .002 to .006	± .005
U =	± .003 to .010	± .005

I.C. SIZE

Number of 1/8" for the insert I.C.

The bottom number is for metric equivalent.

INCH   **C N M G - 4 3 3**
METRIC **C N M G - 12 04 08**

RELIEF ANGLE

N - 0°
A - 3°
B - 5°
C - 7°
P - 10°
D - 15°
E - 20°
F - 25°
G - 30°

TYPE (not always used)

A - With a hole
B - With a hole and countersink (one side)
C - With a hole and countersink (both sides)
D - Smaller than 1/4" I.C. with a hole
E - Smaller than 1/4" I.C. without a hole
F - Clamp-on type with chipbreaker
G - With a hole and a chipbreaker
H - With a hole, countersink (one side), and a chipbreaker
J - With a hole, countersink (both sides) and a chipbreaker
P - 10° Positive, with hole and chipbreaker
S - 20° Positive, with hole and chipbreaker

THICKNESS

Number of 1/32" on inserts smaller than 1/4" I.C. Number of 1/16" on inserts 1/4" I.C. and larger.

Bottom number is metric:
03 = 3.18 mm
T3 = 3.97 mm
04 = 4.76 mm

CORNER RADIUS

1 = 1/64"
2 = 1/32"
3 = 3/64"
4 = 1/16"
6 = 3/32"
8 = 1/8"

Metric:
01 = 0.1 mm
02 = 0.2 mm
04 = 0.4 mm
08 = 0.8 mm
12 = 1.2 mm

*Figure 3-22 **Carbide insert identification number system***

All cutting tools are basically the same except for the type of material from which they are manufactured. They all have geometric characteristics and properties that determine how they can be applied in CNC machining situations. The most important characteristic of cutting tool geometry is "rake." Tool rake, which can range from negative to positive, has an effect on the formation of the chip and on the surface finish. Neutral and negative rake tools are stronger and have a longer cutting life than positive rake tools. Negative rake tools produce rough finishes at low cutting speeds, but give a good finish at high speeds. Positive rake tools are more free-cutting and can produce good finishes in most situations. Negative rake tools require more machine horsepower than positive rake tools. Higher cutting speeds generally produce better surface finishes and are less likely to disturb the grain structure of the workpiece.

Cutting Fluids (Coolant)

Coolant is usually applied to the insert or tool as it cuts metal in order to dissipate the heat and lubricate the cutting action. There is a great difference in tool life and surface finish between cutting with coolant and cutting without coolant (cutting dry). This is due to the cooling effect of the coolant and the lubricating action that reduces friction between the insert or tool and the chip. Additionally, when coolants are applied, pressure welding is reduced. This is especially important when machining softer metals such as aluminum, copper, bronze, and brass, which have a lower melting point and tend to build up on the cutting edge of the insert or tool.

A built-up edge can cause a rough surface finish and tearing of the surface on the workpiece. It should also be noted that a built-up edge is also caused by slow cutting speeds and excessive tool pressure (negative rake cutting). Conversely, at higher cutting speeds, cratering begins to form on the top surface of the insert or tool. This is due mainly to wear from the chips sliding against the insert or tool as it cuts. Therefore, it is important to consider that when coolant is not used, tool performance will substantially decrease. The advantages derived from using coolant are listed below:

- Cools and lubricates the tool and workpiece as it cuts.

- Reduces built-up edge and tool wear.

- Will flush chips away from the cutting edge.

- Produces better surface finishes.

- Enables the tool to cut at optimum cutting speeds and *depths of cut.*

Carbide Insert Features and Identification Number System

Figure 3-23 describes the features and number system for an indexable insert. Each insert is assigned a letter and number comibination that identifies its characteristics.

Figure 3-24 describes the process for identifying indexable inserts by their specific identification number.

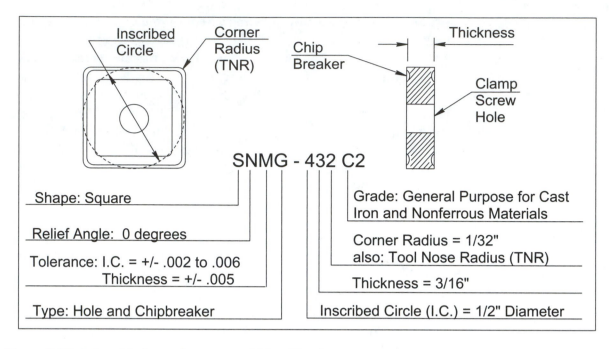

Figure 3-23 *Indexable insert features and identification system*

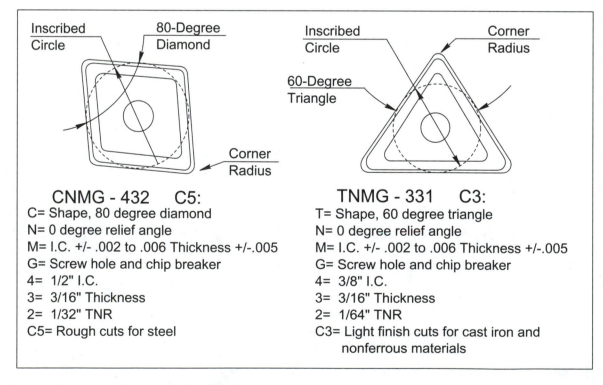

Figure 3-24 *Indexable insert identification process*

■___*Table 3-1* **Standard International Standards Organization (ISO) carbide grades**

ISO GRADE	APPLICATION	WORKPIECE MATERIAL
C-1	Roughing cuts	*Cast iron and nonferrous materials*
C-2	General purpose	Cast iron and nonferrous materials
C-3	Light finishing	Cast iron and nonferrous materials
C-4	Precision boring	Cast iron and nonferrous materials
C-5	Roughing cuts	Steel
C-6	General purpose	Steel
C-7	Finishing cuts	Steel
C-8	Precision boring	Steel

Insert Identification and Grades

Indexable carbide inserts have a standardized system for identification (see Figure 3-20). They also are identified by various grades, which are organized according to their application type and workpiece material as described in Table 3-1.

Carbide inserts are manufactured by various companies, which have their own unique grade specification numbering system. These companies also have developed a variety of other grades to increase performance for specific machining situations. A specific manufacturer's catalog should be used when seeking application information about a specific manufacturer's grade of carbide.

Other Insert Cutting Materials

Indexable inserts can be manufactured from a variety of other materials. Generally, inserts are uncoated, coated, or other. The uncoated inserts are classified into the C1–C8 tungsten carbide group. The coated inserts consist of chemical vapor deposition (CVD) or physical vapor deposition (PVD). The cermet insert contains mostly titanium carbide (TiC) and titanium nitride (TiN) with a metalic binder. Ceramic inserts can be classified into two groups: alumina-base (aluminum oxide) and silicon nitride-base (sialon) compositions. Diamond inserts can also be classified into two groups: *polycrystalline diamond* (PCD) and thin film diamond coatings.

Insert Nose Radius Selection

Selecting the TNR is an important factor because it can affect the tool strength, the surface finish, or perhaps the forming of a fillet on the workpiece. Generally, a 1/32 inch TNR or larger will produce better surface finishes and tool life. A smaller TNR (1/64 inch) is usually best if the setup, tool, or workpiece are not rigid.

■ *Table 3-2 Machining center applications*

MACHINING TYPE	OPERATION DESCRIPTION	TOOL TYPES USED
Milling		
1	Face milling	Face mill cutter, shell mill cutter
2	End milling	End mill cutter
3	Contour milling	End mill cutter
4	Side milling	Side mill cutter
5	Slot milling	Slot mill cutter, End mill cutter
6	Saw milling	Saw cutter
7	T-slot milling	T-slot cutter
8	Dovetail milling	Dovetail cutter
9	Woodruff key milling	Woodruff cutter
10	Keyway milling	End mill cutter
11	Chamfer milling	Chamfer mill cutter
12	Thread milling	Thread mill cutter
13	Cavity mold milling	Ballnose end mill cutter, end mill cutter
14	Pocket milling	End mill cutter, ballnose end mill cutter
15	Groove milling	Slot mill cutter, T-slot cutter
Drilling		
1	Spot drilling	Spot drill, center drill
2	Drilling	Twist drill, insert drill, spade drill, step drill, core drill, coolant-fed drill
3	Deep-hole drilling	Extra-long drill, gun drill
Chamfering		
1	Chamfering	Chamfer tool, spot drill
Spotfacing		
1	Spotfacing	Spotface tool, end mill
2	Back spotfacing	Back spotfacer
Countersink		
1	Countersink	Countersink tool
Reaming		
1	Reaming	Reamer, taper reamer
Boring		
1	Boring	Boring bar
2	Counterboring	Boring bar, end mill
Tapping		
1	Tapping	Tap, taper pipe tap

Insert Geometry Selection

The round insert geometry has the best strength factor, but it is limited to applications where a large radius does not affect the workpiece configuration. The 80° diamond insert geometry, which has the best versatility and strength combination factor, has become very popular geometry for CNC machining. Generally, the first consideration when selecting a geometry is the application. Other items include the type of tool, the workpiece material, the depth of cut, the finish requirement, the tolerances, the rigidity of *part* setup, and the machine. Therefore, the selected geometry must be both capable of machining the workpiece and satisfying all other factors.

CNC MACHINING CENTER APPLICATIONS

The CNC machining center is capable of performing an endless variety of operations, which can range from milling to drilling. Table 3-2 describes some of the types of operations, applications, and tool cutters.

CHAPTER SUMMARY

The following key concepts presented in this chapter include the following:

- The names and descriptions of operations performed on CNC machining centers.

- The names and descriptions of machined holes.

- The names and descriptions of toolholders used on CNC machining centers.

- The names and descriptions of cutting tools used on CNC machining centers.

- Carbide-indexable insert technology.

- Tool cutting speed (SFM) selection.

- Tool cutter speed (RPM) calculations.

- Tool cutting feedrate (IPR) selection.

- Tool cutting feedrate (IPM) calculations.

- Proper cutting tool selection.

REVIEW QUESTIONS

1. List the types of operations performed on CNC machining centers.

2. What is the function of the CNC spot drill?

3. List the common types of CNC toolholders.

4. List ten common types of drillings tools.

5. List six common types of tapping tools.

6. List the two main types of milling tools.

7. What materials are cutting tools made from?

8. List the two features used to measure tool length.

9. When is it necessary to ream a hole?

10. When is it necessary to use a boring bar to finish a hole?

11. Which tap is designed for tapping soft materials such as aluminum?

12. When is climb milling recommended?

13. When is conventional milling recommended?

14. List the advantages of using coolant during a machining operation.

15. Identify the insert shape, type, IC size, thickness, and TNR for DNMG-543.

16. What effect does the insert TNR have on machining operations?

chapter 4

CNC MACHINING CENTER CONTROL AND OPERATION

Chapter Objectives

After studying and completing this chapter, the student should have knowledge of the following:

- *CNC machining center CRT and keypad panel functions*

- *CNC machining center operation panel functions*

- *CNC machining center workholding methods*

- *CNC machining center setup procedures*

This chapter identifies the CNC machine control features and operation functions. This includes descriptions and illustrations of the CRT and keypad panel, the operation control panel, CNC workholding, and CNC machining center setup procedure.

CNC MACHINING CENTER CRT AND KEYPAD PANEL FUNCTIONS

The CRT displays the CNC codes and operational information that are active or stored in its memory (see Figures 4-1 and 4-2). The CNC codes and information are executed by a CNC program, defaulted by the control *memory*, or input by the CNC operator. The latter, which is called manual *data input* (MDI) is performed with the keypad and Cycle Start button. Understanding the code functions, machine operations, and controls is an essential part of safely and efficiently operating a CNC machine tool.

The most common items on the CRT and keypad panel include the on and off power for the control, the control *reset*, which clears active CNC operations, and the menu keys, which are used to access various CRT screens,

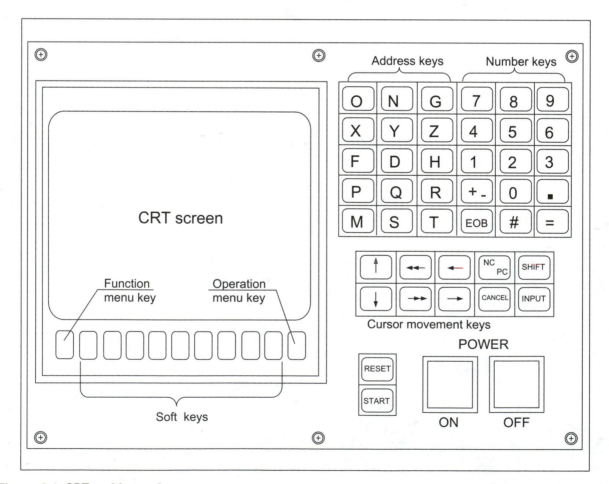

Figure 4-1 **CRT and keypad**

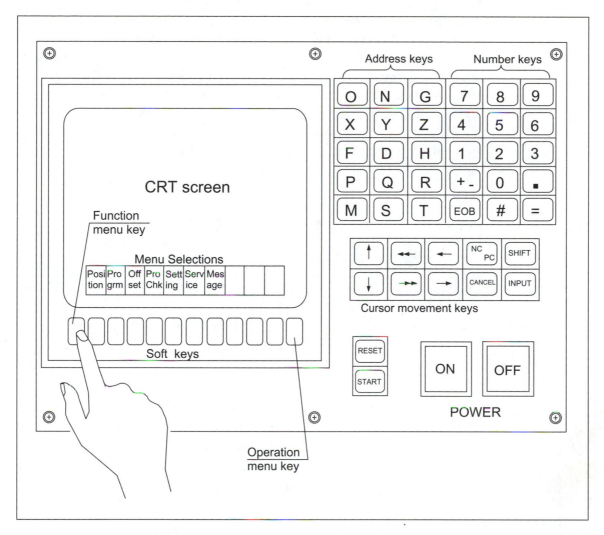

Figure 4-2 **CRT and keypad functions**

such as the *offset* screens, the CNC program directory, the axis *position* screens, and the alarm message screens. The keypad functions include entering or deleting data such as CNC programs, offset values, tool data, MDI commands, and *edit* commands.

CRT and Keypad Descriptions

Figure 4-2 illustrates some common CRT and keypad functions that a CNC operator must identify and understand to safely and effectively operate any CNC machine tool.

When the Function Menu key is pressed (Figure 4-2), a menu selection screen appears on the CRT. The menu items typically displayed on a CRT include

- Position screen
- Offset screen
- Setting screen
- Message screen
- Program screen
- Program check
- Service screen

The menu list may also include other items depending on the machine and control. After selecting and pressing a menu item button, the CRT will display the selected screen. Additionally, each menu selection may have more than one screen to display. The operation menu key enables the operator to select and view other screen options. The CRT and keypad are typically designed with the main feature described. However, to fully understand the CRT and keypad functions, the CNC operator should study the machine manual and receive training before operating the CNC machine.

■ ■ ▪ ▪

CNC MACHINING CENTER OPERATION PANEL FUNCTIONS

The CNC operator must be capable of performing a variety of functions in order to efficiently produce parts on CNC machines. Table 4-1 describes the functions that are mainly performed through the operation control panel that is illustrated in Figure 4-3. Other control features are illustrated in Figures 4-4 to 4-11.

CNC Machining Center Operation Panel

The operation panel example shown in Figure 4-3 illustrates some common features required to operate a CNC machining center.

CNC Machining Center Control Panel Buttons

Figures 4-4 and 4-5 are typical CNC functions that are required knowledge for a CNC machine operator.

When the Emergency Stop button is pushed, all machine functions will stop operating. Usually, the button must be released by turning it clockwise. Therefore, when this button is pressed, the following sequence will occur:

1. The current to the motor is interrupted.

2. The control unit assumes a reset state.

3. The fault causes must be corrected before the button is released.

4. After the button is released, each axis must be homed by the operator.

When the Cycle Start button is pushed, automatic operation starts and the Cycle Start LED (light-emitting diode) is lit.

In the following cases listed below, the Cycle Start button is ignored.

1. When the Feed Hold button is pushed.

2. When the Emergency Stop button is pushed.

3. When the Reset signal is turned ON.

4. When the Mode Select switch is set to a wrong position.

■___*Table 4-1* **Control functions**

CONTROL ITEM	FUNCTION DESCRIPTION
Alarm lamps	When any of these lamps are turned on, the CNC control will not allow the machine to operate until the problem is corrected.
Load meter	This meter indicates the amount of horsepower in percentage load that is being generated by the machining operation.
Spindle Speed meter	This meter indicates the actual RPM of the spindle.
Coolant switch	This switch operates the coolant flow (on or off).
Tool magazine CW/CCW[a]	This switch rotates the tool magazine into position so that the operator can load and unload tools.
Machine lock	This switch locks the machine axis during operation.
Tool Clamp/Unclamp	This push button allows the operator to manually clamp and unclamp tools at the spindle.
Home Position button	When this push button is pressed, the selected axis (table or spindle) will rapid traverse to the Home position limit switch.
Home Position lamps: X Y Z B	When the selected axis (table or spindle) trips the Home position limit switch and completes the move, the corresponding lamp will turn on.
Spindle Orient button	When this push button is pressed, the spindle will rotate to the radial position that aligns the drive keys.
Spindle Jog button	When this push button is pressed, the spindle will rotate at the RPM value stored in memory until the button is released.
Spindle On button	When this push button is pressed, the spindle rotates at the RPM value stored in memory.
Spindle CW switch	When this switch is selected, the spindle will rotate in a clockwise direction, also called forward.
Spindle CCW switch	When this switch is selected, the spindle will rotate in a counterclockwise direction, also called reverse.
Spindle Stop button	When this push button is pressed, the spindle will stop rotating.
Manual Jog + button	When this push button is pressed, the selected axis (table or spindle) will move in a positive direction until the button is released.
Manual Jog − button	When this push button is pressed, the selected axis (table or spindle) will move in a negative direction until the button is released.
Memory Protect	When this switch is in the ON position, the CNC program cannot be altered or deleted.

[a]CW: clockwise; CCW: counterclock.

5. When a sequence number is being searched.

6. When an alarm has occurred.

Mode Selections

The CNC machine is capable of performing various functions, which the CNC operator is required to identify and understand. The Mode Select switch

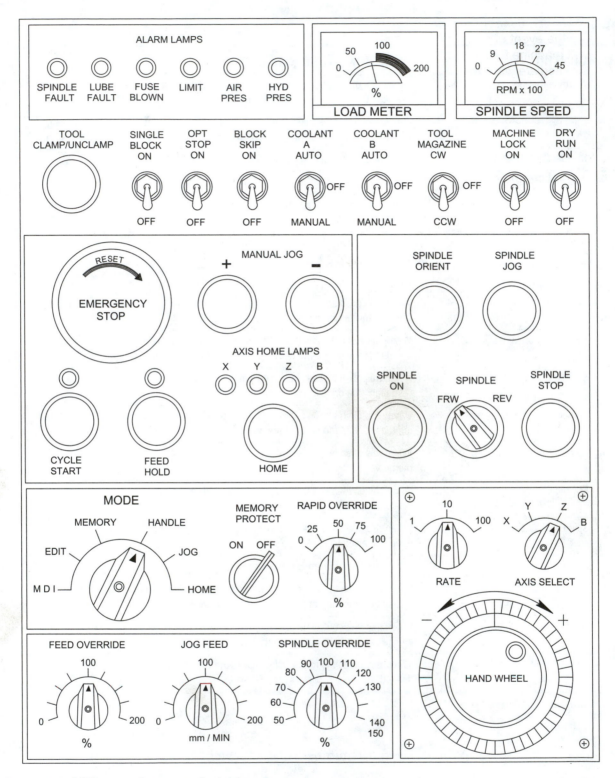

Figure 4-3 **CNC operation control panel**

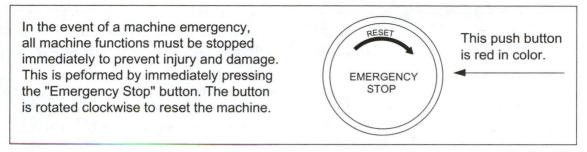

In the event of a machine emergency, all machine functions must be stopped immediately to prevent injury and damage. This is peformed by immediately pressing the "Emergency Stop" button. The button is rotated clockwise to reset the machine.

This push button is red in color.

Figure 4-4 **Emergency Stop button**

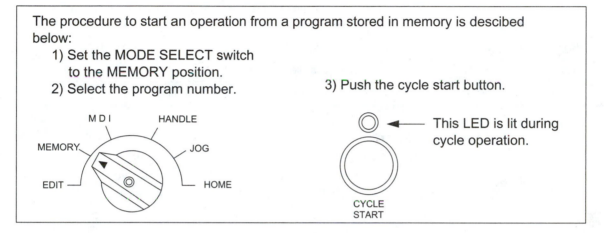

The procedure to start an operation from a program stored in memory is descibed below:

1) Set the MODE SELECT switch to the MEMORY position.
2) Select the program number.

3) Push the cycle start button.

This LED is lit during cycle operation.

Figure 4-5 **Cycle Start button**

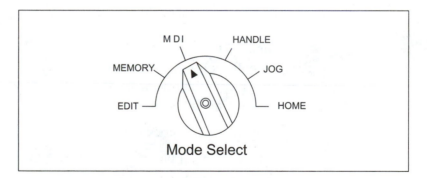

Mode Select

Figure 4-6 **Mode Select switch**

(Figure 4-6) is used to specify the type of function that the control and the machine will perform. Figure 4-6 and Table 4-2 explain the actions typically performed with each mode selection.

Feed Hold, Dry Run, Single Block, and Optional Block Skip

The Feed Hold, *Dry Run*, Single Block, and Block Skip are some other typical operation functions that the CNC operator must identify and fully understand. The control operations described are actions that are usually performed by the CNC operator during the CNC *proveout* phase.

■___*Table 4-2* ***Mode selection descriptions***

MODE SELECTION	FUNCTIONS AND OPERATIONS THAT CAN BE PERFORMED
Edit	(a) Store CNC programs to memory. (b) Modify, add, and delete CNC programs. (c) Retrieve a CNC program from memory.
Memory	(a) A CNC program stored in memory can be executed in auto or single block cycle. (b) Sequence N block number search. (c) CNC program number search.
MDI	Manual data input can be performed via the keypad. The operator enters CNC codes and coordinates, and then executes them using the Cycle Start button.
Handle	Handle feed can be executed.
Jog	Jog feed can be executed.
Home	All machine axes can be reset to their home position.

Feed Hold Button

During automatic operation or any cycle command, when this button is pressed the feed or rapid command moves are stopped. Note that spindle rotation and other miscellaneous functions will remain on.

Dry Run

If this switch is set to ON in the *Cycle* Operation of *Tape*, Memory, or MDI, an "F" function specified in the CNC program is ignored, and the machine tool axes are moved at a faster rate of speed.

Single Block

When Single Block is on, the control will only execute one block of active CNC program information each time the Cycle Start button is pressed. This procedure is described below:

1. Turn on the Single Block switch.

2. Press the Cycle Start button, and a single block of information is executed.

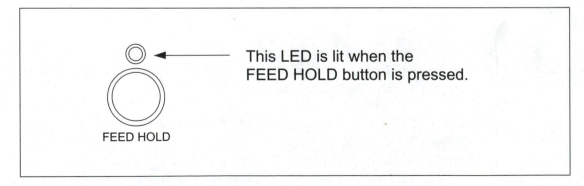

Figure 4-7 *Feed Hold description*

3. When the Cycle Start button is pressed again, another block of the CNC program is read and executed. The cycle repeats until all blocks are executed.

Optional Block Skip

This is the function that allows the control to skip a block or blocks of information in which a slash is programmed before the "N," as the first character in the block. The block would appear as /N005 followed by the codes or coordinates.

Feedrate, Spindle Speed, and Rapid Traverse Overrides

The *Feedrate Override*, Spindle Speed Override, and Rapid Traverse Override dials enable the CNC operator to decrease or increase the programmed values. These operation control features are described in Figures 4-8 through 4-10.

Manual Pulse Generator Hand Wheel

The CNC operator, during the process of setups and operating the CNC machine, will be required to manually move each machine axis. The Manual Pulse Generator Hand Wheel, illustrated in Figure 4-11, is the CNC machine control feature that enables the operator to move the machine in any axis (X, Y, Z, or B). The manual hand wheel operation procedure, which allows fine, rapid, or feed movement, is described in steps 1–4 of Figure 4-11.

CNC MACHINING CENTER WORKHOLDING

Proper location and *orientation* of the workpiece, which ensures both accuracy and *repeatability* of the parts produced, is critical to the overall CNC machining process. In this case, accuracy is defined as machining a part's features to the specified size, form, and location. Repeatability is defined as accurately performing the machining operations for all the parts throughout the production run. Therefore, the workholding method that is selected must be capable of ensuring both the accuracy and repeatability of the CNC production process.

Locating Fundamentals

To ensure accuracy and repeatability, the workpiece must first be properly oriented within the workholding device. To accomplish this, the holding device must perform three primary functions: properly hold, support, and locate the workpiece throughout the CNC machining cycle.

Properly holding the part, a clamp function requires that care be taken to apply sufficient clamping force without distorting the workpiece. The

Feedrate Override Dial

This dial allows the operator to reduce or increase the programmed F-code feedrate. The percent increments are usually 10% but vary depending on the type of control.

This function is active in the following cases:
1) Feed modes by G01, G02, and G03
2) Feed mode during canned cycle execution

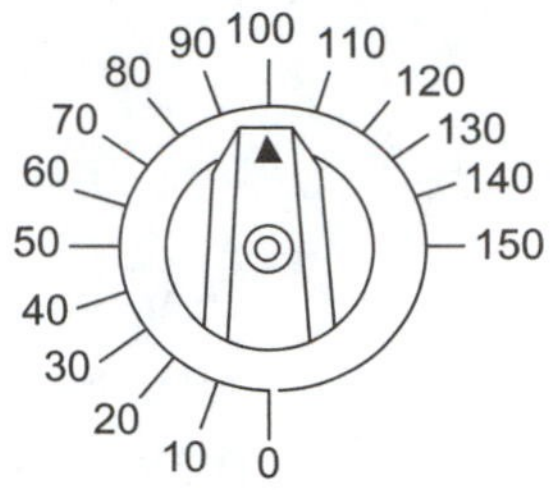

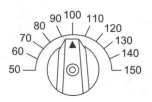

Feed Override

Figure 4-8 Feedrate Override control knob description

Spindle Speed Override Dial

This dial allows the operator to reduce or increase the programmed S-code spindle speed. The percent increments are usually 10% but vary depending on the type of control.

This function is active in the following cases:
1) Spindle is ON by S-code and M03/M04
2) Spindle is ON by MDI mode

RPM Override

Figure 4-9 Spindle Speed Override control knob description

Rapid Traverse Override Dial

The rapid override dial, of 100%, 50%, 25%, and F0 can be provided with the machine operator's panel. When the feedrate is 10 m/min and the switch is set to the position of 50%, the actual feedrate becomes 5 m/min.

This function is active in the following cases:
1) Rapid traverse by G00, G27, G28, G29, and G30
2) Rapid traverse during canned cycle execution
3) Manual rapid traverse operation
4) Rapid traverse Home return

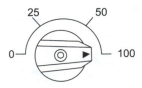

Rapid Override

Figure 4-10 Rapid Traverse Override control knob description

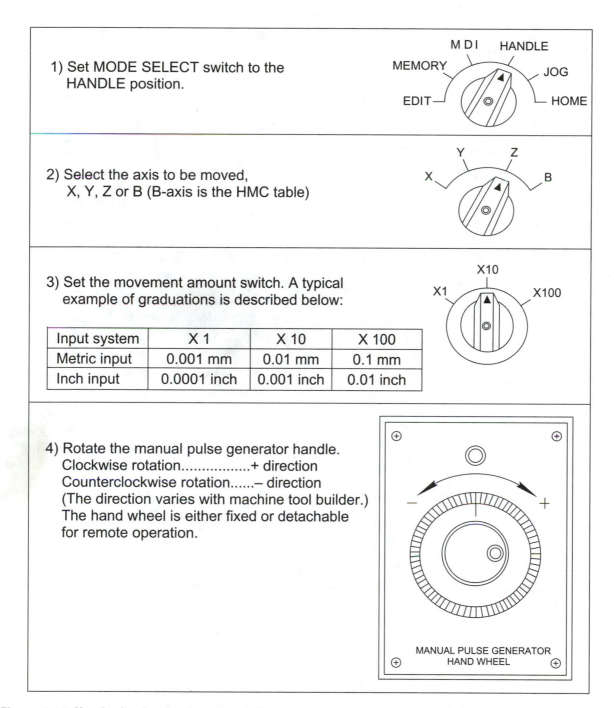

1) Set MODE SELECT switch to the HANDLE position.

2) Select the axis to be moved, X, Y, Z or B (B-axis is the HMC table)

3) Set the movement amount switch. A typical example of graduations is described below:

Input system	X 1	X 10	X 100
Metric input	0.001 mm	0.01 mm	0.1 mm
Inch input	0.0001 inch	0.001 inch	0.01 inch

4) Rotate the manual pulse generator handle.
Clockwise rotation.................+ direction
Counterclockwise rotation......– direction
(The direction varies with machine tool builder.)
The hand wheel is either fixed or detachable
for remote operation.

MANUAL PULSE GENERATOR
HAND WHEEL

Figure 4-11 **Hand wheel operation description**

locators act as positive stops for the workpiece and prevent any movement. Ideally, the workholding device should be designed so that the workpiece is held in place by the locators with clamps directly in line or over the locators.

Every workpiece has three axes that must be considered when designing the locating system. As Figure 4-12 illustrates, these are typically identified as the X-, Y-, and Z-axes. These workpiece surfaces are referred to as datum planes or datum surfaces. They are defined as the primary datum plane, the

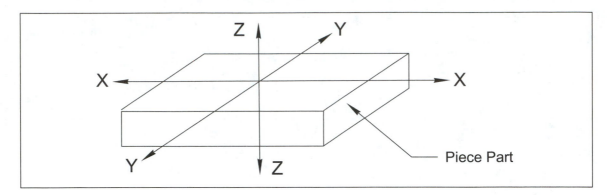

Figure 4-12 ***Cartesian coordinate system***

secondary datum plane, and the tertiary datum plane (see Figure 4-13). The primary datum surface is located on three points. These three locators act as supports (see Figure 4-14). The secondary and tertiary datum planes position the part. The secondary datum plane is positioned with two locators, and the tertiary datum plane is positioned with a single *locator*. In most cases, these datum points are not specified.

Generally, on workpiece drawings, the intended application or function of the part determines the surfaces used for the datum planes. When *geometric dimensioning and tolerancing* (GD&T) is used to specify the part sizes, the datum surfaces are specified directly on the workpiece drawing. In this case, the workholding locators must be on the same surfaces as the part drawing datum system specifies (see Figures 4-13 and 4-14).

CNC Fixture

When a CNC fixture is designed, the positions of the locators are specified on the tool drawing. The locators must be dimensioned the same as the part

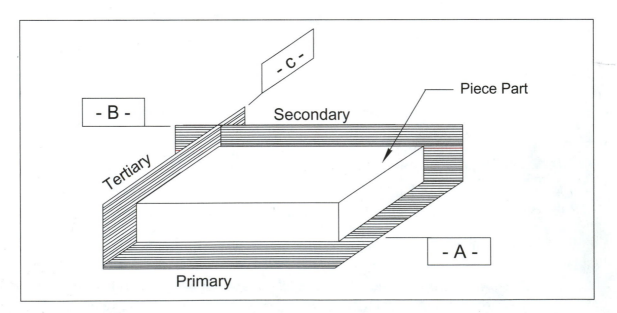

Figure 4-13 ***Datum planes***

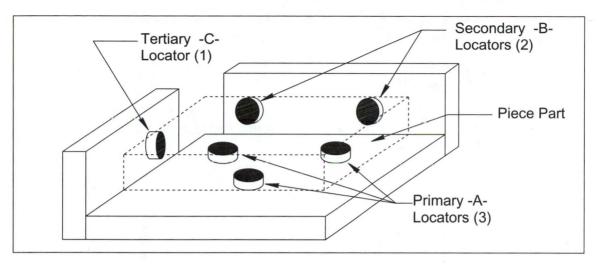

Figure 4-14 Fixture locators

drawing datum system. If necessary, datum targets are used in conjunction with dimensions to precisely position the datum points. Likewise, depending on how these datum targets are represented, they can also show the exact shape or extent of the contact.

Methods of Holding the Part during Machining

The programmer decides how a part is to be held during machining operations and provides documentation (Setup Plan) to the setup person.

For simple shape machining, a setup vise is adequate. If the part requires some peripheral machining, it could be positioned on a plate with three edge locators and held in place with clamps. If the part is complex, the programmer may need to design a fixture. All necessary components such as *support* devices, stops, and clamps, and their locations, must be documented. The programmer must avoid these obstacles when programming tool movements.

Workholding Devices

When a workpiece is setup it must be securely fastened and the setup must be rigid enough to withstand the forces created during the machining operation. The machine operator should be sure that all workholding devices are free from chips and burrs. The workholding devices, specified by the programmer, must be located in the proper position on the machine table. Failure to follow setup instructions may result in operator injury, damage to the machine, or scrapped workpieces. Some common workholding devices include the following:

- *Precision vises* may be manual, air, or hydraulic powered, and are keyed directly to the table slots, which make positioning and clamping of parts fast and accurate (see Figure 4-15).

- **Clamps and studs** are used to fasten parts to the machine table, *angle plate*, *V-blocks*, parallels, or fixtures (see Figure 4-16).

- **Parallels** are used to raise and support a part when held in a vise or clamped to the machine table (see Figure 4-16).

- **3-Jaw chucks** are equipped with hardened steel *jaws* to grip round workpieces, and make positioning and clamping fast and accurate (see Figure 4-17).

- **V-blocks,** which can be used singularly or in pairs, are also used to locate and support round workpieces (see Figure 4-18).

- **Angle plates** are L-shaped devices that are finish machined to a 90° angle (keyed directly to the table slots), and make positioning and clamping of various workpiece setups flexible and accurate (see Figure 4-19).

- **Subplates** are flat plates that are fitted to the machine table to provide quick and accurate location of workpieces, workholding devices, or fixtures (see Figure 4-19).

- **Support jacks** are used to support the workpiece to prevent distortion of the workpiece during clamping or machining. There are various styles available throughout the industry including standard or special designs.

Fixtures

Some workpiece designs are complex, and therefore cannot be located and clamped with standard workholding methods. In these situations, a fixture is required to machine the workpiece (see Figure 4-21). A fixture can be fabricated in a variety of ways. Two common types include modular fixtures and special design fixtures. When designing a fixture to hold the workpiece, the following points must be considered:

1. Positive location of the part.

2. Repeatability to ensure quality.

3. Rigidity for optimum *machinability*.

4. Design as simple as possible.

5. Part loading and unloading should be quick and easy.

6. Part distortion should be minimal.

Modular Fixtures

Modular fixturing has had a dramatic effect on fixture design. In addition to reducing the time required to design and build special fixtures, modular components can be used over and over for many different tools. This makes modular fixturing not only economical and cost effective, but also versatile. There are two basic styles, or types, of modular component fixturing systems. They are either the T-slot system type or the dowel-pin system type.

Tombstone Fixtures

Tombstone fixtures (see Figure 4-20), which may be square, hexagonal, or octagonal, are available in vertical or horizontal models for use on machining centers to hold parts for machining. Parts can be loaded on one side, or any of the available sides of the fixture for machining parts in one setup, if possible. They are especially useful in palletized machining where the workpieces are unloaded and loaded while another pallet is in the machine and cutting the workpiece.

Clamping Hints

- Always place the clamping bolts as close to the workpiece as possible.

- Always support the workpiece to avoid distortion from clamping or tool pressure.

- For finished surfaces, place a piece of soft metal ("shim") between the workpiece surface and the clamp to avoid clamping marks.

- Whenever possible, test for sufficient clamping force before machining.

- Select a bolt length that will engage the entire thread length of the *nut*.

- Never torque a nut to the end of the bolt threads.

Pin Locators and Bushings

Pins and bushings are useful and low-cost methods to locate the workpiece. They are available in standard sizes as well as specials. The two most common types of pins are the round and diamond or relieved *pin*.

Whichever workholding method is selected, the CNC setup person and operator must use proper care to ensure that the workholding device is properly setup. The following is a checklist of items that must be observed and completed:

1. The fixture number must match the setup instructions.

2. All fixture and table surfaces must be clean.

3. The fixture must be mounted and oriented according to the setup instructions.

4. The fixture must be aligned to the table and spindle axes.

5. The fixture should be fastened to the table as rigid as possible.

6. The fixture should locate and clamp the workpiece as rigid as possible.

7. The setup person must find and verify the X-, Y-, and Z-origins.

8. The fixture and workpiece must be checked for interferences.

Workholding Methods

The following setup descriptions (Figures 4-15 to 4-21) illustrate some of the more popular setups used on CNC machining centers. They include precision vise, manual clamping, 3-jaw chuck, V-blocks and clamps, right-angle plate, and the tombstone fixtures.

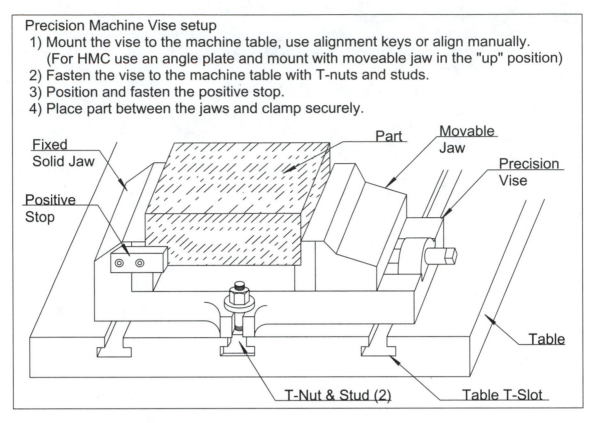

Precision Machine Vise setup
1) Mount the vise to the machine table, use alignment keys or align manually.
   (For HMC use an angle plate and mount with moveable jaw in the "up" position)
2) Fasten the vise to the machine table with T-nuts and studs.
3) Position and fasten the positive stop.
4) Place part between the jaws and clamp securely.

Fixed
Solid Jaw

Positive
Stop

Part

Movable
Jaw

Precision
Vise

Table

T-Nut & Stud (2)

Table T-Slot

Figure 4-15 **Precision vise setup**

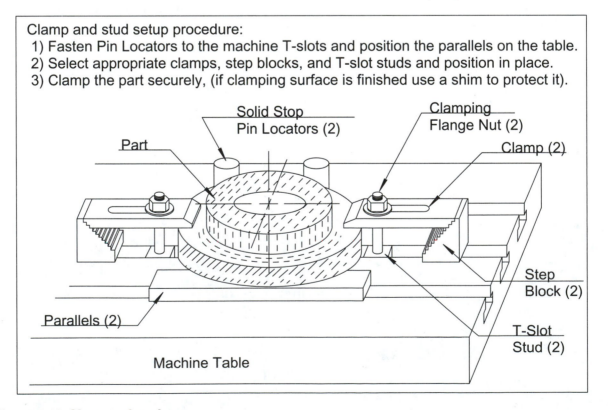

Clamp and stud setup procedure:
1) Fasten Pin Locators to the machine T-slots and position the parallels on the table.
2) Select appropriate clamps, step blocks, and T-slot studs and position in place.
3) Clamp the part securely, (if clamping surface is finished use a shim to protect it).

Solid Stop
Pin Locators (2)

Clamping
Flange Nut (2)

Part

Clamp (2)

Step
Block (2)

Parallels (2)

T-Slot
Stud (2)

Machine Table

Figure 4-16 **Clamp and stud setup**

3-Jaw chuck setup procedure:
 1) Mount the chuck to the machine table with T-Studs (For HMC use an angle plate).
 2) Adjust the jaws to the part diameter, place part between the jaws and clamp
    securely.

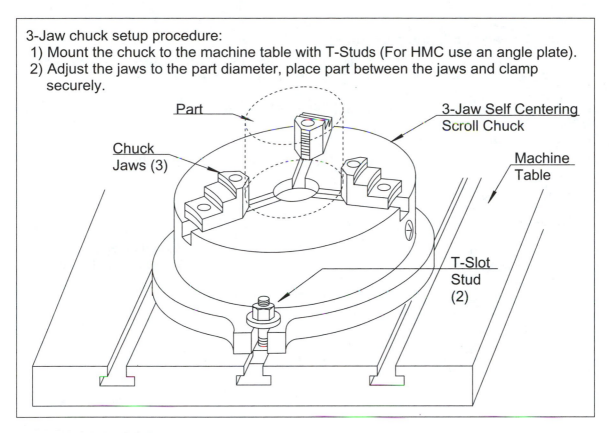

Part

3-Jaw Self Centering
Scroll Chuck

Chuck
Jaws (3)

Machine
Table

T-Slot
Stud
(2)

Figure 4-17 3-Jaw chuck setup

V-block setup procedure:
 1) Align and Clamp the V-Blocks
    to the table.
 2) Place the part on the V-Blocks
    and clamp securely.

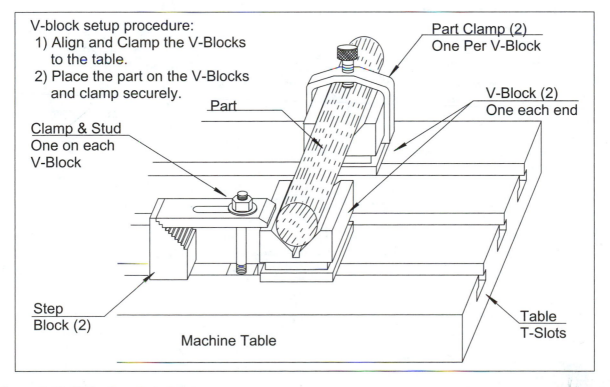

Part Clamp (2)
One Per V-Block

Part

V-Block (2)
One each end

Clamp & Stud
One on each
V-Block

Step
Block (2)

Machine Table

Table
T-Slots

Figure 4-18 V-block and stud setup

Right Angle Plate Setup procedure:

1) Align and Fasten the Right Angle Plate to the table or Subplate (VMC or HMC).
2) Mount and align the workholding device* to the Right Angle Plate.

* (Vise, Chuck, V-Blocks, Fixture, or Modular Fixturing)

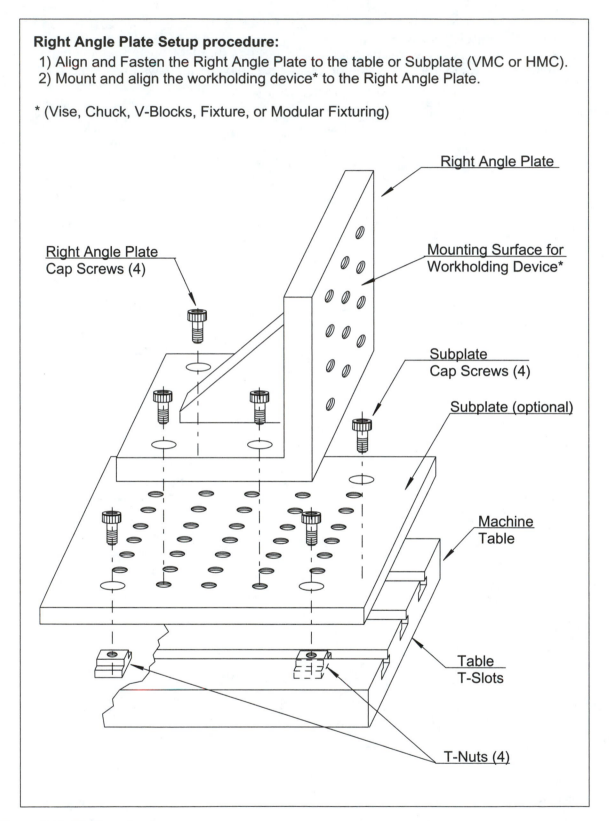

Right Angle Plate

Right Angle Plate
Cap Screws (4)

Mounting Surface for
Workholding Device*

Subplate
Cap Screws (4)

Subplate (optional)

Machine
Table

Table
T-Slots

T-Nuts (4)

Figure 4-19 Right-angle plate setup

Tombstone Setup procedure:

1) Align and Fasten the Tombstone Fixture to the HMC table.
2) Mount and align the workholding device* to the Tombstone Surface.
   * (Vise, Chuck, V-Blocks, Fixture, or Modular Fixturing)
3) Workholding devices can be mounted on all (4) surfaces.

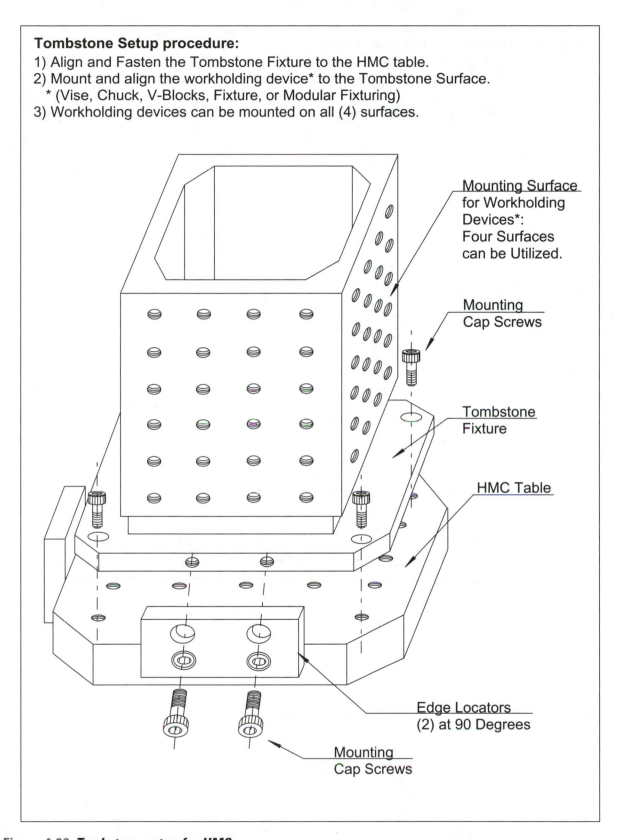

Mounting Surface for Workholding Devices*: Four Surfaces can be Utilized.

Mounting Cap Screws

Tombstone Fixture

HMC Table

Edge Locators (2) at 90 Degrees

Mounting Cap Screws

Figure 4-20 Tombstone setup for HMC

Tombstone Fixture Setup procedure:

1) Align and fasten the Tombstone Fixture to the HMC table.
2) Mount and align the workholding device* to the Tombstone surface.
   * (Vise, Chuck, V-Blocks, Fixture, or Modular Fixturing)
3) Workholding devices can be mounted on all (4) surfaces to improve
   productivity.

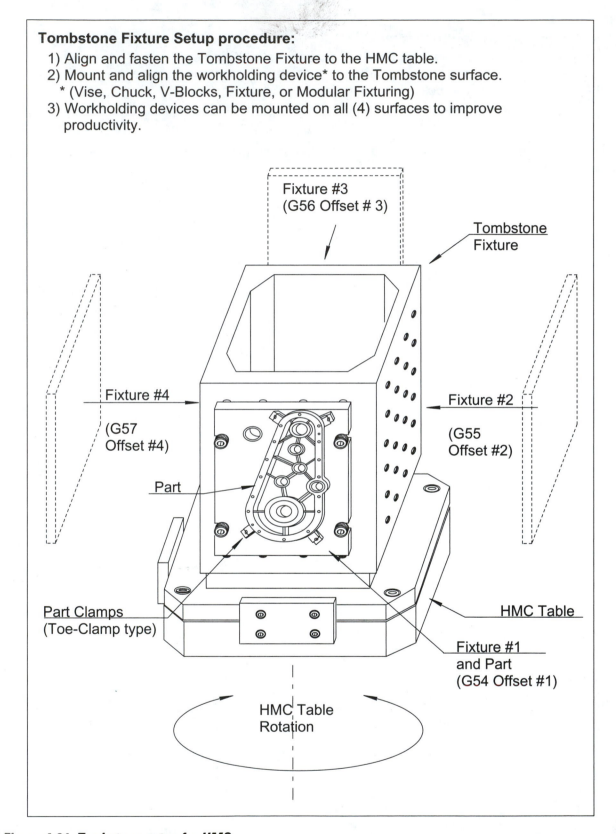

Fixture #3
(G56 Offset # 3)

Tombstone
Fixture

Fixture #4

(G57
Offset #4)

Fixture #2

(G55
Offset #2)

Part

Part Clamps
(Toe-Clamp type)

HMC Table

Fixture #1
and Part
(G54 Offset #1)

HMC Table
Rotation

Figure 4-21 **Tombstone setup for HMC**

■ ■ ■ ■

CNC MACHINING CENTER SETUP PROCEDURE

The setup operation can begin after the workholding device, tooling, CNC program, tool list, setup sheet, and workpiece materials arrive at the CNC machine. Typically, the setup person starts by securing the cutting tools in the toolholders. Then the tooling assemblies are loaded into the tool magazine according to the order outlined in the tool list. Next, the workholding device is put in place on the machine table. It may require clamping and some minor aligning. The workpiece is loaded into the workholding device as specified by the CNC programmer. After loading the CNC program into the MCU, the setup person then finds and enters the location of the part origin (part X0, Y0, Z0), with respect to machine home as indicated on the setup sheet.

The machine home position or machine zero is a location set initially by the machine manufacturer. The CNC machine is homed at the start of the part program setup. When homed, the machine fully retracts the spindle (machine Z0) and positions to its maximum (machine X0, Y0) table preset position.

As previously discussed, the part origin is the zero location of the part X-Y-Z coordinate system. In absolute coordinate programming all tool movements are taken with respect to this origin and not the machine zero. The point from which the dimensions are taken on the part print is usually considered the same as the part origin. This could be the center of a hole, the edge of the part, or a certain predetermined distance from the edge.

The setup X-Y-Z locations of the part origin are from the machine home. One method of setting the part origins is accomplished by placing an edge finder or tool with a known diameter in the spindle and using the MCU's jog buttons to move it to the edge (see Figures 4-23 and 4-24). The X-Y-Z position values are displayed on the CRT screen. There are two main methods of programming this location into the MCU. The first is called the "*fixture* offset method" (G54–G59). This involves using the MCU keypad to enter the X-Y-Z values directly into the MCU unit. The MCU's fixture offset *storage* area is opened and the values entered manually by the operator. Later, when the part program is run, a fixture offset code (G54–G59) in the program directs the MCU to use the X-Y-Z values previously entered in its offset storage area.

The second method requires the setup person to position the machine spindle at X-Y zero and preset the control to read X0.00 and Y0.00 prior to running it. Most CNC shops use these methods; however, some CNC shops allow setup personnel to edit part programs. The preset method and the fixture offset method are safer and more efficient.

Machine Home Position

When a CNC machine is initially turned on, it does not register an accurate position for each axis. The machine needs to be "homed." Homing the machine moves each axis to a known position where each axis is initialized.

Setting the Part Origin

Parts designated for manufacturing on CNC machine tools must first be programmed for the specific CNC machine best suited for machining the part. However, the CNC program is made independent of the CNC machine's coordinate systems. The CNC programmer selects an appropriate location from the part engineering drawing. Most modern engineering drawings specify the origin locations that must be used to program, machine, and inspect a part. This is done with a datum system called Geometric Dimensioning and Tolerancing (GD&T). This location is marked and documented as the X-, Y-, and Z-zeros or part origins. These locations must be setup in a fixed and solid manner so part repeatability is assured. The X-Y origin for square- or rectangular-shaped parts is typically a corner that is accessible and easy for the operator to load and clamp.

The X-Y origin for round-shaped parts is usually the center. The CNC operator homes the machine and then sets the part origin based on the CNC documentation. There are three ways a zero origin point can be set on CNC machines: manually by the operator, by a programmed absolute zero shift (G92), or by using work coordinates (G54).

Manual Setting

When manually setting the part origin, the setup person positions the spindle directly over the specified part zero and zeros out the coordinate systems on the MCU console. Some common methods are described and illustrated in Figures 4-22 through 4-26.

Absolute Zero Shift

The G92 code is used to shift absolute zero. The absolute zero shift is executed by the CNC program. It can be executed as many times as necessary; however, the original zero must be reinstated before ending the CNC program.

Work Coordinates

A work coordinate, sometimes called a "fixture offset," is a modification of the absolute zero shift. This shift is done by manually entering an X-Y-Z value into the control as illustrated in Figure 4-2. Work coordinates are then called when the CNC program reads the appropriate G codes (G54–G59).

Tool Length Offsets

The setup person must next measure and enter the values of the *tool length offsets* for each tool. The different tools used for the machining operations in a CNC program will usually vary in length.

The machine control must be directed to compensate for these variations when moving a tool in the Z-axis direction. It can only do so if it knows the initial distance between the tip of the tool and the part Z0. The setup person sets the spindle to full retract or machine Z0. A tool is loaded, and the distance between the tip of the tool and the part Z0 is measured and recorded. Each tool must be similarly loaded, measured, and then recorded. The values are then entered and stored in the control tool length offset screen as illustrated in Figure 4-2 and under the specified number assigned by the programmer.

In some cases, a tool can have a few tool length offsets. Upon running the CNC program, the control needs to read the *H code* in order to compensate the tool length offset for each tool it places in the spindle.

Edge Finder Setup Method

The CNC process relies on the speed and accuracy of both the machine setup and the machine cycle time to produce parts efficiently. Setting the X-, Y-, and Z-axis origins is part of the machine setup process.

There are various methods that can be used to find these origin locations. Some setup methods for finding the edge locations of square or rectangular parts, include the pin edge finder (Figures 4-22 to 4-24) and the electronic edge finder. In some cases, CNC machines equipped with a probe system can use an electronic probe.

For finding the X-Y origin center of circular features such as bores or turn diameters, other methods may include the coaxial indicator (Figure 4-25) or an electronic probe.

The Z-axis origin registers the distance between the "spindle face" and the "part Z-zero surface." Therefore, a different method must be used to set the Z-axis. The common methods for this include the gage block and spindle touch-off method (Figure 4-26), the tool touch-off method, or a probe. In Figures 4-22 through 4-25, examples illustrate and describe the basic steps for setting the part origins.

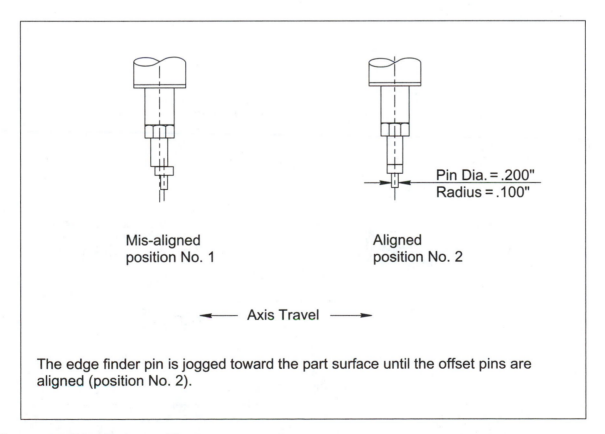

Pin Dia. = .200"
Radius = .100"

Mis-aligned
position No. 1

Aligned
position No. 2

← Axis Travel →

The edge finder pin is jogged toward the part surface until the offset pins are aligned (position No. 2).

*Figure 4-22 **Edge finder positions***

Figure 4-22 describes the two edge finder positions that the CNC operator must identify and understand.

The manual procedure for setting the X-zero origin with an edge finder is described in Figure 4-23. Note: this method is used for straight-square part surfaces.

X-axis Edge Finder Operation Procedure:

1) Clamp toolholder with edge finder into the spindle.
2) Select Jog Mode and jog the edge finder pin near the X-zero part surface.
3) Switch spindle on (slow range 200 RPM).
4) Select the fine jog increment (.001).
5) Jog the edge finder toward the part surface until both diameters are aligned.
6) Select the Z-axis and jog the edge finder above the part.
7) Select the X-axis and jog the spindle .100" toward the part edge.
   (The spindle is at the X-zero position)
8) On the CRT, select the Preset Menu.
9) Using the keypad, enter X0.0; then press the preset key on the CRT.
   (The position screen should read X0.000)

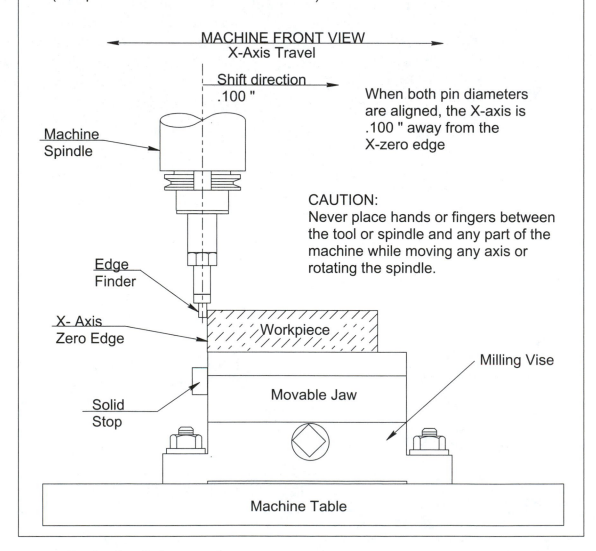

Figure 4-23 **X-axis edge finder operation**

The manual procedure for setting the Y-zero origin with an edge finder is described in Figure 4-24. Note: this method is used for straight-square part surfaces.

Y-axis Edge Finder Operation Procedure:

1) Clamp toolholder with edge finder into the spindle.
2) Select Jog Mode and jog the edge finder pin near the Y-zero part surface.
3) Switch spindle on (slow range 200 RPM).
4) Select the fine jog increment (.001).
5) Jog the edge finder toward the part surface until both diameters are aligned.
6) Select the Z-axis and jog the edge finder above the part.
7) Select the Y-axis and jog the spindle .100" toward the part edge.
   (The spindle is at the Y-zero position)
8) On the CRT, select the Preset Menu.
9) Using the keypad, enter Y0.0; then press the preset key on the CRT.
   (The position screen should read Y0.000)

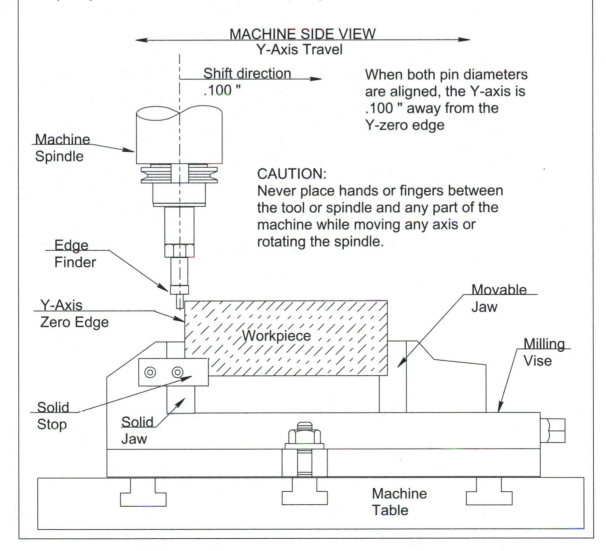

MACHINE SIDE VIEW
Y-Axis Travel

Shift direction
.100"

When both pin diameters are aligned, the Y-axis is .100" away from the Y-zero edge

Machine Spindle

CAUTION:
Never place hands or fingers between the tool or spindle and any part of the machine while moving any axis or rotating the spindle.

Edge Finder

Movable Jaw

Y-Axis Zero Edge

Workpiece

Milling Vise

Solid Stop

Solid Jaw

Machine Table

*Figure 4-24 **Y-axis edge finder operation***

Coaxial Indicator Setup Method

A common method for manually setting the X-Y zero origin of a round feature with a coaxial indicator is described in Figure 4-25.

Coaxial Indicator Setup Procedure:

1) Clamp toolholder with coaxial indicator into the spindle.
2) Select Jog Mode and position the coaxial indicator to the origin location then adjust the centering feeler to contact the bore or turn.
3) Hold the restraint rod, then switch spindle on (slow range 200 RPM).
4) Select the fine jog increment (.001).
5) Jog the X- and/or Y-axis until the indicator needle is stationary.
6) Select the Z-axis and jog the coaxial indicator above the part.
   (The spindle is at X- and Y-zero position)
7) On the CRT, select the Preset Menu.
8) Using the keypad, enter Y0.0; then press the preset key on the CRT.
9) Repeat step 8 for the X-axis.
   (The position screen should read: X0.000 Y0.000)

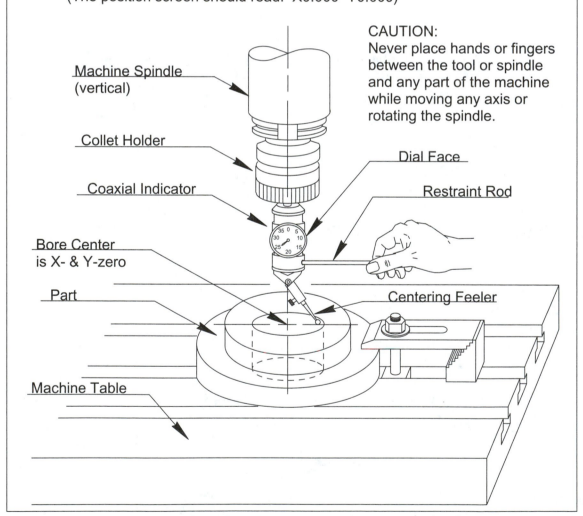

Machine Spindle (vertical)

CAUTION:
Never place hands or fingers between the tool or spindle and any part of the machine while moving any axis or rotating the spindle.

Collet Holder

Dial Face

Coaxial Indicator

Restraint Rod

Bore Center is X- & Y-zero

Part

Centering Feeler

Machine Table

Figure 4-25 Coaxial indicator setup

Gage Block Touch-Off Setup Method

The manual procedure for setting the Z-zero with a *gage block* touch-off to the machine spindle is described in Figure 4-26.

Z-axis Gage Block and spindle touch-off setting procedure:

1) Select a gage block (4.000" for example)
2) Select Jog Mode and Z-axis.
3) Jog spindle to appropriate Z-position over the part surface
   and less than the gage block (approximately 3.500").
4) Place gage block on Z-zero surface and on the spindle diameter.
5) Jog spindle Z-axis up while applying slight pressure toward the
   spindle diameter.
6) Stop the spindle jog when the gage block slides under the spindle face.
7) Remove the gage block.
8) Select the position screen on the CRT.
9) Using the keypad enter the gage block size (Z 4.000" for example),
   then press the Preset key on the CRT.
   (The position should read Z 4.0000)

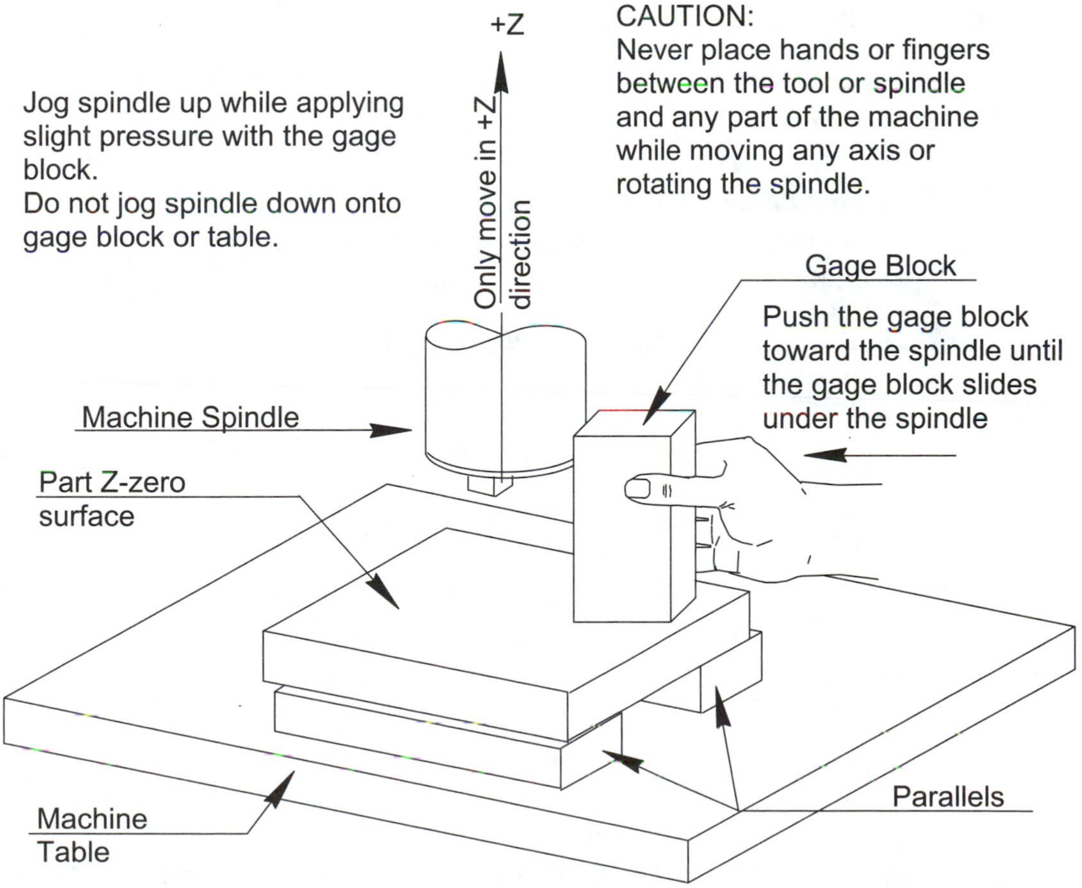

Figure 4-26 **Gage block touch-off setup**

Wiggler Setup Method

The "wiggler" (see Figure 4-27) is another tool used to accurately and easily locate edge surfaces. This tool is used similar to the edge finder described in Figures 4-22 through 4-24.

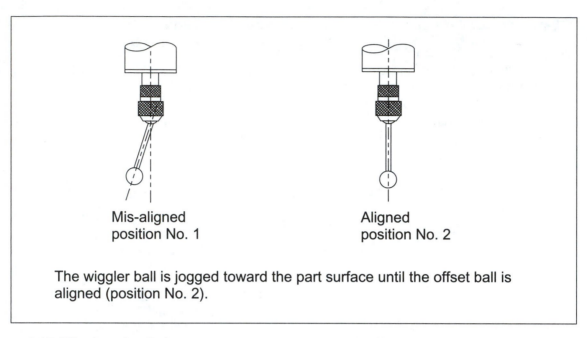

Mis-aligned
position No. 1

Aligned
position No. 2

The wiggler ball is jogged toward the part surface until the offset ball is aligned (position No. 2).

Figure 4-27 Wiggler edge finder

Pointed Contact Setup Method

The illustration in Figure 4-28 describes the pointed contact that can be used to locate center points and scribed lines.

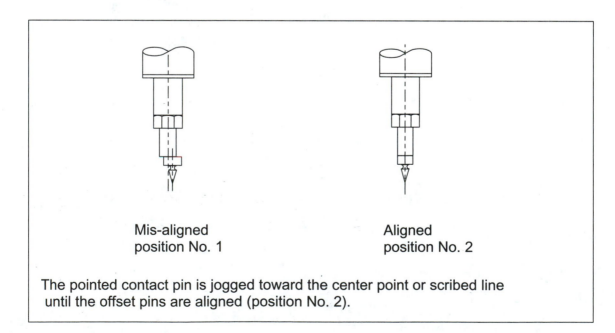

Mis-aligned
position No. 1

Aligned
position No. 2

The pointed contact pin is jogged toward the center point or scribed line until the offset pins are aligned (position No. 2).

Figure 4-28 Pointed contact

■ ■ CHAPTER SUMMARY

The key concepts presented in this chapter include the following:

- The description of the CNC machining center CRT.
- The description of the CNC machining center keypad.
- The names and descriptions of function displays on the CRT.
- The description of the CNC machining center operator panel.
- The names and descriptions of control features on the operator panel.
- The names and descriptions of workholding devices for CNC machining centers.
- The descriptions of setup methods for CNC machining centers.
- The machine home of CNC machining centers.
- The part origin location of CNC machining centers.
- Proper setup guidelines and techniques.

■ ■ REVIEW QUESTIONS

1. List the common items that are displayed on the CNC machining center CRT.
2. Which button when pressed will list the menu items on the CRT?
3. List the functions that can be selected with the mode select knob.
4. What is the function of the manual hand wheel?
5. What happens when the Cycle Start button is pressed?
6. What happens when the Feed Hold button is pressed?
7. How can the operator slow a programmed axis movement?
8. How can the operator slow the programmed spindle speed?
9. When is the Emergency Stop button pressed?
10. How are the machine axes homed?
11. Which control feature indicates that the machine axes are homed?
12. List the common workholding devices and setup methods.
13. List the workholding items that should be completed to perform a proper setup.
14. Explain how the X–Y–Z zero origin setup is performed.
15. Why is it necessary to measure and store the length of each tool?
16. List the items that can be used to find the X-Y surfaces.

chapter 5

CNC MACHINING CENTER TECHNICAL DATA

Chapter Objectives

After studying and completing this chapter, the student should have knowledge of the following:

- *CNC machining calculations*
- *CNC cutting speeds and feedrates*
- *Geometric dimensioning and tolerancing*
- *Surface finish*
- *Material types and hardness ratings*
- *CNC machining center alarm codes*

■ ■ ■ ■

CNC MACHINING CENTER CALCULATIONS

This chapter covers CNC technical calculations and related data. This includes explanations and illustrations that deal with CNC machining calculations, GD&T, surface finish, material types, hardness rating, and basic CNC alarm messages.

Calculation Example 1

The CNC program includes the coordinates that position the spindle and cutting tools. The X-, Y-, and Z-coordinates are mainly derived from the *part drawing* dimensions. However, in some cases where the coordinates are not dimensioned on the part drawing, it becomes necessary to perform calculations. Figures 5-1 through 5-6 illustrate some typical CNC machining calculations for CNC machining center parts.

Calculations for the items labeled A, B, C, and D for Figure 5-1:

A = (2.750 − the 0.250 Radius amount), which equals 2.50.

B = (1.25 − the 0.620 Radius amount), which equals .63.

C = Since the corner specifies a 0.38 × 45° chamfer in two places, the bottom corner is also 0.38 × 45°. The length of the chamfer is 0.38 in both directions because both legs of a 45° triangle are equal. Therefore, (2.750 − 0.38), which equals 2.37.

There are two ways to calculate the slot width D.

Calculation number 1:

D = (0.620 Radius × 2), which equals 1.24.

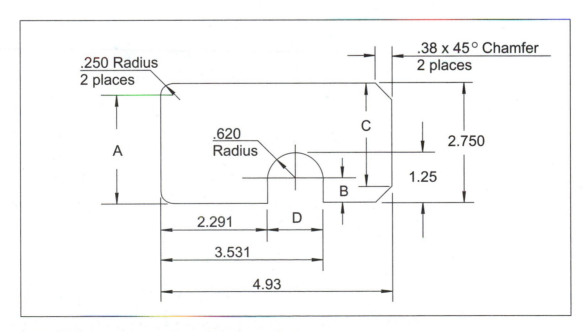

Figure 5-1 **CNC machining math problem example 1**

Calculation number 2:

$$D = (3.531 \text{ Dimension} - 2.291 \text{ Dimension}), \text{ which equals } 1.24.$$

Calculation Example 2

In Figure 5-2, some X-Y coordinates are not dimensioned on the part drawing. Therefore, it is necessary to calculate them based on the given dimensions. Note that the shaded right triangles illustrated are for visualization of the calculations. They are not illustrated on part drawings typically used in industry.

The following process describes how the coordinate points for Figure 5-2 are calculated:

First list the known values: the bolt circle diameter is 3.0 inches; the holes are 72° apart (360 divided by 5); the radius (a, aa) = 1.5 (3.0 divided by 2).

Two right triangles are constructed (shaded areas). To solve triangle 1:

Subtract 72° from 90° = 18°.

To solve triangle 2:

Subtract 18° from 72° = 54°, and 90° minus 54° = 36°. Note that the triangles are labeled a, b, and c. To solve for sides b/bb, apply the trigonometry formula

(b = a × cosine C), and for c/cc, apply the formula (c = a × sine C).

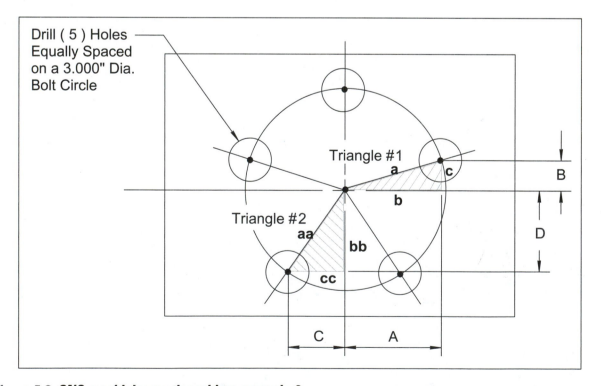

Figure 5-2 **CNC machining math problem example 2**

For b = (1.5 × 0.951056516) = 1.42658, and for c = (1.5 × 0.309016994) = 0.4635; therefore, A = 1.4266, B = 0.4635.

For cc = (1.5 × 0.587785252) = 0.88168, and for bb = (1.5 × 0.809016994) = 1.2135; therefore, C = 0.8817, D = 1.2135.

Calculation Example 3

In Figure 5-3, some coordinates are not dimensioned on the part drawing. Therefore, it is necessary to calculate them based on the given dimensions.

Calculate the items labeled A, B, C, and D for Figure 5-3.

A = Using appendix C: (c = a × sine 30°), a = (5.750 dia./2), and the angle = 360°/3 = 120° (120° − 90°) = 30°; then using a scientific calculator, the sine of 30° = 0.500. Therefore (2.875 × 0.500), which equals 1.4375.

B = The radius of the Bolt Circle diameter; therefore (5.750/2), which equals 2.875.

C = A + B; therefore (1.4375 + 2.875), which equals 4.3125.

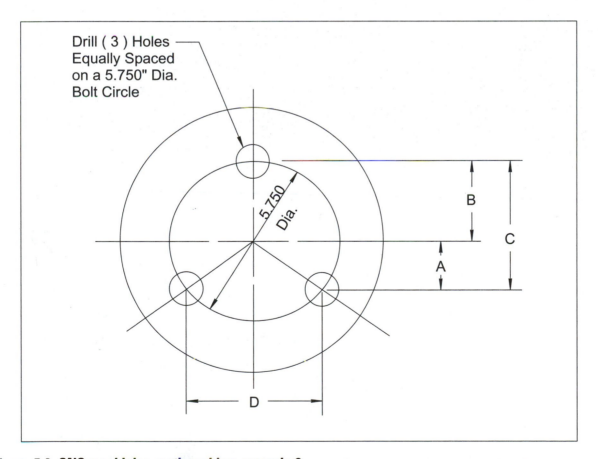

Drill (3) Holes Equally Spaced on a 5.750" Dia. Bolt Circle

Figure 5-3 *CNC machining math problem example 3*

D = Using appendix C: (b = a × cosine 30°), a = (5.750 dia./2), and the angle is the same as previous (30°); then using a scientific calculator, the cosine of 30° = 0.8660. Therefore [(2.875 × 0.8660) × 2], which equals 4.9795.

Calculation Example 4

In Figure 5-4, some coordinates are not dimensioned on the part drawing. Therefore, it is necessary to calculate them based on the given dimensions. Calculate the items labeled A, B, C, and D for Figure 5-4.

A = Using appendix C: (c = a × sine 450°), a = (4.50 dia./2), and the angle = 360°/8 = 45°; then using a scientific calculator, the sine of 45° = 0.7071. Therefore (2.25 × 0.7071), which equals 1.5910.

B = The radius of the Bolt Circle diameter; therefore (4.50/2), which equals 2.25.

C = A + B; therefore (1.5910 + 2.25), which equals 3.841.

D = Since the angle between the holes is 45°, all the distances "A" are equal (1.5910) in both directions (X and Y). Therefore (1.591 × 2), which equals 3.182.

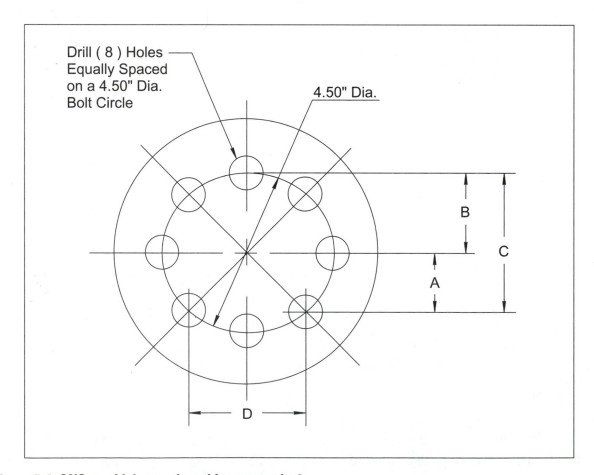

Drill (8) Holes Equally Spaced on a 4.50" Dia. Bolt Circle

4.50" Dia.

Figure 5-4 CNC machining math problem example 4

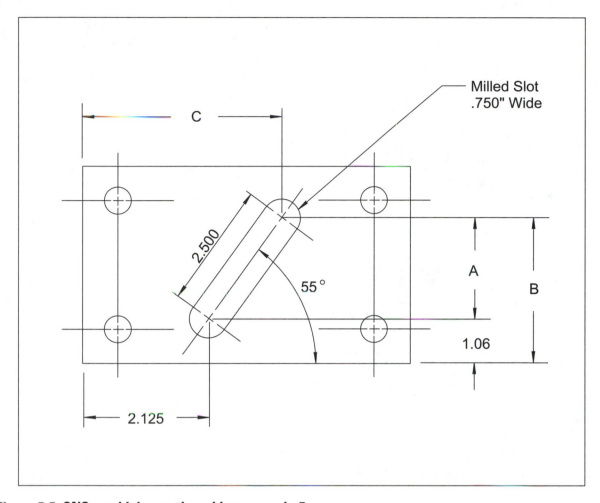

Figure 5-5 **CNC machining math problem example 5**

Calculation Example 5

In Figure 5-5, some coordinates are not dimensioned on the part drawing. Therefore, it is necessary to calculate them based on the given dimensions.

Calculate the items labeled A, B, and C for Figure 5-5.

A = Using appendix R: (b = a × sine 55°), a = (2.500), and using a scientific calculator, the sine of 55° = .8192. Therefore (2.500 × 0.8192), which equals 2.048.

B = 1.06 + distance A: (1.06 + 2.048), which equals 3.108.

C = Using appendix R: (c = a × cosine 55°), a = (2.500), and using a scientific calculator, the cosine of 55° = 0.5736, (2.500 × 0.5736) = 1.434; therefore (2.125 dimension + 1.434), which equals 3.559.

Calculation Example 6

In Figure 5-6, some coordinates are not dimensioned on the part drawing. Therefore, it is necessary to calculate them based on the given dimensions.

Calculate the items labeled A, B, C, and D for Figure 5-6.

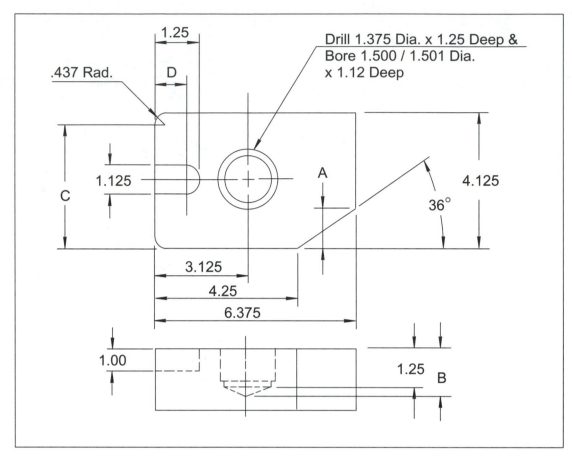

Figure 5-6 **CNC machining math problem example 6**

A = Using appendix R: (c = b × tan 36°), b = (6.375 − 4.25), and using a scientific calculator, the tan of 36° = 0.7265. Therefore (2.125 × 0.7265), which equals 1.5438.

B = [The drill depth + (0.3 × the drill diameter)], because 0.3 is the factor used for 118° drill points; therefore [1.25 + (0.3 × 1.375)], which equals 1.663.

C = (The overall length − the corner radius), since the corner specifies a 0.437 Rad., (4.125 − 0.437), which equals 3.688.

D = (The slot length − the radius of the slot), [1.25 − (1.125/2)], which equals 0.6875.

CNC CUTTING SPEEDS AND FEEDRATES

Carbide tooling is an important aspect of modern CNC tooling. Carbide-insert tools for hole operations have been described and illustrated earlier. Tool speeds and feedrates determine, to a large extent, the performance of these cutting tools. They are defined for the various drilling applications as well as the various milling applications. This section also includes a matrix for calculating the RPM and one for selecting and calculating the feedrate. For a specific diameter tool and tool type, the two most important parameters that must be specified when cutting a particular material are the speed and feedrate of the tool. These values are important because they influence tool life and cutting performance.

Cutting Speed (SFM) and Spindle Speed (RPM) for Hole Operations

The cutting speed is defined as the speed of any point on the circumference of the cutter. It is usually expressed in *surface feet per minute* (SFM). The tool speed and cutter diameter selected will be used to calculate the spindle RPM. These items are detailed and described in appendices B–G.

The cutting speed (SFM) selected for a particular hole operation will depend on several factors, listed below:

- Type of hole operation
- Material type
- Material hardness
- Hole depth
- Type of tool to be used
- Type of coolant system used
- Rigidity of the tool
- Rigidity of the setup

Selecting a cutting speed that is too high can lead to excessive dulling and burning of the tool's cutting edges. Conversely, speeds that are too low may cause the tool to wear excessively or break during operation. Appendix D contains a comprehensive list of drill cutting speeds (SFM).

For reaming, the cutting speeds (SFM) are approximately one-half of what is used for drilling. This is due to the larger cutting edge contact area, which causes more tool pressure. Appendix D includes a listing of suggested SFM values for reaming.

The cutting speeds (SFM) for tapping are usually much slower than for drilling. This is also due to the larger cutting edge contact area, which causes more tool pressure. Appendix D includes a listing of suggested SFM values for tapping.

Counterboring and countersinking speeds are approximately one-fourth of what is used for drilling. Again, the larger cutting edge contact area of these tools typically causes more tool pressure.

Boring operations, which are typically performed with carbide-insert or carbide-tipped type tools, usually require speeds that are approximately five times faster than what is used for drilling. Boring tools usually have minimal cutting edge contact area and consequently cause less tool pressure.

The following is an example that describes the RPM calculation process for a drilling operation of a mild steel plate:

- Part material is mild steel.

- Drill diameter is 0.750 inch.

- Drill material is HSS.

- From appendix D, cutting speed (CS) = (60–100 SFM) = 80 mean range

- Formula from appendix B:

$$RPM = (4 \times CS)/DIA. = (4 \times 80)/0.75 = 427\,RPM.$$

To facilitate this process in a step-by-step manner, a calculation matrix has been designed (see Figure 5-7). This matrix can be used as a guide to ensure that all the data for calculating the RPM is applied correctly. It is important to note that this process will produce a recommended starting RPM value. Due to the various items that affect machining, the RPM values can be adjusted to the optimum level during the actual metal cutting proveout.

Tool Feedrate (IPR and IPM) for Hole Operations

The tool feedrate for hole operations is defined as the amount (inches) at which the tool advances into the work per one revolution. It is expressed as IPR. By multiplying the tool feedrate in IPR by the spindle RPM (IPM = IPR $\times$ RPM), the tool feedrate is converted to IPM. The tool feedrate is the value that is input

Machining Center Speed (RPM) Calculation Matrix		
Step	**Description**	**Input Values**
1	List the **operation type**: milling, drilling, tapping, reaming, boring.	
2	List the **tool material**: uncoated carbide, coated carbide, HSS, ceramic, diamond.	
3	List the **part material**: aluminum, steel, alloy steel, cast iron, stainless steel, brass, copper.	
4	Select the **cutting speed** (SFM). Reference chart:	
5	List the **diameter** (cutter OD)	
6	RPM formula: RPM = SFM $\times$ 4 /Diameter	

Figure 5-7 **RPM calculation matrix**

into the CNC program. The recommended feedrate values for HSS hole operations are listed in appendix F.

The following is an example that describes the IPM calculation process for a drilling operation of a mild steel plate:

- Part material is mild steel.

- Drill diameter is 0.750 inch.

- From appendix F, IPR chart, IPR = 0.014.

- RPM from previous calculation = 427.

- Formula from appendix B.

$$IPM = RPM \times IPR = (427 \times 0.014) = (5,978),\ 6.0\ IPM.$$

Reaming feedrates are approximately 2–3 times greater than what is typically used for drilling. This is due to the material amount per side, which is typically 0.005–0.032 inch, and the material shaving action of the reamer blades.

Boring feedrates, which are usually expressed in IPR, can range from 0.0005 up to 0.030 inch IPR depending a variety of factors. Some factors include surface finish required, tolerance allowed, material hardness, and type of cut (rough or finish) to be performed. Appendix N lists some recommended IPR values for boring.

Tapping feedrates are derived from the "*pitch*" of the thread size to be tapped. Since the tap basically represents the screw thread to be machined, the feedrate of the tap must be the same as the screw thread pitch. Therefore, the feedrate is calculated by dividing one by the number of threads per inch of the screw thread. The tap designation, which is described below, is used to calculate the feedrate (pitch).

The tap designation **1/4-20 UNC-2B**, is translated as follows:

1/4	Represents the major diameter (0.250 inch nominal).
20	Represents the number of threads per inch (20).
UN	Specifies the thread series type (Unified National).
C	Specifies course, F is used for fine.
2	Specifies the class fit: 1 = loose fit, 2 = moderate fit, 3 = tight fit.
B	Specifies internal threads; for external threads an A is used.

Consequently, for a 1/4–20 UNC-2B tap, the feedrate is calculated by dividing one inch by 20 threads per inch (1/20 = 0.05). This tapping feedrate of 0.05 IPR is considered a heavy machining feedrate. To compensate, the RPM of the tool is slower.

Tool Speed and Feedrates for Milling Operations

As previously stated, the tool speed (RPM) and feedrate (IPM) are the two most important parameters to be specified for the cutting tool. This holds true especially for milling processes. The success of an operation in terms of accuracy of the cut, surface finish, and tool wear depends upon the proper specification of both the tool speed (RPM) and the feedrate (IPM).

Tool Speed (RPM) for Milling Operations

The tool speed (RPM) calculation process defined previously for hole operations is similar for calculating milling-type operations. Appendix C also contains a comprehensive list of recommended cutting speeds (SFM) for the various milling operations that are typically performed on CNC machining centers.

The following is an example that describes the RPM calculation process for a face milling operation of a mild steel plate:

- Part material is mild steel.
- Use a 4 inch cutter with 6 inserts (coated carbide).
- From appendix C, cutting speed = 500–900 SFM = 700 mean range.
- Formula from appendix B.

$$RPM = 4 \times CS/D = 4 \times 700/4 = 700 \, RPM.$$

Tool Feedrate (IPM) for Milling Operations

The tool feedrate, which is expressed in IPM, is defined as the rate at which the cutter advances into the work. The milling feedrate formula is **IPM = RPM × T × N**, where **RPM** is the speed of the milling cutter, **T** is the feed per tooth (FPT), and **N** is the number of inserts on the milling cutter.

The following is an example that describes the IPM calculation process for a face milling operation of a mild steel plate. Also, a calculation matrix has been designed (see Figure 5-8). This matrix can be used as a guide to ensure that all the data for calculating the IPM is applied correctly:

Machining Center Feedrate Selection and Calculation Matrix		
Step	**Description**	**Input Values**
1	List the **operation type**: milling, drilling, tapping, reaming, boring.	
2	List the **tool material**: uncoated carbide, coated carbide, HSS, ceramic, diamond.	
3	List the **part material**: aluminum, steel, *alloy steel,* cast iron, stainless steel, brass, copper.	
4	Select the **feed rate** (IPR or FPT), Reference chart:	
5	Select the **feed mode**: (IPR or IPM)	
6	If IPM use formula: IPM: (IPM = RPM × IPR)	

*Figure 5-8 **Feedrate selection and calculation matrix***

- Part material is mild steel.

- Use a 4 inch cutter with 6 inserts (coated carbide).

- From appendix E, feed per tooth = 0.004 − 0.016 = 0.01 mean range.

- Formula from appendix B.

$$IPM = RPM \times T \times N = (700 \times 0.01 \times 6) = 42 \text{ IPM}.$$

■ ■ ■ ■

GEOMETRIC DIMENSIONING AND TOLERANCING

Most manufacturers that require engineering drawings for machining purposes apply the American Society of Mechanical Engineers (ASME) Y14.5M tolerancing methods to their drawings. This tolerancing method uses various symbols for geometric feature control instead of written notes. These symbols are easily learned because they generally reflect the geometric characteristic under consideration. The more common geometric symbols are illustrated in Figures 5-9 through 5-11 and 15-11 through 15-14.

This method of tolerancing is referred to as GD&T and applies a datum system to establish how the part must be located for manufacturing and inspection purposes (see Figures 5-12 and 5-13). This is the type of tolerancing system that is applied to each of the project drawings in chapters 6 through 10 of this textbook.

GD&T

The part drawing is the most common method of illustrating the workpiece features and the associated dimensions and tolerances. These features are functional requirements of the end product and usually require some form of machining. Therefore, each feature, which requires machining, must be dimensioned in a manner that clearly defines the functional requirements and eliminates errors from incorrect or dual interpretations.

The GD&T system clearly defines the part requirements through the use of standardized symbols that are easy to interpret and facilitates *inspection* with functional *gauges*. The GD&T datum system also defines the surfaces important to the functionality of the part. The feature control frame includes the geometric form of the feature, the tolerance, and the surfaces to which they are related.

Datum Dimensioning

The *datum dimensioning* method requires that all dimensions on a drawing are placed in reference to one fixed zero point origin. Datum dimensioning is ideally suited for absolute positioning applications and machines. The GD&T method that matches the drawing dimensions with the CNC program coordinates allows effective communications between the CNC operator, the *coordinate measuring machine* (CMM) inspectors, and any other personnel.

GD&T Symbol	Application Description
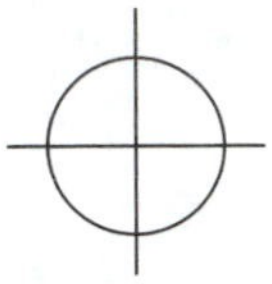 ⊕ **POSITION**	This symbol specifies the location tolerance for a machined hole to a datum system, such as: a drill hole, a ream hole, a bore or other type of machined diameter.
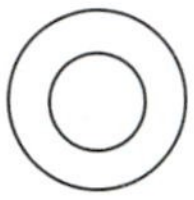 ◎ **CONCENTRICITY**	This symbol specifies the axial alignment tolerance for machining a diameter relative to another diameter. For outside (turn) or inside (bore) diameters.
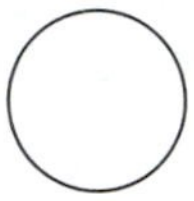 ○ **CIRCULARITY (ROUNDNESS)**	This symbol specifies the roundness tolerance for machining an outside (turn) or inside (bore) diameter.
 ⌀ **CYLINDRICITY**	This symbol specifies the roundness tolerance for the length of the feature to be machined, either an outside (turn) or inside (bore) diameter.
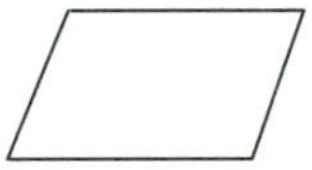 ▱ **FLATNESS**	This symbol specifies the flatness tolerance of a feature to itself, such as a milled surface, a face, or a shoulder.
 ⊥ **PERPENDICULARITY**	This symbol specifies the squareness tolerance of one feature to another, such as two milled surfaces, a face to a turn, or a shoulder to a bore.
 // **PARALLELISM**	This symbol specifies the parallel tolerance of one feature to another, such as two milled surfaces, two faces, or two other features.
— **STRAIGHTNESS**	This symbol specifies the straightness tolerance of a feature to itself, such as a shaft or bore.

Figure 5-9 **GD&T symbols and descriptions**

GD&T Symbol	Application Description
 CIRCULAR RUNOUT	This symbol specifies the dial indicator reading amount at any given area when the part is rotated relative to a datum axis, usually two turn diameters or between centers.
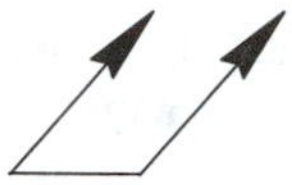 TOTAL RUNOUT	This symbol specifies the dial indicator reading amount across the entire area when the part is rotated relative to a datum axis, usually two turn diameters or between centers.
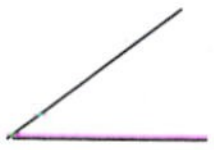 ANGULARITY	This symbol specifies the tolerance zone that an anglular surface can deviate, usually two flat surfaces or features.
 PROFILE OF A LINE	This symbol specifies the tolerance zone that a profile surface can deviate at a any given area of the part profile relative to a datum system.
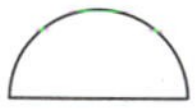 PROFILE OF A SURFACE	This symbol specifies the tolerance zone that a profile surface can deviate across the entire part profile relative to a datum system.
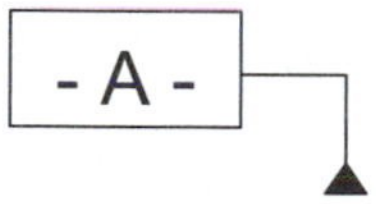 - A - DATUM	This symbol specifies the exact feature, plane, axis, or point which is established as a datum.
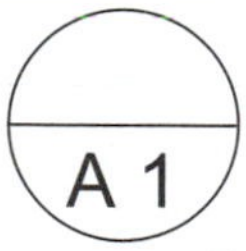 A 1 DATUM TARGET	This symbol specifies the exact location point of the datum feature which has been established.
1.500 BASIC DIMENSION	This symbol specifies the theoretical exact size, profile, orientation or location of a feature or datum target.
(1.500) REFERENCE DIMENSION	The parenthesis are used to specify dimensions that are only used as reference and are not to be machined or measured.

Figure 5-10 **GD&T symbols and descriptions**

GD&T Symbol	Application Description
(M) MAXIMUM MATERIAL CONDITION	This symbol specifies that a geometric tolerance or datum reference applies when the condition of a feature is at the maximum size allowed by the feature tolerance, (turn at high limit, hole at low limit).
(L) LEAST MATERIAL CONDITION	This symbol specifies that a geometric tolerance or datum reference applies when the condition of a feature is at the minimum size allowed by the feature tolerance, (turn at low limit, hole at high limit).
(P) PROJECTED TOLERANCE ZONE	This symbol specifies the height of the tolerance zone that a geometric tolerance is applied. Usually a hole to insure alignment of the mating part.
∅ DIAMETER TOLERANCE ZONE OR FEATURE	This symbol is used to specify a diameter. It is used to specify the tolerance value of a geometric tolerance or a feature dimension, such as a turn diameter or hole size.
[▱ | .005] FEATURE CONTROL FRAME - UNRELATED	This frame is used to specify the geometric condition and tolerance of a feature that is to be machined. This surface must be milled flat within .005 inches. Because this geometric condition is not related a datum is not specified.
[⊥ | .005 (M) | A] FEATURE CONTROL FRAME - RELATED	This frame is used to specify the geometric condition and tolerance of a feature that is to be machined. A feature may include: one or more milled surfaces, a turn diameter, a face, a bore, a set of drilled holes, or screw threads, and so on. This feature condition also specifies a related surface (datum -A-).

Figure 5-11 **GD&T symbols and descriptions**

Datum Reference

A datum is a specific surface, line, plane, or feature assumed to be perfect and used as a reference point for dimensions or features. Datums are identified by letters, and are shown on the part with datum feature symbols that refer to the datum reference in the *feature control symbol*, (see Figure 5-12). The letter A in the feature control symbol refers to datum A on the part (see Figure 5-10). The datum feature symbol is shown as a letter with a dash on either side contained in a box. All letters in the alphabet except the letters O, Q, and I may be used to identify datums. These letters are not used since they closely resemble numbers. On complex part prints, where more than 23 datums are required, double letters (AA, AB, AC, etc.) may also be used.

Datum references can be specified as a single datum (A) or as multiple datums (A, B, C). When only one letter is specified in the feature control symbol, it means the feature is related only to that single datum. Datums are arranged in their order of precedence.

GD&T Symbols and Descriptions

The symbols illustrated in Figures 5-9 through 5-11 are applied to engineering drawings in order to specify the part manufacturing requirements.

GD&T Datum System and Fixturing

The datum system illustrated in Figure 5-12 must be determined and specified in order to facilitate the entire manufacturing process. These surfaces are used as the locating surfaces for CNC machining and inspection.

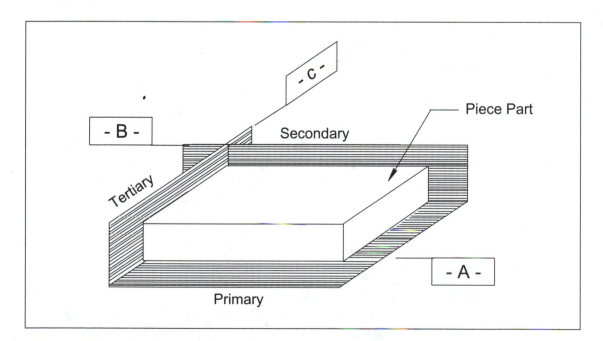

Figure 5-12 **GD&T datum plane system**

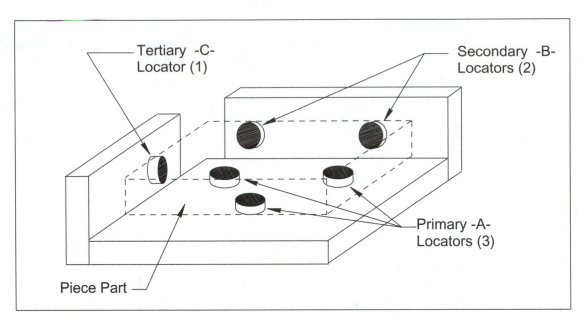

Figure 5-13 *GD&T fixture locating*

Figure 5-13 is an example of a holding fixture design that has appropriately applied the datum system illustrated in Figure 5-12.

Consequently, a primary datum is specified first, a secondary datum is specified next, and a tertiary datum is specified last. Supplementary, or modifying, symbols are used in addition to the basic elements in the feature control symbol to further define and clarify the meaning and intent of the basic feature control symbol. The symbols commonly used with geometric dimensioning and tolerancing are illustrated in Figures 5-9 through 5-11.

Feature Control Frame

The feature control frame is a rectangular frame used to specify the geometric condition, tolerance, and relationship of a feature (see Figure 5-13). The first item specified in the feature control frame is a geometric condition in the form of a symbol (illustrated in Figures 5-9 and 5-10). The second item specified is the tolerance amount, usually in inches or millimeters. The next item specified is the datum or datums related to the geometric condition. In cases where the geometric condition is not related to a datum, none would be specified. An example of this is a *flatness* requirement of a surface because it can only be flat to itself. Conversely, a perpendicular or a parallel surface always requires another surface to measure from. Therefore, the feature control frame may be related or unrelated.

Maximum Material Condition

The maximum material condition, or *MMC*, is the condition in which the part feature has its maximum amount of material within the specified limits of size. This is the largest size of an external feature and the smallest size of an internal feature. The MMC modifier is shown as a circled M (see Figure 5-11), and is only used on part features that vary in size.

This modifier may be used with the tolerance value, the datum reference, or both. In use, it means the specified tolerance value only applies at the MMC size of the feature.

Least Material Condition

The least material condition, or *LMC*, is the opposite of MMC. The LMC shows the least material within the specified limits of size. This value is the smallest size of an external feature or the largest size of an internal feature.

The LMC modifier is shown as a circled L (see Figure 5-11), and like the MMC modifier, is used only for features that can vary in size. This modifier may also be used with the tolerance value, the datum reference, or both. When used, this modifier means the specified tolerance value only applies at the LMC size of the feature.

Projected Tolerance Zone

This symbol, shown as a circled P (see Figure 5-11), is used to define the extension of a feature beyond the actual part surface to be inspected. This is usually applied to express the mating part alignment or function.

This modifier may be used with the tolerance value, the datum reference, or both. In use, it means the specified tolerance value only applies at the MMC size of the feature.

The machining process determines and controls the surface finish characteristics that are produced during a CNC machining cycle. The machining process includes the type of operation or operations that are performed such as drilling, milling, boring, and reaming. Also included in the machining process is the number of cuts required for roughing as well as for finishing. The cutting speed (SFM) and the feedrate are also components of the machining process that can be varied. Another factor includes the type of tool material used to machine the part, such as HSS, coated carbide, uncoated carbide, ceramics, polycrystalline cubic boron nitride (PCBN), and diamond. The tool corner radius, which can vary, is another factor that affects the surface finish produced. The rigidity of both the tool and the workholding device are also contributing factors in the surface finish that is produced.

SURFACE FINISH

The two primary design reasons that surface finish controls are specified on engineering drawings are to reduce friction and to control wear of a product. Overfinishing and underfinishing the machined surface adds unnecessarily to the production cost and can result in the rejection of a part or an entire production run. Surface texture measurement, which calls for the use of accurate measuring equipment, is covered by ANSI. Surface finish or texture measurements are typically measured with either a digital surface gage or visual comparison gages within a degree of reasonable accuracy.

All machined surfaces consist of a series of peaks and valleys that when magnified resemble the surface of a phonograph record (see Figure 5-14). The characteristics of surface texture, or finish, are *roughness*, waviness, lay, and flaw. Surface roughness usually caused by the cutting action of the tool edges, the cutting pattern of a grinding wheel, or by the speed and feed of the machine tool is measured over a specific distance. The measurement is typically in *microinches* (0.0000001 inches). Waviness, measured in inches, is similar to roughness but it occurs over a wider spacing than that used for determining roughness. Roughness may be considered as superimposed on a "wavy" surface. Lay indicates the direction of the predominant surface pattern. Flaws are irregularities, or defects, which occur at one place, or relatively infrequently on a surface.

Surface finish is only one of the items affected whenever the machining process is altered. Other items such as part size dimensions, premature tool wear, change in cycle time, and quality may also be affected. Therefore, it is important that the process is documented and adhered to during each production run. Surface finish requirements are typically specified on engineering drawings with a symbol and a value (see Figure 5-15).

There are various process methods that produce surface finishes. They range from very fine to very rough, or 2 microinches to 1000 microinches. The surface finish requirement must be considered whenever a production part is processed for CNC machining.

Chart 5-1 lists some of the more common manufacturing processes that can be selected for the manufacture of a production part. As Chart 5-1 illustrates, selecting the appropriate process type is very important because it affects the surface finish, which in turn affects the quality of the parts produced and the production cost.

Surface Finish Characteristics and Finish Notes

Figure 5-14 illustrates what a machined surface appears like when it is viewed under high magnification. These characteristics are used to measure and control the quality of production parts produced by any of the various machining processes. The CNC machine operator can alter the surface finish produced mainly by changing the cutting feedrate. However, there are other tool and machining features that may also alter the surface finish.

Figure 5-15 illustrates some typical surface finish notes that are applied to engineering drawings.

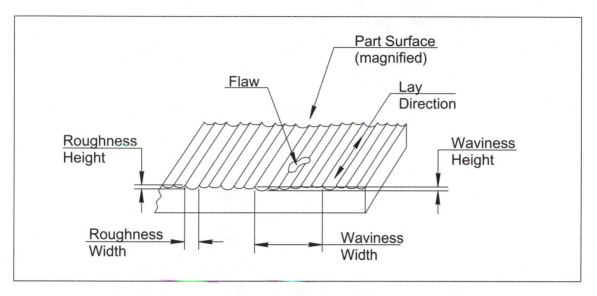

*Figure 5-14 **Surface finish characteristics***

Surface Finish Produced by Process Type

Chart 5-1 describes some typical surface finish values that can be produced by machining, casting, and other various processes. Each range shown is typical of each listed process type. However, higher or lower values can be obtained when the process variables are nontypical. The values are expressed in microinches ranging from 1000 microinches to 1 microinch. These values are typically specified on the part features that are illustrated on the engineering drawing. It should be noted that because of the many variables that can be altered in the process, the values in this chart are for general consideration only.

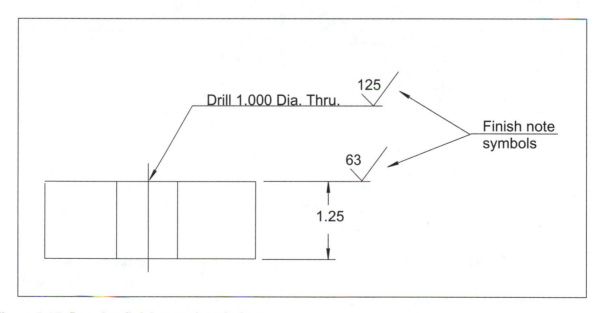

*Figure 5-15 **Drawing finish note descriptions***

Chart 5-1 Surface finish range description

PROCESS TYPE	SURFACE FINISH PRODUCED (Average Application Range) Roughness Height Rating (Microinches, Arithmetical Average)										
	1000	500	250	125	63	32	16	8	4	2	1
Drilling			■	■	■						
Milling			■	■	■	■					
Reaming				■	■	■					
Boring, Turning			■	■	■	■	■				
EDM				■							
Laser			■	■	■	■					
Broaching				■	■	■					
Sawing	■	■	■	■							
Grinding						■	■	■	■		
Honing						■	■	■	■	■	
Roller Burnishing								■			
Polishing							■	■	■		
Lapping								■	■	■	
Superfinishing								■	■	■	
Flame Cutting	■										
Hot Rolling	■										
Cold Rolling				■	■	■					
Forging		■	■	■							
Sand Casting	■										
Investment Casting				■	■						
Die Casting					■						

■ Indicates the Average Application Range

MATERIAL TYPES AND HARDNESS RATING

There are various types of materials that are used to manufacture products. These products or components of a product usually require some form of CNC machining. Therefore, it is necessary to become familiar with the various types of materials and their characteristics. Metals, which are the most common materials that are machined on CNC machines, are identified as either ferrous or nonferrous. Iron and steel metals are considered ferrous, and

Table 5-1 *Material types and characteristics*

MATERIAL TYPE	COLOR	WEIGHT	HARDNESS	MACHINABILITY
Aluminum	Silver-white	Lightweight	Soft	Built-up edge
Brass	Gold	Heavy	Medium	Abrasive
Copper	Reddish	Medium	Soft	Built-up edge
Bronze	Reddish-gold	Heavy	Soft	Abrasive
Plastic PVC	Black or gray	Lightweight	Soft	Light
Cast Iron	Gray	Heavy	Medium-hard	Abrasive
Steel	Silver-gray	Heavy	Medium-hard	Average
Stainless steel	Silver	Heavy	Medium-hard	Abrasive or tough
Alloy Steel	Silver	Heavy	Hard	Tough

any metal other than iron or steel is considered nonferrous. Table 5-1 identifies some common materials and some identifying characteristics.

Types of Steel

There are many different types of steels available for wide-ranging uses such as paper clips, bridges, springs for automobiles, and cutting tools. All of these steels have one thing in common: they all contain iron and carbon, the main elements in steel. Carbon is the element that may have the greatest effect on the steel's properties because it is the hardening agent. As the percentage of carbon in steel increases over 0.83%, the hardness, hardenability, tensile strength, and the wear resistance of steel are increased.

There are four distinct categories of steels: low-carbon, medium-carbon, high-carbon, and alloy steels.

- *Low-carbon steel* contains from 0.02 to 0.30% carbon, and because of the low carbon content, cannot be hardened. The surface of this steel can be case-hardened by heating and increasing its carbon content. Low-carbon steel is used for shafts, nuts, washers, sheet metal, and so on.

- *Medium-carbon steel* contains from 0.30 to 0.60% carbon, and can be hardened (toughened), which increases its tensile strength. It is widely used for forgings, wrenches, hammers, screwdrivers, and so on.

- *High-carbon steel,* also called tool steel, contains over 0.60% and as high as 1.7% carbon. It can be fully hardened and is used for cutting tools such as drills, taps, reamers, milling cutters, and lathe toolbits.

- *Alloy steels* are generally some form of high-carbon steel to which certain alloying elements have been added to provide it with qualities such as increased hardness, toughness, tensile strength, wear and corrosion resistance.

Tool Steels

There are many types of tool steels produced by manufacturers under their own trade names. These steels are generally classified as water-hardening, oil-hardening, air-hardening, or high-speed steels to suit a variety of uses. To avoid confusion with the trade names, it is wise to order steel by its AISI (American Iron and Steel Institute) or SAE (Society of Automotive Engineers) number.

- *High-speed steels* may have a tungsten or molybdenum base, and are used primarily for cutting tools such as drills, reamers, milling cutters, and lathe tools. They provide good wear resistance and hardness.

- *Hot-work steels* may have a chromium, tungsten, or molybdenum base, and provide good wear resistance, toughness, and excellent hardness. These steels resist cracking during *heat treatment*, and provide good life in forging tools, piercing dies, and die-casting applications.

- *Cold-work steels* generally a high-carbon, high-chromium base, provide deep hardenability, size stability, excellent wear resistance, and good life in tool and die applications and cutting tools.

Hardness Rating Methods

The hardness of a material has a direct impact on the manufacturing process. The hardness level, as well as other factors, affects the spindle speed (RPM), the feedrate (IPR), the depth of cut (DOC), the insert geometry, the insert grade, the rake angle, the insert TNR, and the surface finish that is produced. The hardness of a part material is usually specified on the engineering drawing. The most common methods of rating a material hardness include the four listed.

Brinell Hardness Number (BHN)

BHN is determined by applying a known load for a given length of time, indenting the test surface with a hardened steel or carbide ball of known diameter. The diameter of the indentation is microscopically measured and converted to the BHN by standard tables.

Rockwell Hardness Test (Rc or Rb)

The Rockwell tester measures the depth of residual penetration by a steel ball or a diamond cone under given conditions of load. The "C" scale, Rc, is utilized for hardened steels and some case-hardened parts. The "B" scale, Rb, is used for medium-hard and/or some annealed metals such as medium carbon steel or cast iron.

Scleroscope Hardness Test

For this test a diamond-tipped hammer falls from a specific height through a glass tube with a graduated scale. The distance of rebound is observed visually or recorded by indicator to determine the hardness. The scleroscope tester is portable, and therefore ideal and easy to apply on large heavy parts.

Vickers Diamond Pyramid Number (DPN)

The DPN test is similar to the BHN test, except the load is applied to the test surface for a specific time through a square-based pyramid with 136° between opposing faces. The number (DPN) is in kilograms divided by the square millimeters of the indentation area.

CNC MACHINE ALARM CODES

The CNC machine continually diagnoses itself to determine if any error is present. When an error is detected, the machine will stop automatically and the CRT will display the word "ALARM." Some errors are simple while others are complex and time-consuming. Pressing the Alarm Message button on the control CRT can identify the specific error that has occurred. Table 5-2 describes a sample of typical error codes that may be generated. Each control manufacturer provides a specific list for their control. Most CNC machines have diagnostic feedback capability that enables the operator and maintenance personnel to correct any system errors.

When an error occurs, the CNC machine will not operate until the error has been corrected and the control is reset. Additionally, there are other situations that can cause the CNC machine to stop operation. For example, most CNC machines are designed to include safety features that will inhibit certain operations when a condition is determined to be unsafe. Some of these situations may include but are not limited to

- Guard doors open when Cycle Start is pressed.

- Guard doors open when Spindle Start is pressed.

- High-Pressure-Coolant inactive when guard doors are open.

- Tool not in clamped position.

When these types of situations occur, the CRT may or may not display a message. Usually it will display a "command rejected" message. In either case, the CNC operator should be adequately trained to correct and minimize these errors from occurring.

In situations where the error is due to a machine mechanical or electrical malfunction, the CNC operator should contact the appropriate repair personnel. In situations where the error is due to a CNC program function, the CNC operator should contact the appropriate CNC programmer. The CNC operator should never attempt to correct any dangerous malfunctions that may cause injury.

■____*Table 5-2 Alarm code messages and descriptions*

ERROR CODE	MESSAGE DISPLAYED ON CRT	ERROR DESCRIPTION
1	TOO MANY DIGITS	Data exceeding the allowable number of digits was input.
2	ILLEGAL USE OF NEGATIVE VALUE	A − sign was specified in an address where no − is allowed.
3	ILLEGAL USE OF DECIMAL POINT	A decimal point · was specified in an address where no decimal point is allowed.
4	IMPROPER G CODE	An illegal G code was specified.
5	IMPROPER NC ADDRESS	An illegal address was specified.
6	INVALID BREAK POINT OF WORD	NC word delimiter is in error.
7	ILLEGAL PROGRAM NO. POSITION	Address 0 or N was specified at an illegal position (after a macro, etc.).
8	ILLEGAL PROGRAM NO. FORMAT	Address 0 or N was not followed by a number (numeral).
9	TOO MANY WORDS IN ONE BLOCK	The number of words in a block exceeded an allowable range.
10	EOB NOT FOUND	EOB code is missing at the end of a program, which was input in MDI mode.
11	ZERO DIVIDE	Division was specified with a divisor "0."
12	S COMMAND OUT OF RANGE	A specified spindle speed exceeded the maximum spindle motor speed.
13	SEQUENCE NUMBER NOT FOUND	A specified sequence number was not found in sequence number search.
14	PROGRAM NOT FOUND	The program whose program number was called by M98 code cannot be found.
15	PROGRAM IN USE	A program being edited in the background was called to execute.
16	DUPLICATE NC-WORD & M99	An address other than O, N, P, and L was specified in the same block as M99 under the macromodal call mode.
17	ILLEGAL LOOP	DO−END No. exceeded allowed range.
18	SEQUENCE NO. OUT OF RANGE	A sequence number is other than 1–9999.
19	ZERO RETURN NOT FINISHED	After turning on the power supply, a travel command was specified for an axis with which reference point return has not been finished yet.
20	CIRCLE CUT IN RAPID	F0 (rapid traverse of F1-digit feed or inverse feed) was specified in the circular mode (G02/G03).
21	TOO MANY SIMULTANEOUS CONTROL AXES	A travel command was specified for axes exceeding the simultaneously controllable axes.

■___*Table 5-2 Continued*

ERROR CODE	MESSAGE DISPLAYED ON CRT	ERROR DESCRIPTION
22	TOO LARGE DISTANCE	A travel amount exceeded the maximum command value as a result of offset, and so on.
23	ILLEGAL PLANE SELECT	A plane select command is erroneous. (Parallel axes were specified simultaneously.)
24	FEED ZONE (COMMAND)	Cutting feedrate command (F code) was 0.
25	OVER TOLERANCE OF RADIUS	An arc, where the difference of radius values at start point and end point is larger than the parameter set value, was specified.
26	OFFSET C START UP CANCEL BY CIRCLE	Tool nose radius compensation is being started up or canceled in the circular mode.
27	OFFSET C ILLEGAL PLANE	Plane is being changed during tool nose radius compensation.
28	OFFSET C INTERFERENCE	Too much cutting is caused in tool nose radius compensation.
29	OFFSET C NO SOLUTION	There is no intersection point in tool nose radius compensation.
30	ILLEGAL ADDRESS	An illegal address was specified when loading parameters.
31	TOO MANY DIGITS	The number of digits of data exceeded an allowable value when loading parameters.
32	PROGRAM DOES NOT MATCH	A tape program does not match a program in the memory.
33	TOO MANY ADDRESS CODES	Two or more were commanded in the same block.
34	MEMORY ACCESS OVER RANGE	An attempt was made to write data to or read data from an address, which is outside the NC memory area.
35	NO P, Q COMMAND AT G70–G72	Neither P nor Q was specified in the block of multiple repetitive cycle G70, G71, or G72 for turning. (P: sequence no. for a cycle start block; Q: sequence no. for a cycle end block.)
36	HYDRAULIC OIL FAILURE	Hydraulic oil reservoir is low or system is not operating.
37	ILL COMMAND IN G70–G76	A command in the block of multiple repetitive cycle (G70 G76) for turning is improper.
38	P, Q BLOCK NOT FOUND	A sequence no. block for P or Q specified in the block of multiple repetitive cycle G70, G71, or G72 for turning cannot be found.
39	ILLEGAL COMMAND IN P-BLOCK	A command in the sequence no. block for P specified in the block of multiple repetitive cycle G70, G71, or G72 for turning is improper.

CHAPTER SUMMARY

The key concepts presented in this chapter include the following:

- Calculations for CNC machining center coordinates.
- Tool cutting speed (SFM) selection.
- Tool cutter speed (RPM) calculations.
- Tool cutting feedrate (IPR) selection.
- Tool cutting feedrate (IPM) calculations.
- The GD&T symbols datum system features and applications.
- The names and descriptions of surface finish characteristics.
- The descriptions of finish notes on engineering drawings.
- The average surface finish produced by specific process types.
- The names and descriptions of material types.
- The hardness rating method types and descriptions.
- The CNC machining center alarm code types and descriptions.

REVIEW QUESTIONS

1. Why is it important to select the proper cutting speed and feedrate?
2. Determine the feedrate (IPR) for a 0.750 inch diameter HSS drill, to drill mild steel.
3. Determine the feedrate (FPT) range for a HSS end mill, to mill cast iron.
4. Determine the cutting speed (SFM) range for a HSS tap, to tap mild steel.
5. Calculate the RPM for a 5.0 inch carbide face mill, to face mill aluminum.
6. Calculate the feedrate (IPR) for a 5/8-11 UNC-2B tap.
7. What is the purpose for applying GD&T to engineering drawings?
8. Why are datums required?
9. Sketch the symbol for perpendicularity, parallelism, flatness, position, straightness, and concentricity.
10. What determines the surface finish that is produced by machining?
11. List the surface finish range produced by milling, drilling, and reaming.
12. List the various types of materials that are machined on CNC machines.
13. List the machinability characteristic for aluminum, steel, and alloy steel.
14. List the types of material hardness testing methods.
15. What should the CNC operator do if the CRT displays an alarm message?

CNC MACHINING CENTER RAPID AND FEED MOVES

Chapter Objectives

After studying and completing this chapter, the student should have knowledge of the following:

- *G00 and G01 codes*

- *G00 rapid movement in X-, Y-, and Z-axes*

- *G01 feed movement in X-, Y-, and Z-axes*

- *HMC five-axis machining*

- *Calculating X-Y coordinates*

RAPID TRAVERSE AND FEED MOVES

The first set of illustrations in this chapter, Figures 6-1 through 6-6, describe the G00 rapid movements for both VMCs and HMCs. The second set of illustrations, Figures 6-7 through 6-9, describe the G01 feed movements for VMCs. These figures are designed to illustrate the differences between rapid moves and feed moves. Additionally, the chapter includes a coordinate calculation worksheet and a sample program to identify some basic CNC program format.

G00 RAPID MOVEMENT

The G00 rapid code is primarily used to move the X-, Y-, or Z-axis as fast as possible. Therefore, after the G00 code an axis or combination of axes follows. The axis letter (X, Y, or Z) include a positive or negative number, which represents the coordinate position of the machine. The positive sign is the machine default sign and therefore not required.

The G00 code is modal and will remain active for each subsequent line until it is cancelled. Therefore, any axis letter present in a subsequent line of the CNC program will move in rapid traverse mode whether the G00 code is present or not.

When the G00 code along with an axis coordinate position is read (Cycle Start), the specified machine axis or axes will move automatically at a very rapid rate. This is only recommended when it is absolutely certain that a collision will not occur. Typically, this is after the CNC program has been Dry Run or proved out. In cases where the rapid moves have not been proved out, the CNC operator should override the rapid traverse rate to the lowest possible setting. This will allow the CNC operator time to react to a possible collision and stop the movement before the collision occurs.

G00 X-axis Rapid Traverse Description for VMC

The rapid traverse mode initiated with the G00 code is illustrated for a VMC in Figure 6-1. The rapid traverse code is modal, and moves the spindle and table at a high rate of motion, usually 200 IPM or greater. The X- and Y-axes move the table and the Z-axis moves the spindle. Whenever rapid moves are executed, caution should be taken to avoid collisions. The G00 code is canceled by either the G01, G02, or G03 code.

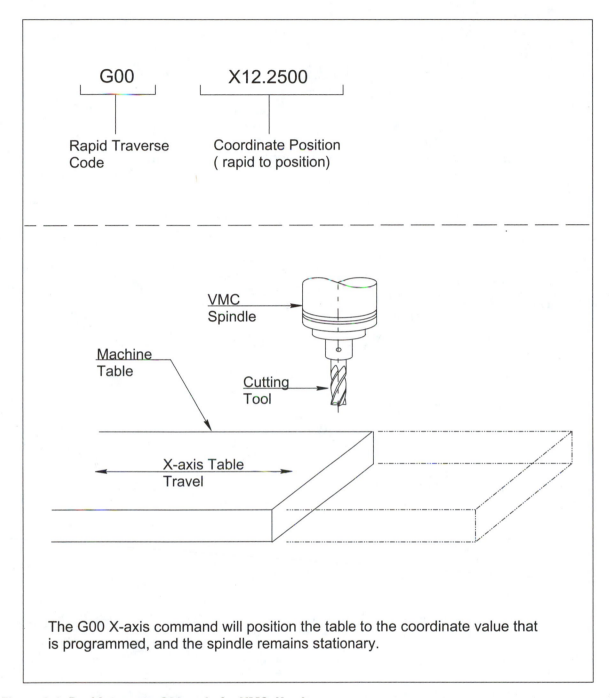

G00 — Rapid Traverse Code

X12.2500 — Coordinate Position (rapid to position)

VMC Spindle

Machine Table

Cutting Tool

X-axis Table Travel

The G00 X-axis command will position the table to the coordinate value that is programmed, and the spindle remains stationary.

Figure 6-1 Rapid traverse G00 code for VMC, X-axis

G00 Y-axis Rapid Traverse Description for VMC

The rapid traverse mode initiated with the G00 code is illustrated for a VMC in Figure 6-2. The rapid traverse code is modal, and moves the spindle and table at a high rate of motion, usually 200 IPM or greater. The X- and Y-axes move the table and the Z-axis moves the spindle. Whenever rapid moves are

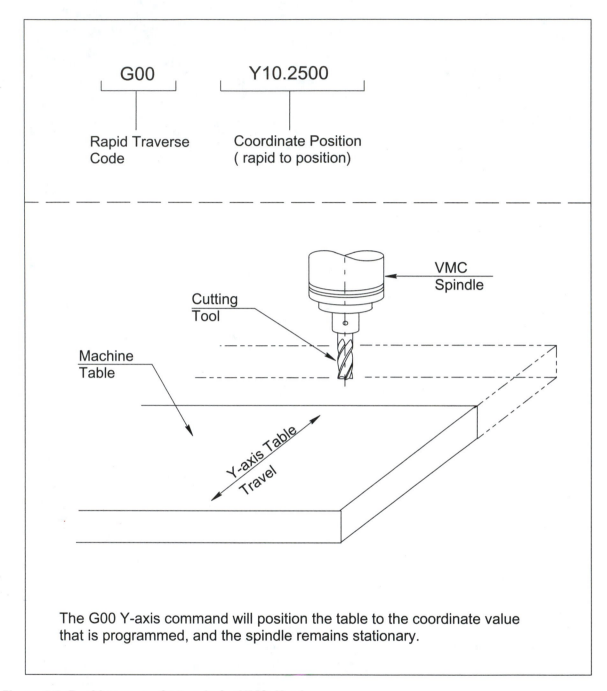

G00   Y10.2500

Rapid Traverse   Coordinate Position
Code   (rapid to position)

VMC
Spindle

Cutting
Tool

Machine
Table

Y-axis Table
Travel

The G00 Y-axis command will position the table to the coordinate value
that is programmed, and the spindle remains stationary.

Figure 6-2 *Rapid traverse G00 code for VMC, Y-axis*

executed, caution should be taken to avoid collisions. The G00 code is canceled by either the G01, G02, or G03 code.

G00 Z-axis Rapid Traverse Description for VMC

The rapid traverse mode initiated with the G00 code is illustrated for a
VMC in Figure 6-3. The rapid traverse code is modal, and moves the

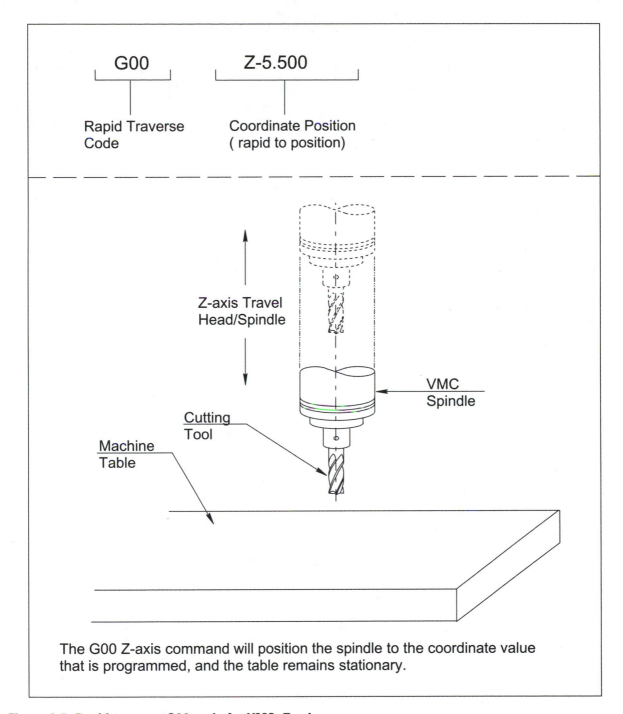

Figure 6-3 *Rapid traverse G00 code for VMC, Z-axis*

spindle and table at a high rate of motion, usually 200 IPM or greater. The X- and Y-axes move the table and the Z-axis moves the spindle. Whenever rapid moves are executed, caution should be taken to avoid collisions. The G00 code is canceled by either the G01, G02, or G03 code.

G00 X-axis Rapid Traverse Description for HMC

The rapid traverse mode initiated with the G00 code is illustrated for a HMC in Figure 6-4. The rapid traverse code is modal, and moves the spindle and table at a high rate of motion, usually 200 IPM or greater. The X- and Y-axes move the table and the Z-axis moves the spindle. Whenever rapid moves are executed, caution should be taken to avoid collisions. The G00 code is canceled by either the G01, G02, or G03 code.

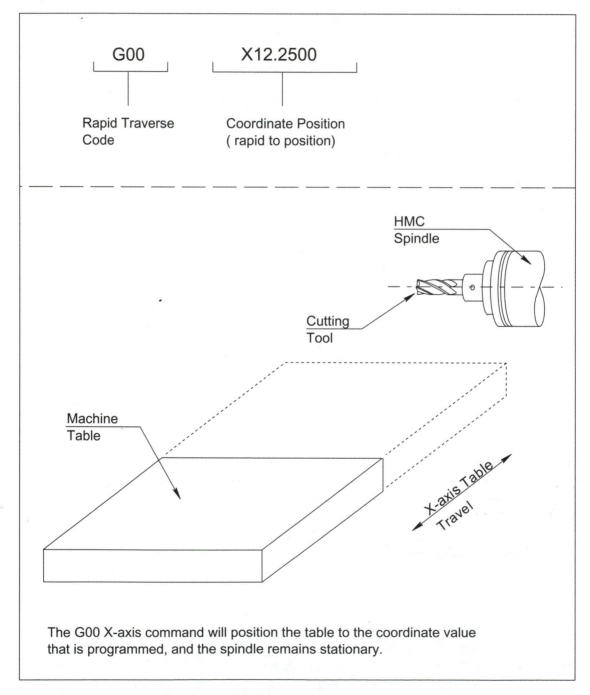

G00   X12.2500

Rapid Traverse   Coordinate Position
Code   (rapid to position)

HMC
Spindle

Cutting
Tool

Machine
Table

X-axis Table
Travel

The G00 X-axis command will position the table to the coordinate value that is programmed, and the spindle remains stationary.

Figure 6-4 Rapid traverse G00 code for HMC, X-axis

G00 Y-axis Rapid Traverse Description for HMC

The rapid traverse mode initiated with the G00 code is illustrated for a HMC in Figure 6-5. The rapid traverse code is modal, and moves the spindle and table at a high rate of motion, usually 200 IPM or greater. The X- and Y-axes move the table and the Z-axis moves the spindle. Whenever rapid moves are executed, caution should be taken to avoid collisions. The G00 code is canceled by either the G01, G02, or G03 code.

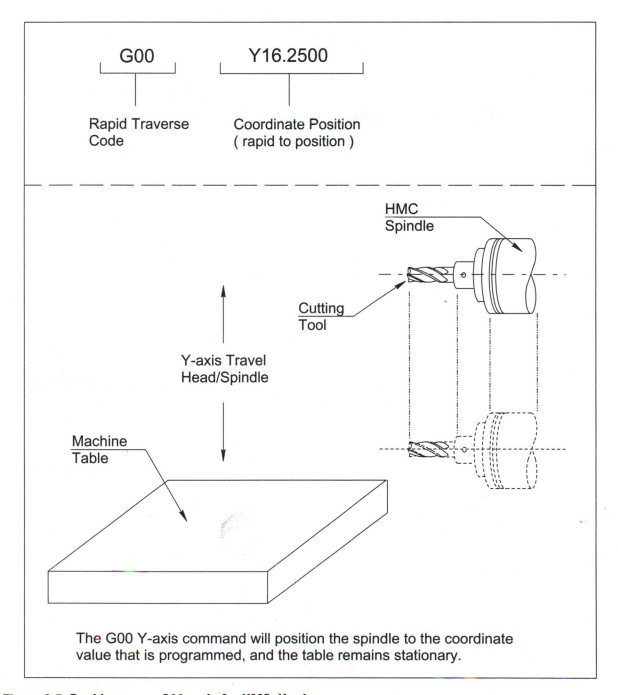

The G00 Y-axis command will position the spindle to the coordinate value that is programmed, and the table remains stationary.

Figure 6-5 **Rapid traverse G00 code for HMC, Y-axis**

G00 Z-axis Rapid Traverse Description for HMC

The rapid traverse mode initiated with the G00 code is illustrated for a HMC in Figure 6-6. The rapid traverse code is modal, and moves the spindle and table at a high rate of motion, usually 200 IPM or greater. The X- and Y-axes move the table and the Z-axis moves the spindle. Whenever rapid moves are executed, caution should be taken to avoid collisions. The G00 code is canceled by either the G01, G02, or G03 code.

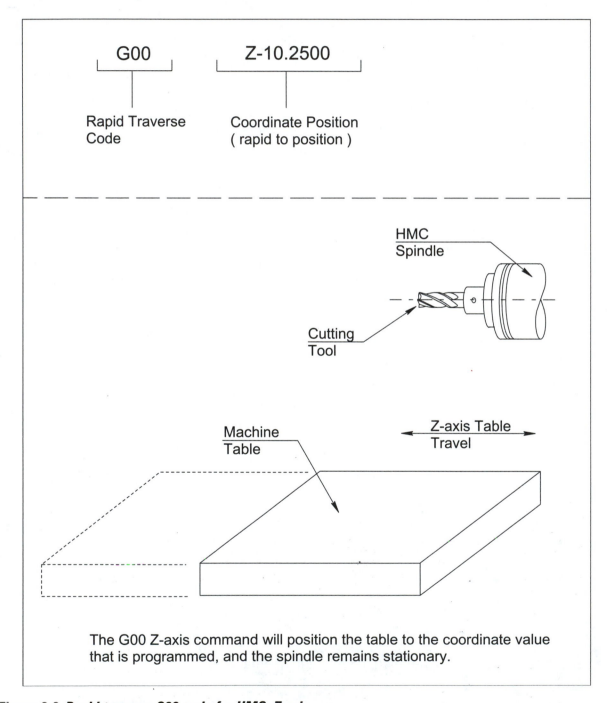

G00

Rapid Traverse
Code

Z-10.2500

Coordinate Position
(rapid to position)

HMC
Spindle

Cutting
Tool

Machine
Table

Z-axis Table
Travel

The G00 Z-axis command will position the table to the coordinate value that is programmed, and the spindle remains stationary.

Figure 6-6 Rapid traverse G00 code for HMC, Z-axis

■ ■ ■ ■
G01 FEED MOVEMENT

The G01 feed code is primarily used to move the X-, Y-, or Z-axis at a predetermined feedrate to cut the workpiece. Therefore, after the G01 code an axis or combination of axes will follow. The axis letter (X, Y, or Z) will include a positive or negative number, which represents the coordinate position of the machine. Additionally, for the G01 code a feedrate value is required. It should also be noted that cutting requires the tool to be rotating. The G01 code and the *"F" code* are also modal and will remain active for each subsequent line until they are changed or cancelled.

When the G01 code along with an axis coordinate position and a feedrate is read (Cycle Start), the specified machine axis or axes will move at a relatively slow rate and cut the workpiece. This is only recommended when it is absolutely certain that a collision will not occur. Typically this is after the CNC program has been Dry Run or proved out. In cases where the feed moves have not been proved out, the CNC operator can override the feedrate to adjust or completely stop the feed movement.

The CNC program format common to most industrial applications is described in the CNC program exercises in the following chapters. The CNC programs in an industrial application should be created and maintained in a format consistent and as "operator-friendly" as possible. This will increase the efficiency of all related CNC operations throughout the plant.

G01 X-axis Feed Move Description for VMC

The feed (cutting) mode initiated with the G01 (linear interpolation) code is illustrated for a VMC in Figure 6-7. The feed code is modal, and moves the spindle and table at the feedrate value programmed, usually 1–100 IPM. The X- and Y-axes move the table and the Z-axis moves the spindle. Whenever a feed move is executed, caution should be taken to avoid tool breakage from an excessive feedrate or a collision. Feed moves are mainly executed for cutting the part material, and therefore the spindle must be rotating (on) prior to reading the G01 command code. The G01 code is a modal command, and is canceled by either the G02, G03, and G00 code.

G01 Y-axis Feed Move Description for VMC

The feed (cutting) mode initiated with the G01 (linear interpolation) code is illustrated for a VMC in Figure 6-8. The feed code is modal, and moves the spindle and table at the feedrate value programmed, usually 1–100 IPM. The X- and Y-axes move the table and the Z-axis moves the spindle. Whenever a feed move is executed, caution should be taken to avoid tool breakage from an excessive feedrate or a collision. Feed moves are mainly executed for cutting the part material, and therefore the spindle must be rotating (on) prior to reading the G01 command code. The G01 code is a modal command, and is canceled by either the G02, G03, and G00 code.

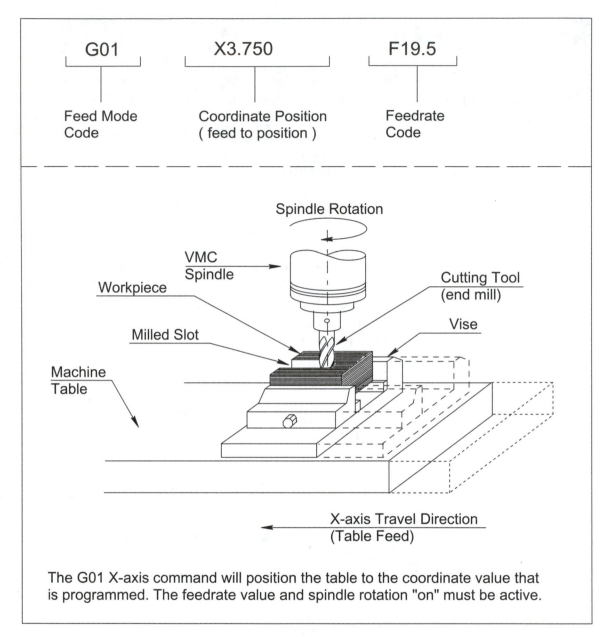

Figure 6-7 *Feed G01 code for VMC, X-axis*

G01 Z-axis Feed Move Description for VMC

The feed (cutting) mode initiated with the G01 (linear interpolation) code is illustrated for a VMC in Figure 6-9. The feed code is modal, and moves the spindle and table at the feedrate value programmed, usually 1–100 IPM. The X- and Y-axes move the table and the Z-axis moves the spindle. Whenever a feed move is executed, caution should be taken to avoid tool breakage from an excessive feedrate or a collision. Feed moves are mainly executed for cutting the part material, and therefore the spindle must be rotating (on) prior to reading the G01 command code. The

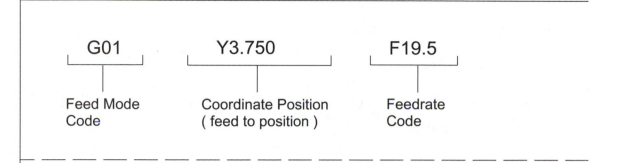

G01	Y3.750	F19.5
Feed Mode Code	Coordinate Position (feed to position)	Feedrate Code

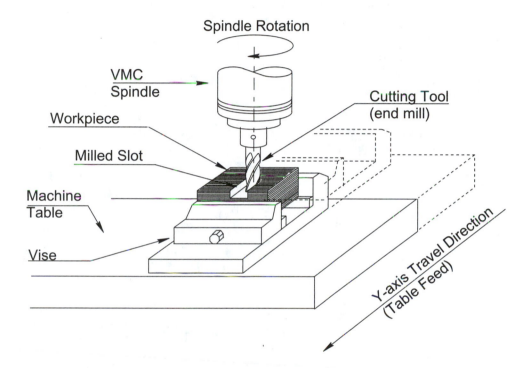

The G01 Y-axis command will position the table to the coordinate value that is programmed. The feedrate value and spindle rotation "on" must be active.

Figure 6-8 Feed G01 code for VMC, Y-axis

G01 code is a modal command, and is canceled by either the G02, G03, and G00 code.

HMC FIVE-AXIS MACHINING

The previous examples of feed moves have illustrated the spindle in fixed positions. However, in cases where a workpiece feature requires machining at an old angle, a CNC machine designed with a C-axis head (see Figure 6-10) may

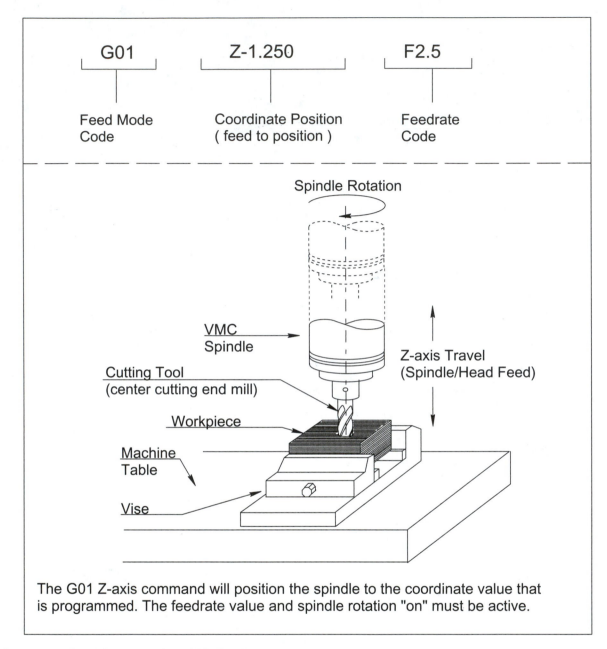

The G01 Z-axis command will position the spindle to the coordinate value that is programmed. The feedrate value and spindle rotation "on" must be active.

Figure 6-9 **Feed G01 code for VMC, Z-axis**

be used. Additionally, the spindle requires a right-angle attachment type holder as illustrated. It is also possible to machine features at odd angles using another setup or special fixturing.

In Position 1, the illustration shows the tool cutter at the typical angle of zero degrees. In Position 2, the spindle C-axis is rotated to the required angle (12°) of the workpiece. This is a five-axis operation because the table (B-axis) is rotated, the X-Y are positioned, the spindle (C-axis) is rotated, and the Z-axis feeds the cutter.

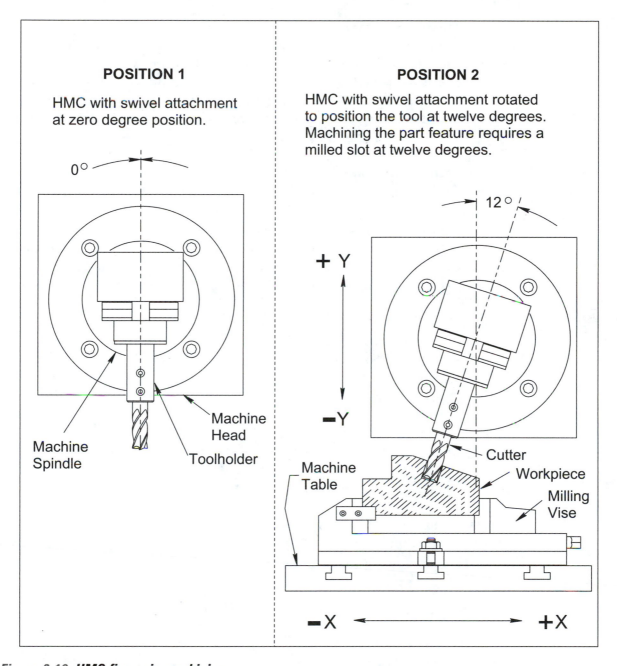

Figure 6-10 **HMC five-axis machining**

INSTRUCTIONS FOR CNC PROGRAM EXAMPLE 1000

This section includes a coordinate worksheet, CNC documentation, and a CNC program for a vertical machining center. Each of these items is designed to illustrate the CNC setup and operation process. Therefore, it is important to understand the purpose and role of each of the following items:

Coordinate Worksheet

The coordinate worksheet is intended to be an instructional aide. Its main purpose is to determine how the X-Y coordinates of the CNC program are developed. It is also designed to help the student identify and read the coordinates in the CNC program example. It is typically not used as a manufacturing document. Obtaining the dimensions from the engineering drawing will be required to complete the coordinate worksheet.

Engineering Drawing

The engineering drawing is primarily used to determine that the workpiece is being manufactured as required. It is used to verify the configuration, dimensions, tolerances, finish requirements, material, and datums of the workpiece.

A *process drawing* is sometimes substituted for the engineering drawing. It is used to define the workpiece requirements more clearly and eliminate items that are not included in that particular operation.

CNC Machine Setup Plan

The setup plan is used to specify what device will be used to hold the workpiece, how the workpiece will be orientated, and where the origin is located. The setup plan can be used to visualize and anticipate the machine moves from the CNC program.

CNC Tool List

The tool list is required to assemble and load the tools in the same order as the CNC program tool change commands. Therefore, the CNC operator can and should reference the tool list to verify the CNC program.

Instructions for CNC Program Example 1000

For this CNC program, the student is required to identify and note the main CNC program items that are listed below:

1. Program number

2. Information notes

3. Tool changes

4. Default code setting blocks

5. Rapid moves

6. Feed moves

The example below illustrates how to identify and note the CNC program items above.

N130 T1 M06	*Auto Tool Change Tool 1*
N140 M01	
N150 G0 G90 X-1.5 Y-.6 S1200 M3	*Rapid, Spindle On CW @ 1200 RPM*
N160 G43 H1 Z.3 M08	*Activate Offset 1, rapid to Z.3, Coolant On*

COORDINATE WORKSHEET 1000

Instructions: Mark the X-Y zero origin, then calculate the coordinates for each point. Record the calculations in the chart below. For the dimensions, use part drawing 1000 on the following page (Figure 6-11).

Use the standard origin symbol at right to mark the X-Y zero.

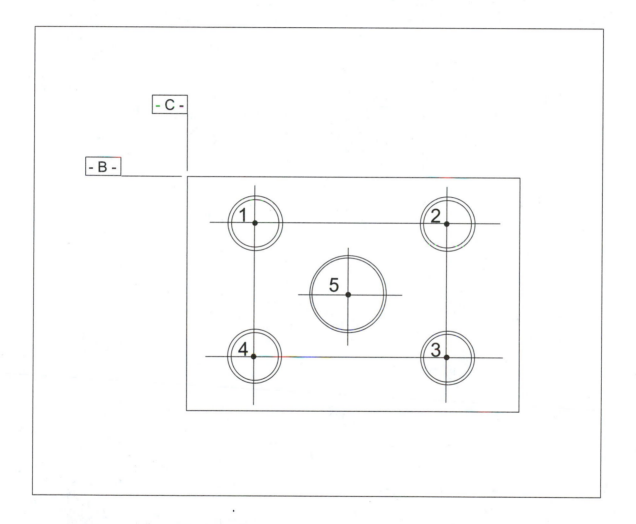

PART NO.	X-COORDINATE	Y-COORDINATE
1		
2		
3		
4		
5		

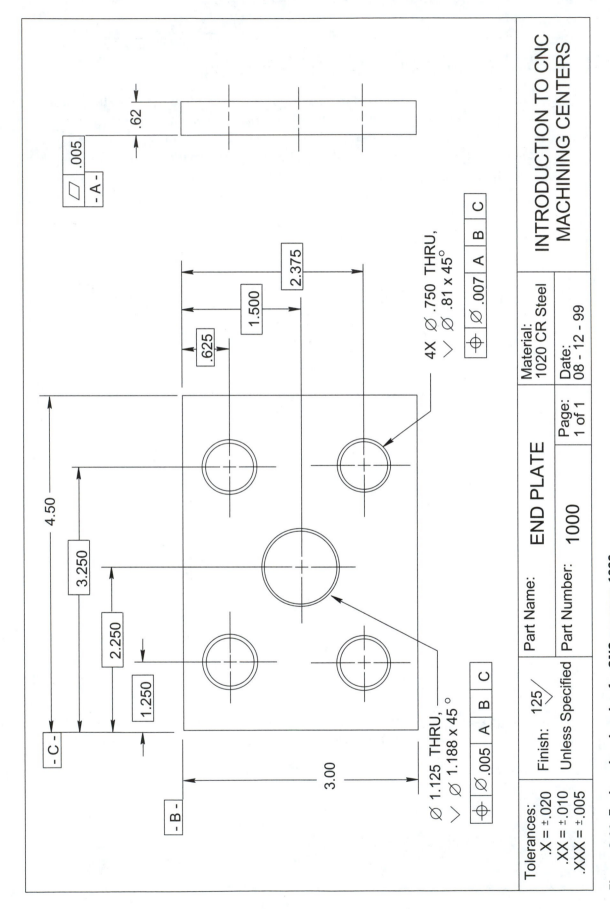

Figure 6-11 Engineering drawing for CNC program 1000

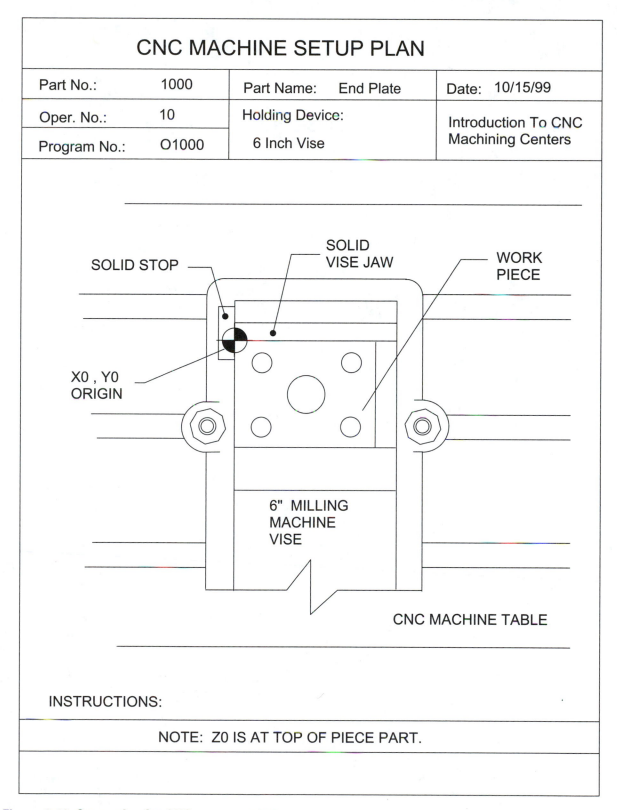

Figure 6-12 Setup plan for CNC program 1000

CNC TOOL LIST AND OPERATION

Part No.: 1000	Prog. No.: O1000	Oper. No.: 10
Part Name: End Plate	Date: 03/08/00	Page 1 of 1

Pocket No.	Length Offset	Radius Offset	Tool No.	Tool Dia.	Tool Description and Operation
T01	H01	D	8001	2.0	2.0" Dia. Face Mill Cutter
T02	H02	D	3002	1.500	90 Degree Spot Drill spot (5) holes
T03	H03	D	1015	.750	.750" Dia Drill Drill (4) holes
T04	H04	D	1007	1.125	1.125 Dia. Drill Drill (1) hole

Figure 6-13 Tool list for CNC program 1000

CNC PROGRAM EXAMPLE 1000

%
O1000
(END PLATE - PROGRAM NO. 1000)
(DATE, 08/12/99)
(TOOL - 01 2.00 DIA. - FACE MILL)
(TOOL - 02 1.500 DIA. - SPOTDRILL)
(TOOL - 03 .750 DIA. - DRILL)
(TOOL - 04 1.125 DIA. - DRILL)

N100 G00 G17 G40 G49 G80 G90
N110 G91 G28 Z0.
N120 G28 X0. Y0.

(TOOL - 01 2.00 DIA. - FACE MILL)
N130 T1 M06
N140 M01
N150 G0 G90 X-1.5 Y-.6 S1200 M3
N160 G43 H1 Z.3 M08
N170 G1 Z0. F50.
N180 X5.75 F24.
N190 Y-2.4
N200 X-1.5
N210 G0 Z1. M09
N220 M05
N230 G91 G49 G28 Z0.

(TOOL - 02 1.50 DIA. - SPOTDRILL)
N240 G00 G17 G40 G49 G80 G90
N250 T2 M06
N260 M01
N270 G0 G90 X1.25 Y-.625 S0535 M3
N280 G43 H2 Z1. M08
N290 G99 G81 Z-.41 R.1 F.01
N300 X3.25
N310 Y-2.375
N320 X1.25
N330 G99 G81 X2.25 Y-1.5 Z-.59 R.1 F.01
N340 G80 M09
N350 M05
N360 G91 G49 G28 Z0.

(TOOL - 03 1.50 DIA. - SPOTDRILL)
N370 G00 G17 G40 G49 G80 G90

CNC PROGRAM EXAMPLE 1000 (*continued*)

```
N380 T3 M06
N390 M01
N400 G0 G90 X1.25 Y-.625 S0535 M3
N410 G43 H3 Z1. M08

N420 G99 G81 Z-.905 R.1 F.01
N430 X3.25
N440 Y-2.375
N450 X1.25
N460 G80 M09
N470 M05
N480 G91 G49 G28 Z0.

(TOOL - 04 1.125 DIA. - DRILL)
N490 G00 G17 G40 G49 G80 G90
N500 T4 M06
N510 M01
N520 G0 G90 X2.25 Y-1.5 S0355 M3
N530 G43 H4 Z1. M08
N540 G99 G81 Z-1.013 R.1 F.01
N550 G80 M09
N560 M05
N570 G91 G49 G28 Z0.
N580 M30
%
```

CHAPTER SUMMARY

The key concepts presented in this chapter include the following:

- The G00 rapid traverse code and machine movement descriptions.

- The G01 feed (linear interpolation) code and machine movement descriptions.

- The descriptions of X-, Y-, and Z-axes movements.

- The description of the C-axis operation and right-angle attachment.

- Calculations for CNC machining center coordinates.

- The description of an engineering drawing for CNC machining centers.

- The description of a setup plan for CNC machining centers.
- The description of a tool list for CNC machining centers.
- The format description of a CNC program for CNC machining centers.
- The basic codes and structure required of a CNC program for CNC machining centers.

REVIEW QUESTIONS

1. Why is a Dry Run and proveout required before running the CNC machine in automatic mode?

2. When moving in rapid G00 mode what must the CNC operator be careful to avoid?

3. When moving in rapid G00 mode, how can the CNC operator slow and stop the machine movement?

4. When moving in feed G01 mode, what must the CNC operator be careful to avoid?

5. When moving in feed G01 mode, how can the CNC operator slow and stop the machine movement?

6. Which VMC machine component moves when the control reads G00 X 2.000? When the control reads G00 Y 2.000? When the control reads G00 Z 2.000?

7. Which HMC machine component moves when the control reads G00 X 2.000? When the control reads G00 Y 2.000? When the control reads G00 Z 2.000?

8. Which HMC machine component moves when the control reads G00 B90.000?

9. For a VMC and an HMC, which codes are required to change a tool?

10. Discuss and identify the CNC program format structure and codes that were covered in this chapter. The items that should be identified include. CNC program number, tool information notes, tool change blocks, default code setting blocks, spindle starts, coolant on, rapid and feed move blocks, machine home command blocks, length offset activation blocks, and end of program block.

CNC PROGRAM EXERCISE FOR VMC

Using the sketch in Figure 6-14, calculate the G00 rapid and G01 feed moves for the X-, Y-, and Z-axes. Enter the coordinates in the CNC program provided on the following page.

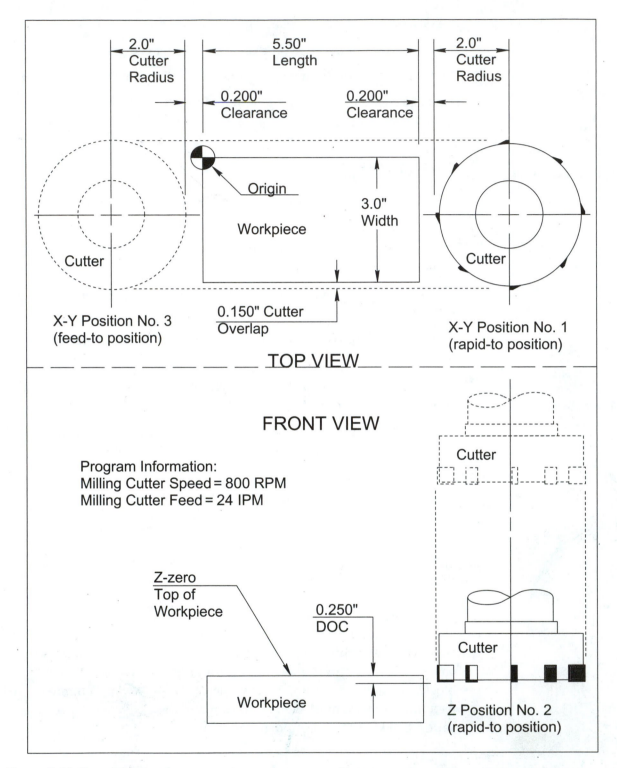

Figure 6-14 ***Program exercise***

■ ■ CNC PROGRAM EXERCISE 1001

Fill in the missing program information of CNC program 1001 as indicated below:

Block N150: enter the G code, the X-Y coordinate values, and RPM value.

Block N170: enter the G code, and Z-coordinate value.

Block N180: enter the G code, the X-coordinate value, and IPM value.

Block N190: enter the G code and the Z-coordinate value to move the cutter 1.0 inch above the top surface of the workpiece.

```
%
O1001
(PLATE - PROGRAM NO. 1001)
(DATE, 05/12/03)
(TOOL - 01 4.00 DIA. - FACE MILL)

N100 G00 G17 G40 G49 G80 G90
N110 G91 G28 Z0.
N120 G28 X0. Y0.

(TOOL - 01 4.00 DIA. - FACE MILL)
N130 T1 M06
N140 M01

N150 _____ X _____ Y _____ S _____ M3
N160 G43 H1 Z.3 M08
N170 _____ Z _____
N180 _____ X _____ F_____
N190 _____ Z _____

N200 M05
N210 G91 G49 G28 Z0. M09
N220 M30
%
```

chapter 7

CNC MACHINING CENTER CIRCULAR INTERPOLATION

Chapter Objectives

After studying and completing this chapter, the student should have knowledge of the following:

- *G02 and G03 circular interpolation codes*
- *M03, M04, M05, and "S" spindle codes*
- *CW radius contour feed movement*
- *CCW radius contour feed movement*
- *Radius contouring features*
- *Radius contouring methods*
- *Calculating X–Y coordinates*
- *CNC program format descriptions*

CIRCULAR INTERPOLATION

This chapter introduces circular interpolation, which is the CNC method used to cut circular features on a workpiece. To perform circular interpolation basically requires a code (G02 or G03), the *arc* radius size, and the starting and ending coordinates of the arc. For CNC machining centers, circular interpolation is typically performed using an end mill cutter to machine the workpiece. Therefore, the examples in this chapter describe and illustrate circular interpolation as it is performed using an end mill.

The first set of illustrations in this chapter, Figures 7-1 through 7-4, describes the radius contouring features and methods that are identical for both VMCs and HMCs. The set of illustrations that follows, Figures 7-5 through 7-8, describes the codes, coordinate values, and movements associated with radius contour milling. These figures are designed to illustrate the differences between CW and CCW radius milling. Additionally, this chapter includes a coordinate calculation worksheet and a sample program to identify basic CNC program format.

As previously mentioned, an arc or radius cut is performed on CNC machining centers by the process called "circular interpolation." The G02 and G03 codes are designated to perform this function. Both codes are modal and each will remain active for all subsequent lines until a G00 or G01 cancels either code. Therefore, any axis coordinate present in a subsequent line of the CNC program will move in radial feed mode whether the G02 or G03 code is present in the subsequent line or not.

G02 CIRCULAR INTERPOLATION CW

The G02 code is used to cut a radius in a clockwise direction (see Figure 7-1). It basically moves the X- and Y-axes simultaneously CW at a predetermined feedrate to cut a specific radius on the workpiece. After the G02 code, a combination of two axis coordinates follows.

The coordinate position of the machine, which is represented by the axis letters (X, Y, or Z), will include a positive- or negative-valued number. Additionally, the G02 code requires a code for the radius value. This value, which is represented by the letter "R" or the letters "I" and "J," is typically derived from the engineering drawing specifications. Because the circular interpolation code is intended for cutting, the spindle or tool must be rotating and a feedrate value (F12.0 for example) is also required.

The CNC radius machining can be performed in the X-Y axes, the X-Z axes, or the Y-Z axes. Each axis combination requires a code that specifies which axes are selected for cutting the radius. The G17 code is used for the X-Y axes. The G18 code is used for the X-Z axes. The G19 code is used for the Y-Z axes. The required code must be read by the control before the G02 code is read.

When cutting a radius in the X-Y axes, a standard end mill with a flat end is used. However, when cutting a radius in the X-Z or Y-Z axes, it is necessary to use a ballnose-style end mill. This type of end mill is necessary because it is designed to cut both radially and axially. Another code commonly used to compensate for the diameter of the end mill is the (G41 or G42) "tool diameter compensation" code.

When the G02 code along with an axis coordinate position is read (Cycle Start), the specified machine axes will move automatically at a relatively slow feedrate and cut the radius on the workpiece. Initiating a Cycle Start is only recommended when it is absolutely certain that a collision will not occur. Typically, this is after the CNC program has been Dry Run or proved out. In cases where the radius feed moves have not been proved out, the CNC operator can override the feedrate to adjust or completely stop the radius feed movement. This will allow the CNC operator time to react to a possible collision, heavy cuts, or an excessive feedrate, and stop the movement before the damage occurs.

CNC machining often requires contour milling, which may include arcs and circles. This requires that the CNC machine generate a circular tool path. This tool path is accomplished by the circular interpolation method. Circular interpolation can be generated in two directions. One direction is CW, the other is CCW.

G02 Circular Interpolation VMC Description

Figure 7-1 describes the items required to create a G02 CW circular tool path. The command requires that a feedrate be specified (F10.5, for example). The

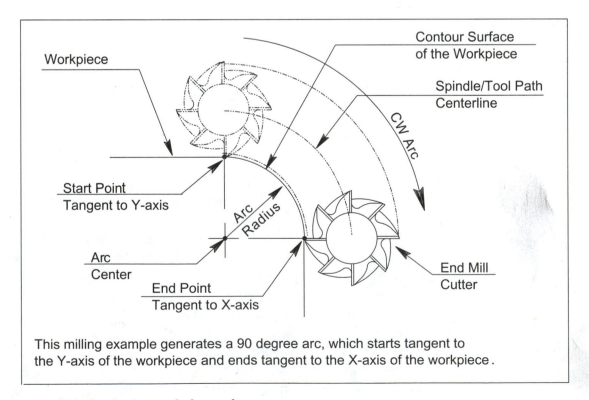

Figure 7-1 G02 circular interpolation code

feedrate can be specified on the same line as the G02 command, or it must be specified prior to the G02 command. The G02 command is a modal command that is canceled by either G00, G01, or G03. The example in Figure 7-1 is an arc that starts and ends tangent to the X-axis and Y-axis.

■ ■ ■ ■
G03 CIRCULAR INTERPOLATION CCW

The G03 code is used to cut a radius in a CCW direction (see Figure 7-2). Similar to the G02 code, it basically moves the X- and Y-axes simultaneously CCW at a predetermined feedrate to cut a specific radius on the workpiece. The G03 and the G02 codes are similar in all respects except direction of cut.

G03 Circular Interpolation VMC Description

Figure 7-2 illustrates the G03 CCW circular interpolation radius machining characteristics. They are similar to CW circular interpolation except that the arc is generated in a CCW direction.

This illustration describes the items required to create a G03 CCW circular tool path. The command requires that a feedrate be specified (F15.5, for example). The feedrate can be specified on the same line as the G03 command, or it must be specified prior to the G03 command. The G03 command is a modal command canceled by either G00, G01, or G02. The example in Figure 7-2 is an arc that starts and ends tangent to the X-axis and Y-axis.

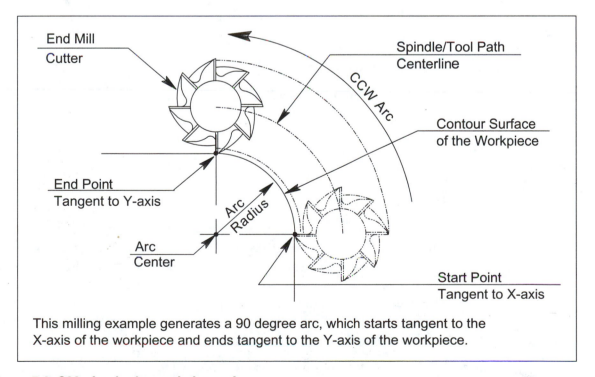

This milling example generates a 90 degree arc, which starts tangent to the X-axis of the workpiece and ends tangent to the Y-axis of the workpiece.

Figure 7-2 G03 circular interpolation code

SPINDLE FUNCTION CODES

As previously stated, metal cutting on a CNC machine requires the tool to be rotating as it feeds into the workpiece. For this reason, the machine spindle is primarily used to rotate the cutting tool at a predetermined speed (RPM). The RPM is first calculated and then programmed by using the "S" *code* along with the numerical RPM value.

M03, M04, and "S" Spindle Function Codes

The M03 code is used to rotate the spindle in a CW direction. The M04 code is used to rotate the spindle in the CCW direction. To start the spindle rotation, the CNC program must read the S code and RPM value (S1525, for example) either before or at the same time that the "M" code is read. It is typically programmed as follows:

S1525 M03

These spindle rotation code descriptions are applied to all cutting situations. The M03, M04, and S codes are required for all types of machining operations. They are programmed and executed in the same manner for all machining situations.

M05 Spindle Stop Code

The M05 code is used in situations when the spindle rotation needs to be stopped during the automatic execution of the CNC program. The M05 spindle stop code is required in situations such as the following:

- Before tool changes
- To inspect a machined feature size
- To remove machining metal chips
- To relieve clamping pressure
- To move a workholding clamp
- To relocate the workpiece
- To unload and load a workpiece
- To manually change a tool

After the M05 code is executed, the spindle will remain stopped until the CNC program restarts it. The control can read these CNC program codes in various format configurations. However, the codes should be structured in a consistent format to increase the efficiency of reading CNC programs.

There are also situations when the spindle rotation is stopped without an M05 code. The spindle rotation is stopped internally by the MCU in situations such as the following:

- When the Emergency Stop button is pressed
- During the execution of an automatic tool change
- During the execution of some canned cycles

In the first two situations the spindle will remain stopped until it is restarted manually, by the CNC program or by MDI. In the case of the canned cycles, the spindle is restarted automatically at the appropriate time.

Contour Milling Description for VMC

CNC machining often requires contour milling, which may include arcs greater than 90°. This circular tool path can be generated by both vertical or horizontal CNC machining centers. The circular interpolation can be generated in either direction, G02 CW or G03 CCW.

Figure 7-3 describes what is required to generate a G02 clockwise circular tool path that is greater than 90°. For this situation, the coordinate points usually require trigonometry calculations. The example in Figure 7-3 is an arc greater than 90°.

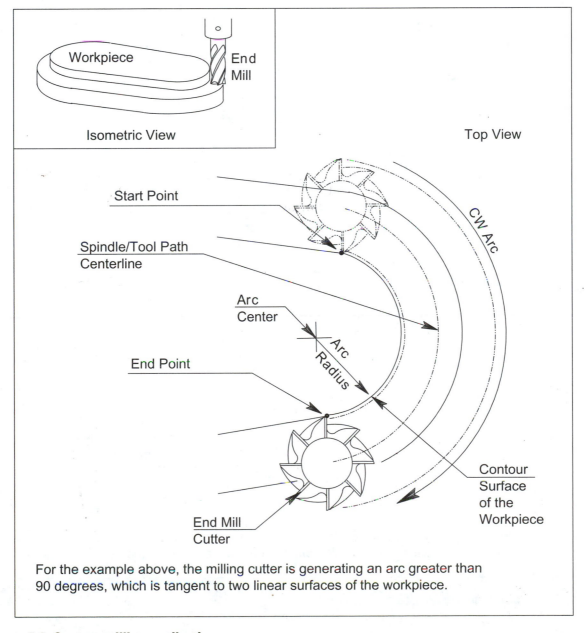

For the example above, the milling cutter is generating an arc greater than 90 degrees, which is tangent to two linear surfaces of the workpiece.

Figure 7-3 Contour milling application

360° Contour Milling Description for VMC

CNC machining often requires contour milling, which may include 360° circles. This circular tool path can also be generated by a CNC vertical or horizontal machine. The circular interpolation can also be generated in either direction, G02 CW or G03 CCW.

Figure 7-4 describes the items required to create a 360° circular tool path. For this situation, the tool starts and ends at the same location.

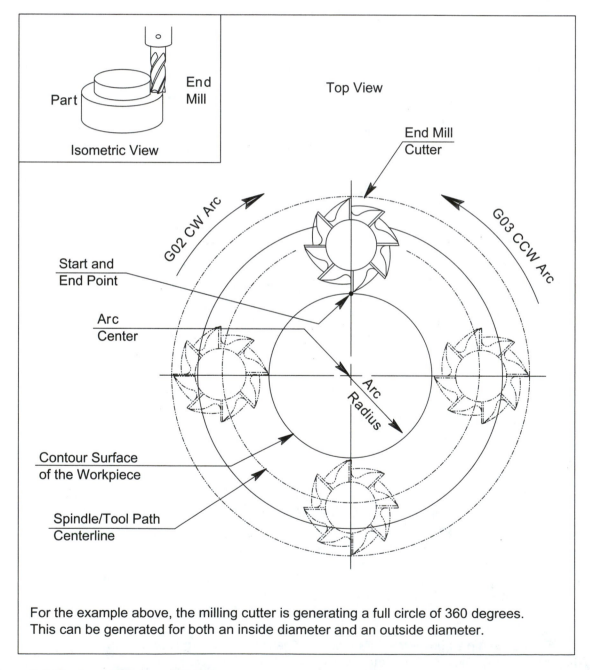

For the example above, the milling cutter is generating a full circle of 360 degrees. This can be generated for both an inside diameter and an outside diameter.

Figure 7-4 Contour milling application

G02 Code Description

A circular tool path can be generated by either of two methods: the "Center Point Method" or the "Radius Method." The example in Figure 7-5, illustrates the center point method in G90 absolute mode. For this example, the feedrate is not specified on the same line as the G02 command, it is specified on the prior block.

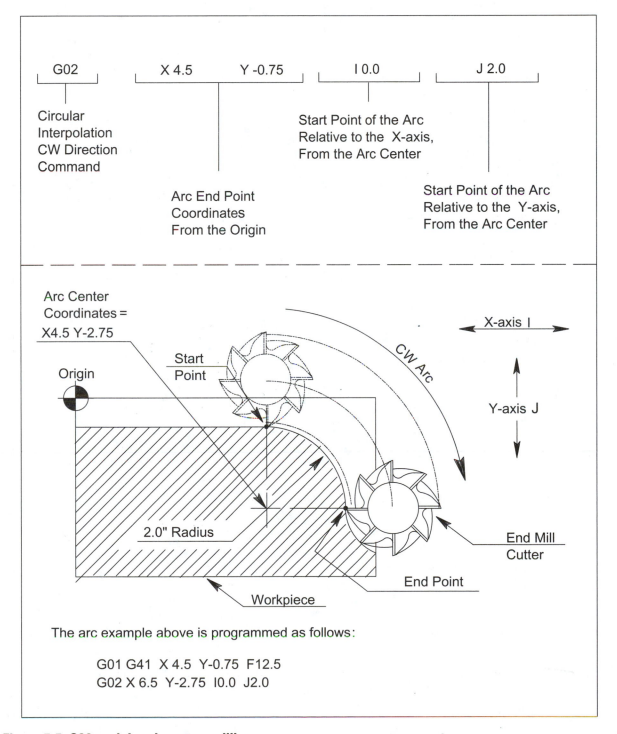

*Figure 7-5 **G02 peripheral contour milling***

G03 Code Description

A circular tool path can be generated by either of two methods: the Center Point Method or the Radius Method. The example in Figure 7-6 illustrates the center point method in G90 absolute mode. For this example, the feedrate is not specified on the same line as the G03 command, it is specified on the prior block.

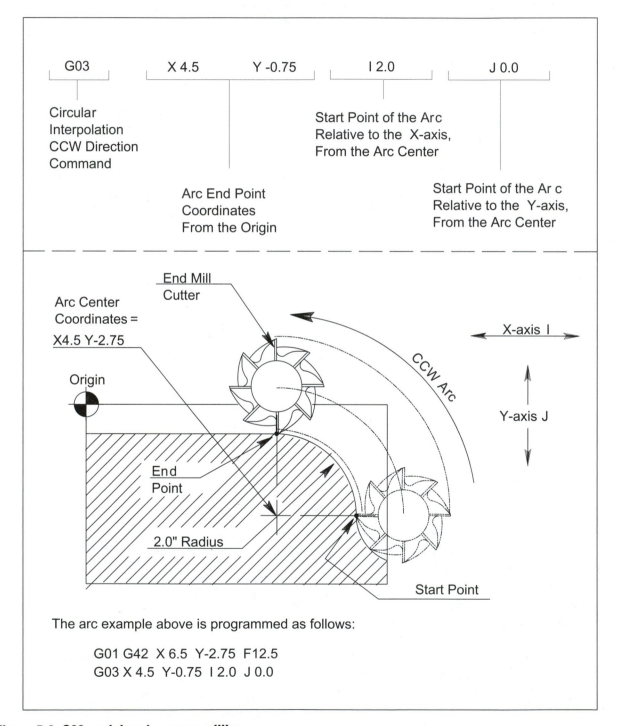

Figure 7-6 *G03 peripheral contour milling*

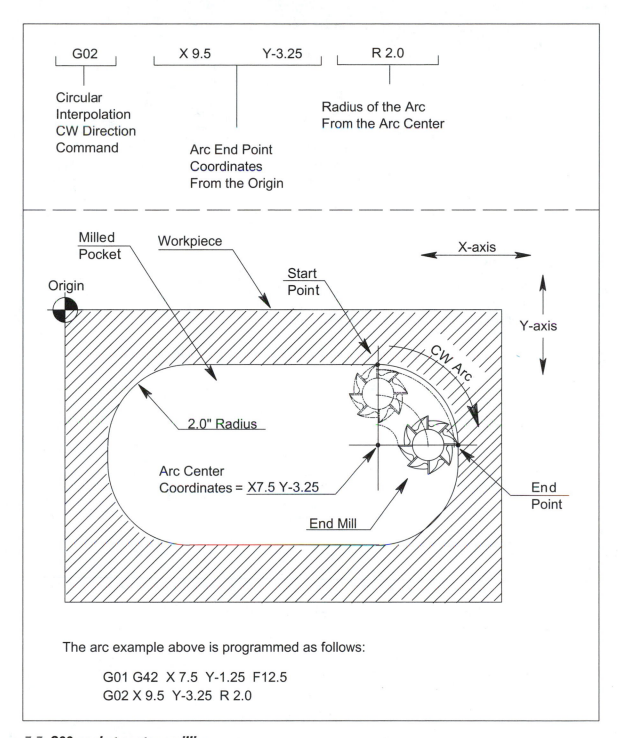

Figure 7-7 ***G02 pocket contour milling***

The example in Figure 7-7 illustrates the radius method in G90 absolute mode. For this example, the feedrate is not specified on the same line as the G02 command, it is specified on the prior block.

The example in Figure 7-8 illustrates the radius method in G90 absolute mode. For this example, the feedrate is not specified on the same line as the G03 command, it is specified on the prior block.

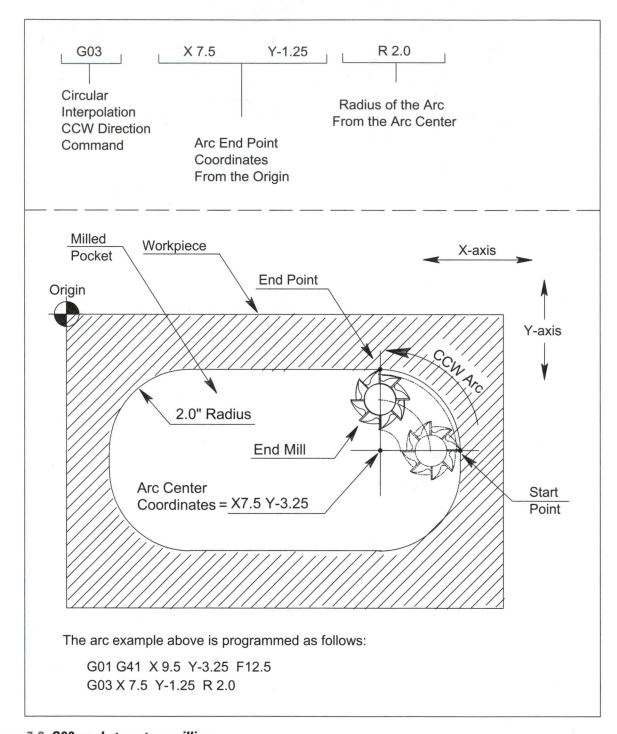

Figure 7-8 G03 pocket contour milling

INSTRUCTIONS FOR CNC PROGRAM EXAMPLE 2000

This section includes a coordinate worksheet, CNC documentation, and a CNC program for a VMC center. As explained in chapter 6, each of these items is designed to illustrate the CNC setup and operation process.

Therefore, it is important to review and understand the purpose and role of each CNC item. These items are listed below:

- Coordinate worksheet (for instructional purposes)
- Engineering drawing/Process drawing
- CNC machine setup plan
- CNC tool list

This program is similar in format to the other program examples. The similarities that should be noted include

- Program number
- Information notes
- Tool changes
- Default code setting
- Spindle starts
- Coolant on
- Machine home command
- Length offset activation
- End of program

Instructions for CNC Program 2000

For this CNC program, the student is required to identify and note the main CNC program items that are listed below:

1. Program number
2. Information notes
3. Tool changes
4. Default code setting blocks
5. Rapid moves
6. Feed moves
7. Radius feed moves
8. End of program

The example below illustrates how to identify and note the CNC program items above.

(TOOL - 01 2.00 DIA. - FACE MILL)	*Tool Information Note*
N20 G00 G90 G17 G40 G49 G80	*Default Code Setting Block*
N30 T5 M06	*Auto Tool Change Tool 5*
N40 M01	*Optional Stop*

COORDINATE WORKSHEET 2000

Instructions: Mark the X-Y zero origin, then calculate the coordinates for each point. Record the calculations in the chart below. For the dimensions, use part drawing 2000 on the following page (Figure 7-9).

Use the standard origin symbol at right to mark the X-Y zero.

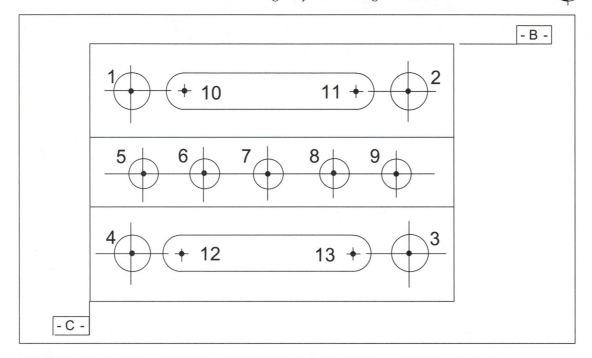

PART NO.	X-COORDINATE	Y-COORDINATE
1		
2		
3		
4		
5		
6		
7		
8		
9		
10		
11		
12		
13		

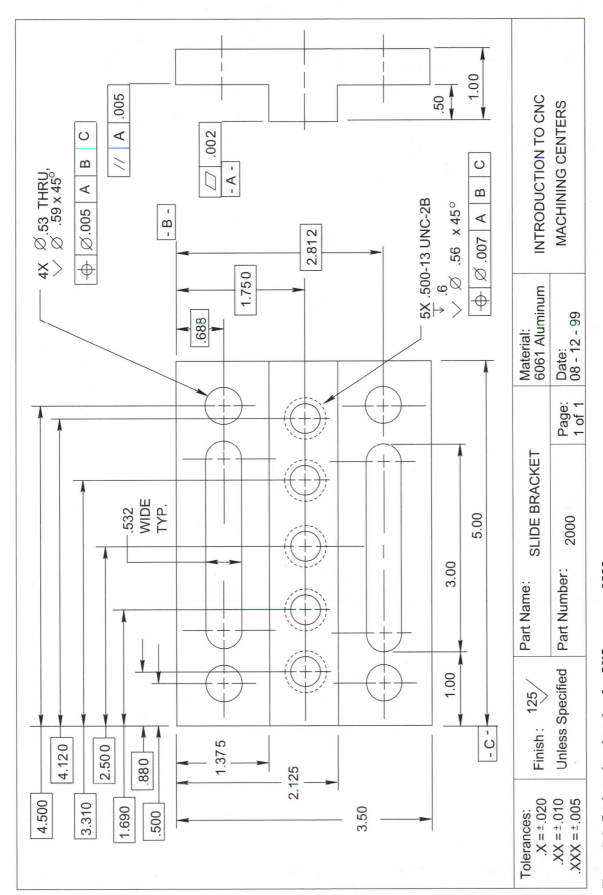

Figure 7-9 Engineering drawing for CNC program 2000

165

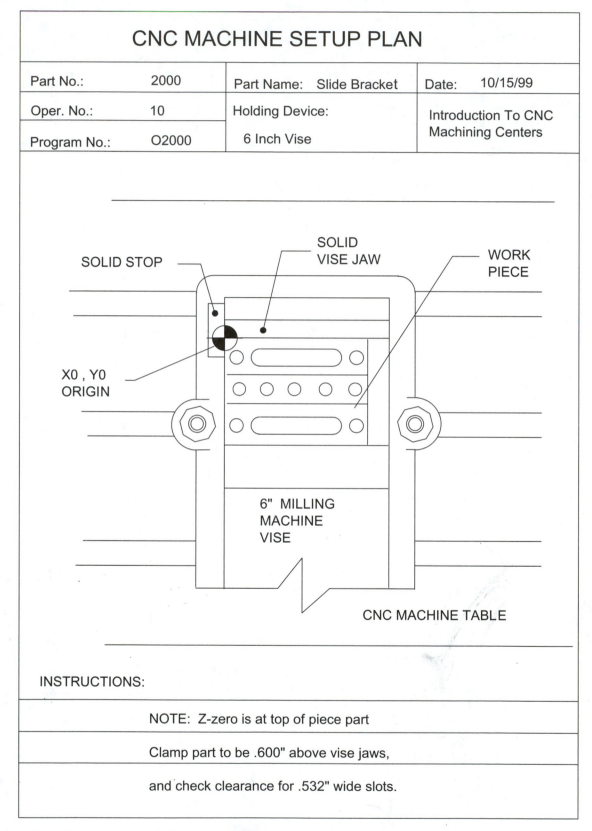

CNC MACHINE SETUP PLAN

Part No.:	2000	Part Name: Slide Bracket	Date: 10/15/99
Oper. No.:	10	Holding Device:	Introduction To CNC Machining Centers
Program No.:	O2000	6 Inch Vise	

SOLID STOP

SOLID VISE JAW

WORK PIECE

X0 , Y0 ORIGIN

6" MILLING MACHINE VISE

CNC MACHINE TABLE

INSTRUCTIONS:

NOTE: Z-zero is at top of piece part

Clamp part to be .600" above vise jaws,

and check clearance for .532" wide slots.

Figure 7-10 Set-up plan for CNC program 2000

CNC TOOL LIST AND OPERATION

Part No.: 2000	Prog. No.: O2000	Oper. No.: 10
Part Name: Slide Bracket	Date: 10/15/99	Page 1 of 1

Pocket No.	Length Offset	Radius Offset	Tool No.	Tool Dia.	Tool Description and Operation
T01	H01	D	8001	2.0	2.0" Dia. End Mill Cutter, 4-Flute, HSS. Mill (2) Steps
T02	H02	D	3002	1.0	90 Degree Spot Drill spot (8) holes
T03	H03	D	1007	.437	.437 Dia. Drill Drill (5) holes
T04	H04	D	1015	.531	.53" Dia Drill Drill (4) holes
T05	H05	D	5021	.500	.500-13 UNC Tap Tap (5) holes
T06	H06	D	7012	.375	.375 Dia. End Mill, 2-Flute, CC, HSS. Mill (2) .532 Slots

Figure 7-11 **Tool list for CNC program 2000**

■ ■ ■ ■

CNC PROGRAM EXAMPLE 2000

```
%
O2000
(SLIDE BRACKET - PROGRAM 2000)
(DATE, 08/12/99)
(TOOL - 01 2.00 DIA. - END MILL)
(TOOL - 02 1.000 DIA. - SPOTDRILL)
(TOOL - 03  .437 DIA. - DRILL)
(TOOL - 04  .531 DIA. - DRILL)
(TOOL - 05  .500-13 UNC - TAP)
(TOOL - 06  .375 DIA. - END MILL)
N010 G00 G40 G49 G80 G90
N015 G91 G28 Z0.
N020 G28 X0. Y0.

(TOOL - 01 2.00 DIA. - END MILL)
N025 T1 M06
N030 G0 G90 X6.1 Y-.875 S1145 M3
N035 G43 H1 Z.2 M08
N040 G1 Z0. F5.73
N045 X-1.1 F11.45
N050 Y-2.625
N055 X6.1
N060 G0 Z.2
N065 Y-.375
N070 G1 Z-.24 F5.73
N075 X-1.1 F11.45
N080 G0 Z.2
N085 X6.1
N090 G1 Z-.48 F5.73
N095 X-1.1 F11.45
N100 G0 Z.2
N105 X6.1
N110 G1 Z-.5 F5.73
N115 X-1.1 F11.45
N120 G0 Z.2
N125 Y-3.125
N130 G1 Z-.24 F5.73
N135 X6.1 F11.45
N140 G0 Z.2
N145 X-1.1
N150 G1 Z-.48 F5.73
N155 X6.1 F11.45
N160 G0 Z.2
```

■ ■ ■ ■

CNC PROGRAM EXAMPLE 2000 (*continued*)

N165 X-1.1
N170 G1 Z-.5 F5.73
N175 X6.1 F11.45
N180 G0 Z.2 M09
N185 G91 G49 G28 Z0. M05

(TOOL — 02 1.000 DIA. — SPOTDRILL)
N190 G00 G40 G49 G80 G90
N195 T2 M06
N200 M01
N205 G0 G90 X.5 Y-.688 S0500 M3
N210 G43 H2 Z1. M08
N215 G98 G81 Z-.795 R-.4 F12.
N220 X4.5
N225 Y-2.812
N230 X.5
N235 G80
N240 G0
N245 G99 G81 X.88 Y-1.75 Z-.28 R.1 F12.
N250 X1.69
N255 X2.5
N260 X3.31
N265 X4.12
N270 G80 M09
N275 G91 G49 G28 Z0. M05

(TOOL - 03 .437 DIA. - DRILL)
N280 G00 G40 G49 G80 G90
N285 T3 M06
N290 M01
N295 G0 G90 X.88 Y-1.75 S0915 M3
N300 G43 H3 Z1. M08
N305 G99 G81 Z-.85 R.1 F10.
N310 X1.69
N315 X2.5
N320 X3.31
N325 X4.12
N330 G80 M09
N335 G91 G49 G28 Z0. M05

(TOOL - 04 .531 DIA. - DRILL)
N340 G00 G40 G49 G80 G90
N345 T4 M06
N350 M01

CNC PROGRAM EXAMPLE 2000 (*continued*)

```
N355 G0 G90 X.5 Y-.688 S0900 M3
N360 G43 H4 Z1. M08
N365 G98 G81 Z-1.209 R-.4 F15.
N370 X4.5
N375 Y-2.812
N380 X.5
N385 G80 M09
N390 G91 G49 G28 Z0. M05

(TOOL - 05  .500-13 UNC - TAP)
N395 G00 G40 G49 G80 G90
N400 T5 M06
N405 M01
N410 G0 G90 X.88 Y-1.75 S0400 M3
N415 G43 H5 Z1. M08
N420 G99 G84 Z-.6 R.2 F30.77
N425 X1.69
N430 X2.5
N435 X3.31
N440 X4.12
N445 G80 M09
N450 G91 G49 G28 Z0. M05

(TOOL - 06  .375 DIA. - END MILL)
N455 G00 G40 G49 G80 G90
N460 T6 M06
N465 M01
N470 G0 G90 X1.266 Y-.688 S1055 M3
N475 G43 H6 Z1. M08
N480 G0 Z-.4
N485 G1 Z-.76 F5.0
N490 Y-.6095 F22.
N495 X3.734
N500 G2 Y-.7665 I0. J-.0785
N505 G1 X1.266
N510 G2 Y-.6095 I0. J.0785
N515 G1 Z-1.02 F5.0
N520 G1 X3.734 F22.
N525 G2 Y-.7665 I0. J-.0785
N530 G1 X1.266
N535 G2 Y-.6095 I0. J.0785
N540 G0 Z1.
N545 Y-2.812
```

CNC PROGRAM EXAMPLE 2000 (*continued*)

```
N550 Z-.4
N555 G1 Z-.76 F5.0
N560 Y-2.7335 F22.
N565 G1 X3.734
N570 G2 Y-2.8905 I0. J-.0785
N575 G1 X1.266
N580 G2 Y-2.7335 I0. J.0785
N585 G1 Z-1.02 F5.0
N590 G1 X3.734 F22.
N595 G2 Y-2.8905 I0. J-.0785
N600 G1 X1.266
N605 G2 Y-2.7335 I0. J.0785
N610 G0 Z.1M09
N615 G91 G49 G28 Z0. M05
N620 G90
N625 M30
%
```

CHAPTER SUMMARY

The key concepts presented in this chapter include the following:

- The G02 CW circular interpolation code and machine movement descriptions.

- The G03 CCW circular interpolation code and machine movement descriptions.

- The Center Point Method circular interpolation code descriptions.

- The Radius Method circular interpolation code descriptions.

- The M03 spindle CW rotation code description.

- The M04 spindle CCW rotation code description.

- The S spindle RPM speed code description.

- The M05 spindle stop code description.

- The description of contour milling on CNC machining centers.

- Calculations for CNC machining center coordinates.

- The format description of a CNC program for CNC machining centers.

- The basic codes required of a CNC program for CNC machining centers.

■ ■ REVIEW QUESTIONS

1. Which type of tool is used for cutting a radius (G02 or G03) on CNC machining centers?

2. What machine feature does the operator use to optimize the programmed feed moves during the proveout?

3. Which VMC components move when the control reads G02 X2.0 Y1.5 R.5?

4. Which HMC components move when the control reads G02 X2.0 Y1.5 R.5?

5. What is the difference between the Center Point Method and the Radius Method?

6. Which VMC component is activated when the control reads an M03 or M04?

7. Which CNC code controls the spindle speed for machining centers?

8. Which CNC code is used to stop the spindle?

9. Which code in a CNC program signifies that the program is finished?

10. Discuss and identify the CNC program format structure and codes covered in this chapter. The items that should be identified include CNC program number, tool information notes, tool change blocks, default code setting blocks, spindle starts, coolant on and off, machine home command blocks, length offset activation blocks, circular interpolation blocks, and end of program block.

■ ■ CNC PROGRAM EXERCISE FOR VMC

Using the Figure 7-12, calculate the G02 and G03 circular feed moves for the X- and Y-axes. Enter the coordinates in the CNC program provided on the following page.

■ ■ CNC PROGRAM EXERCISE 2001

Fill in the missing program information of CNC program 2001 as indicated below:

Block N180: enter the x-coordinate, and IPM value.

Block N190: enter the G code, the X-Y coordinate value, and R value.

Block N200: enter the G code, the X-Y coordinate value, and R value.

Block N210: enter the G code, and the X-coordinate value.

Block N220: enter the G code, the X-Y coordinate value, and R value.

Block N230: enter the G code, and the Y-coordinate value.

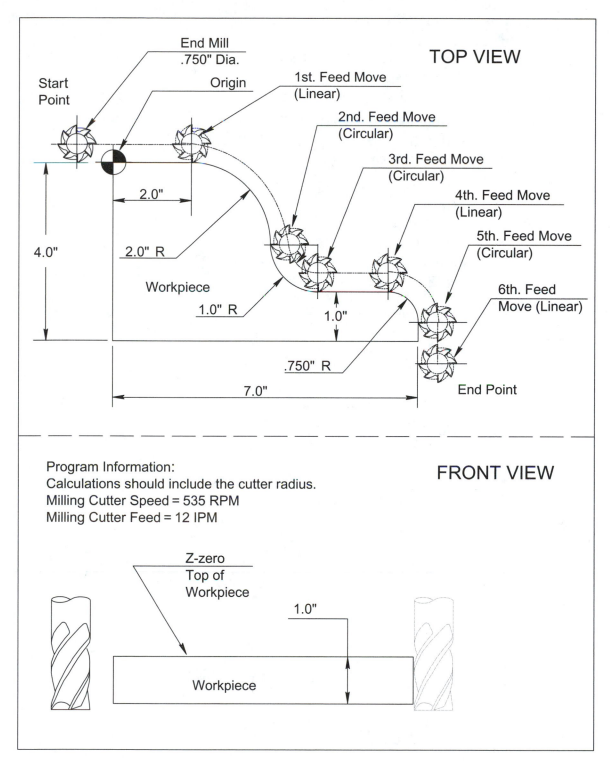

Figure 7-12 *Program exercise*

```
%
O2001
(CAM - PROGRAM NO. 2001)
(DATE, 06/22/03)
(TOOL - 01 .750 DIA. - END MILL)

N100 G00 G17 G40 G49 G80 G90
N110 G91 G28 Z0.
N120 G28 X0. Y0.

(TOOL - 01 .750 DIA. - END MILL)
N130 T1 M06
N140 M01
N150 G0 X-0.9 Y0.375 S535 M3
N160 G43 H1 Z.3 M08
N170 G1 Z-1.1 F25.

N180 X _____ F _____
N190 G _____ X _____ Y _____ R _____
N200 G _____ X _____ Y _____ R _____
N210 G _____ X _____
N220 G _____ X _____ Y _____ R _____
N230 G _____ Y _____
N240 M05
N250 G91 G49 G28 Z0. M09
N260 M30
%
```

chapter 8

CNC CUTTER DIAMETER COMPENSATION

Chapter Objectives

After studying and completing this chapter, the student should have knowledge of the following:

- *G41 cutter diameter compensation left*

- *G42 cutter diameter compensation right*

- *G43 tool length compensation*

- *G40 and G49 cancel codes*

- *G28 machine home return*

- *Calculating X-Y coordinates*

- *CNC program format descriptions*

CUTTER DIAMETER COMPENSATION

The set of illustrations in this chapter (Figures 8-1 to 8-6) describe the *cutter diameter compensation* features, codes, and methods. Cutter compensation is typical for both VMCs and HMCs. CNC program 3000 applies diameter and length compensation codes, coordinate values, and movements as it contour mills the part. Additionally, this chapter includes other miscellaneous codes, a coordinate calculation worksheet, and a sample program to familiarize the reader with the basic CNC program format structure.

Cutter Diameter Compensation Codes

The typical application for cutter diameter compensation is contour end milling. This cutter compensation can be calculated into the CNC program coordinate moves, or the G41 and G42 codes can be used to compensate for the radius of the end mill. These codes, once activated, will command the CNC machine to compensate appropriately.

G41 Cutter Diameter Compensation Left

The G41 cutter diameter compensation code is used to shift the axes (X,Y) by a specified radius amount, which is half of the cutter diameter. The CNC operator or setup person is responsible for entering this value into the offset memory. The letter "D" offset number of the CNC program must match the cutter specified on the tool list and the offset memory number. When the control reads the G41 code, any axis letter (X or Y) will shift by the amount stored in the memory. It is important to note that while cutter diameter compensation is active, the cutter can change direction with another axis (X to Y, or Y to X), but if the same axis (X to X, or Y to Y) changes direction 180°, an alarm will occur. Additionally, because this code is intended for cutting applications, the cutter must be rotating and a feedrate value is required.

When the G41 code along with an axis coordinate position is read (Cycle Start), the specified machine axes will shift and move automatically to cut the workpiece. This is only recommended when it is absolutely certain that alarms or collisions will not occur. Typically this is after the CNC program has been Dry Run or prove out. In cases where the compensation moves have not been proved out, the CNC operator can use the override features to slow the axis movement or completely stop the axis movement. This will allow the CNC operator time to react to a possible collision and stop the movement before any damage occurs.

G42 Cutter Diameter Compensation Right

The G42 diameter compensation code is similar to the G41 except that it is used to shift the cutter in the opposite direction (right). All the items described for the G41 code also apply to the G42 code.

G40 Cutter Diameter Compensation Cancel

The G41 and G42 diameter compensation codes are both modal, and therefore require the G40 cancel code when the cutting is finished or when a change of direction is required.

G43 Tool Length Compensation

The G43 tool length compensation code is used to shift the Z-axis by a specified length amount.

The amount is the actual tool length measurement (Figure 3-9) that the CNC operator or setup person is responsible for entering into the length offset memory. The letter "H" offset number of the CNC program must match the tool specified on the tool list and the offset memory number. When the control reads the G43 code, it will shift the spindle (Z-axis) by the amount stored in the memory. It is important to note that the G43 tool length compensation is used to shift the spindle and cutter in a negative direction toward the table and workpiece.

When the G43 code along with a Z-axis coordinate position is read (Cycle Start), the spindle axis will shift and move automatically to cut the workpiece. This is only recommended when it is absolutely certain that a collision will not occur. Typically, this is after the CNC program has been Dry Run or proved out. In cases where the compensation moves have not been proved out, the CNC operator can use the override features to slow the axis movement or completely stop the axis movement. This will allow the CNC operator time to react to a possible collision and stop the movement before any damage occurs. This is required for each tool length offset in the CNC program.

In most cases, the cutting tool controls the depth of the feature that it machines. Some depths may require a higher degree of accuracy. This is determined by the tolerance on the engineering drawing. In these situations, the CNC operator or setup person must first measure the depth of the feature, determine if an adjustment is required, then adjust the depth by adding or subtracting the required amount from the length offset. It is important to note that if the adjustment amount exceeds 0.100 inch, the CNC program or tool length offset should be checked for a possible error.

G49 Tool Length Compensation Cancel

The G43 length compensation code is modal, and therefore it requires the G49 cancel code when the cutting is finished. This is required for every tool that is in the CNC program.

G28 Machine Home Return

The CNC machine must be homed in all axes (X, Y, Z, and B for HMC) before the CNC program can be cycled. The CNC machine axes can be homed manually with the control, or automatically with the G28 code. The G28 code can be cycled either by MDI mode or by the CNC program. In either case, each axis (X, Y, Z, or B) must be specified when the home command is initiated.

The command to home only the Z-axis would read **G28 Z0.0;** to home the X- and Y-axes, it would read **G28 X0.0 Y0.0;** and so on. Additionally, the tool length compensation must be canceled (G49) for the Z-axis prior to the G28 command. A G28 Z-axis machine home command is required prior to each manual or automatic tool change command. In order to avoid collisions, the Z-axis is usually the first axis to be homed. The CNC operator can use the override features to slow the axis movement or completely stop it while visually ensuring that a collision will not occur when homing any machine axis.

These CNC program codes discussed are common in format and application to most CNC programs used in industrial applications. The format should be consistent in order to increase the efficiency of all related CNC operations.

G41 and G42 Code Description: External Cutting

Cutter compensation is used on milling applications. As previously discussed, there are two G codes that activate cutter compensation. They are G41 for left cutter compensation and G42 for right cutter compensation. When cutter compensation is activated, the cutter shifts by an amount equal to the radius of the cutter diameter and tangent to the machined surface. To determine which code to use, the cutter is viewed from behind as it feeds away. If the cutter is to the left of the workpiece, surface G41 is used. If it is to the right of the workpiece surface, then G42 is used.

In Figure 8-1, various end mill cutting and compensation direction examples are illustrated. However, the spindle only cuts with one tool and on one surface as illustrated in isometric view example.

G41 and G42 Code Description: Pocket Cutting

Cutter compensation is used for both external and internal milling applications. When cutter compensation is activated for internal surfaces such as the milled pocket in Figure 8-2, the cutter shifts in the same manner as an external surface. Likewise, to determine which code to use, the cutter is also viewed from behind as it feeds away. If the cutter is to the left of the workpiece surface, G41 is used. If it is to the right of the workpiece surface, then G42 is used.

In Figure 8-2, various end mill cutting and compensation direction examples are illustrated. However, the spindle only cuts with one tool and on one surface as illustrated in the isometric view example.

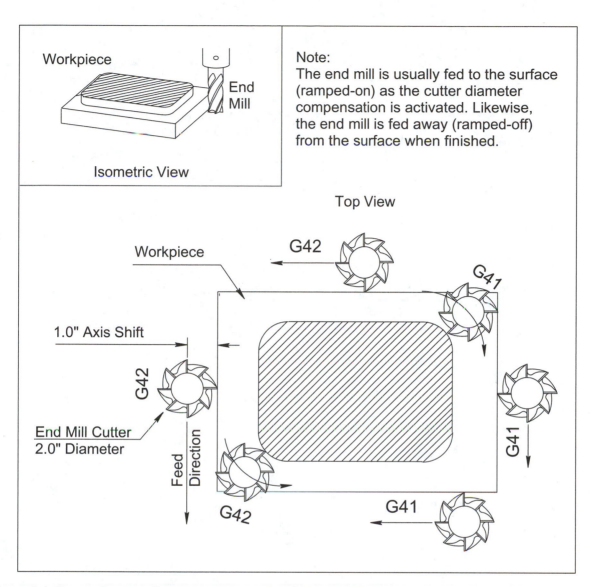

Figure 8-1 *External cutter diameter compensation G41 and G42 codes*

■ ■ ■ ▪

G41 CUTTER DIAMETER COMPENSATION VMC DESCRIPTION

The G41 cutter diameter compensation left code is illustrated in Figures 8-3 and 8-4. These illustrations describe the features associated with cutter diameter compensation left. The G41 code is executed before cutting the workpiece surface. The command is usually activated as described and illustrated with an axis command move.

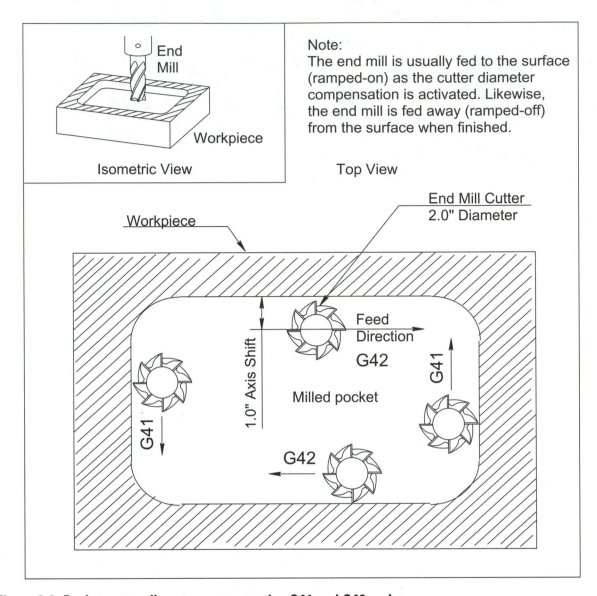

Figure 8-2 Pocket cutter diameter compensation G41 and G42 codes

■ ■ ■ ■

G42 CUTTER DIAMETER COMPENSATION
VMC DESCRIPTION

The G42 cutter diameter compensation right code is illustrated in Figures 8-5 and 8-6. These illustrations describe the features associated with cutter diameter compensation right. The G42 code is executed before cutting the workpiece surface. The command is usually activated as described in the figure with an axis command move.

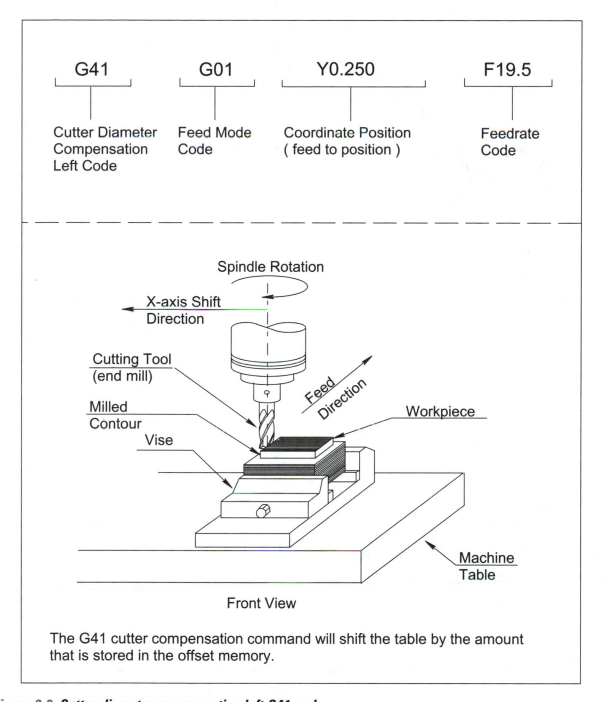

G41	G01	Y0.250	F19.5
Cutter Diameter Compensation Left Code	Feed Mode Code	Coordinate Position (feed to position)	Feedrate Code

Spindle Rotation

X-axis Shift Direction

Cutting Tool (end mill)

Feed Direction

Milled Contour

Workpiece

Vise

Machine Table

Front View

The G41 cutter compensation command will shift the table by the amount that is stored in the offset memory.

Figure 8-3 **Cutter diameter compensation left G41 code**

INSTRUCTIONS FOR CNC PROGRAM EXAMPLE 3000

This section includes a coordinate worksheet, CNC documentation, and a CNC program for a VMC. As explained previously, each of these items is designed to illustrate the CNC setup and operation process. Therefore, it is

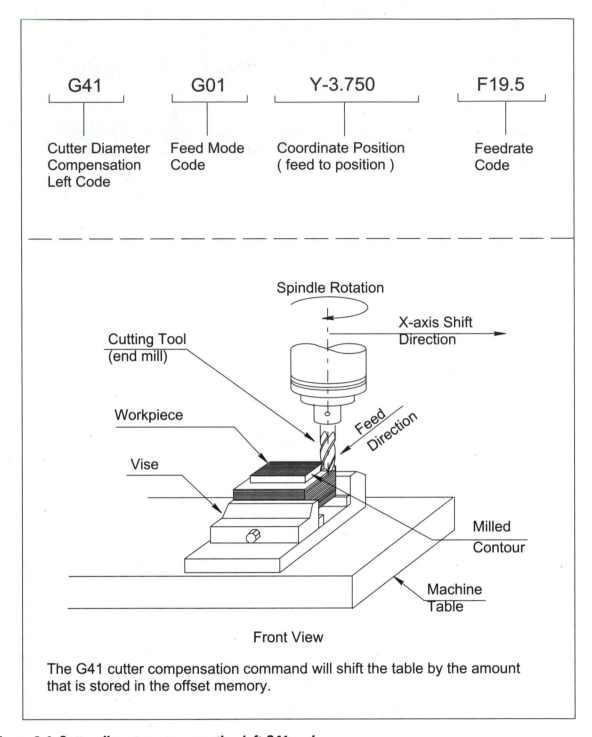

G41 — Cutter Diameter Compensation Left Code

G01 — Feed Mode Code

Y-3.750 — Coordinate Position (feed to position)

F19.5 — Feedrate Code

Spindle Rotation

X-axis Shift Direction

Cutting Tool (end mill)

Workpiece

Vise

Feed Direction

Milled Contour

Machine Table

Front View

The G41 cutter compensation command will shift the table by the amount that is stored in the offset memory.

Figure 8-4 Cutter diameter compensation left G41 code

important to review and understand the purpose and role of each CNC item. These items are listed below:

- Coordinate worksheet (for instructional purposes)

- Engineering drawing or process drawing

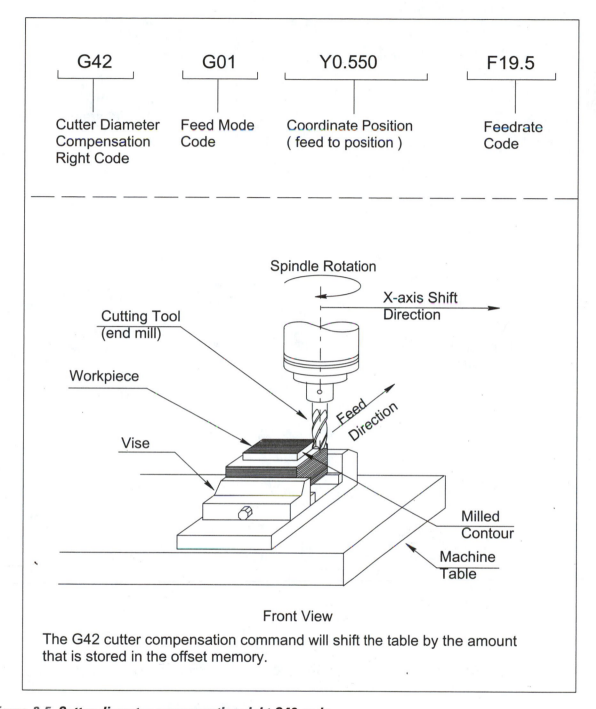

Figure 8-5 *Cutter diameter compensation right G42 code*

- CNC machine setup plan
- CNC tool list

This program is similar in format to the other program examples. The similarities that should be noted include

- Program number
- Information notes

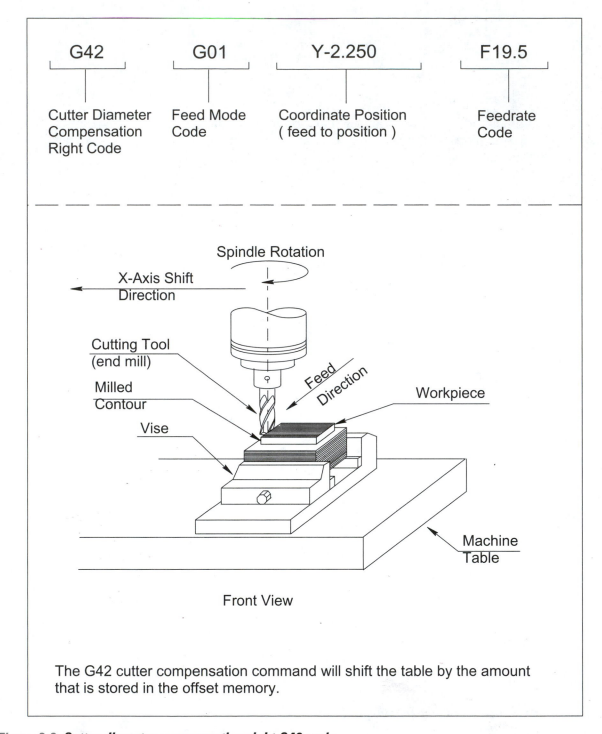

G42	G01	Y-2.250	F19.5
Cutter Diameter Compensation Right Code	Feed Mode Code	Coordinate Position (feed to position)	Feedrate Code

Spindle Rotation

X-Axis Shift Direction

Cutting Tool (end mill)

Milled Contour

Vise

Feed Direction

Workpiece

Machine Table

Front View

The G42 cutter compensation command will shift the table by the amount that is stored in the offset memory.

Figure 8-6 *Cutter diameter compensation right G42 code*

- Tool changes

- Default code setting

- Spindle starts

- Coolant on

- Machine home command

- Length offset activation

- End of program

Instructions for CNC Program Example 3000

For this CNC program, the student is required to identify and note the main CNC program items that are listed below:

- Program number

- Information notes

- Tool changes

- Default code setting blocks

- Rapid moves

- Feed moves

- Radius feed moves

- Tool radius compensation codes

- End of program

The example below illustrates how to identify and note the CNC program items above.

N600 G1 X1.266	*Feed Move X1.266*
N605 G2 Y-2.7335 I0. J.0785	*Radius CW Feed Move*
N610 G0 Z.1M09	*Rapid Move Z.1, Coolant Off*
N615 G91 G49 G28 Z0. M05	*Machine Home Z0, Length Offset Cancel, Spindle Stop*
N620 G90	*Absolute Mode*
N625 M30	*End of Program*

COORDINATE WORKSHEET 3000

Instructions: Mark the X-Y zero origin, then calculate the coordinates for each point. Record the calculations in the chart. For the dimensions, use part drawing 3000 illustrated in Figure 8-7.

Use the standard origin symbol at right to mark the X-Y zero.

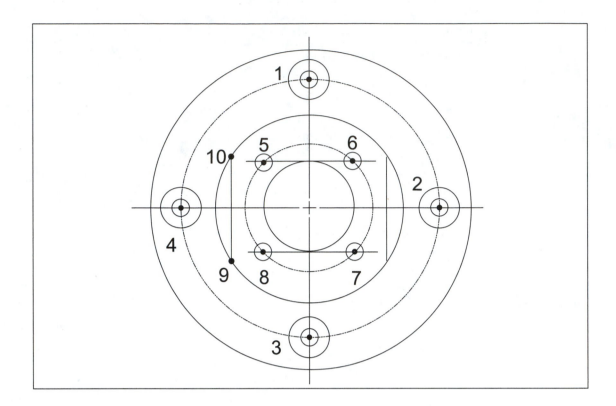

PART NO.	X-COORDINATE	Y-COORDINATE
1		
2		
3		
4		
5		
6		
7		
8		
9		
10		

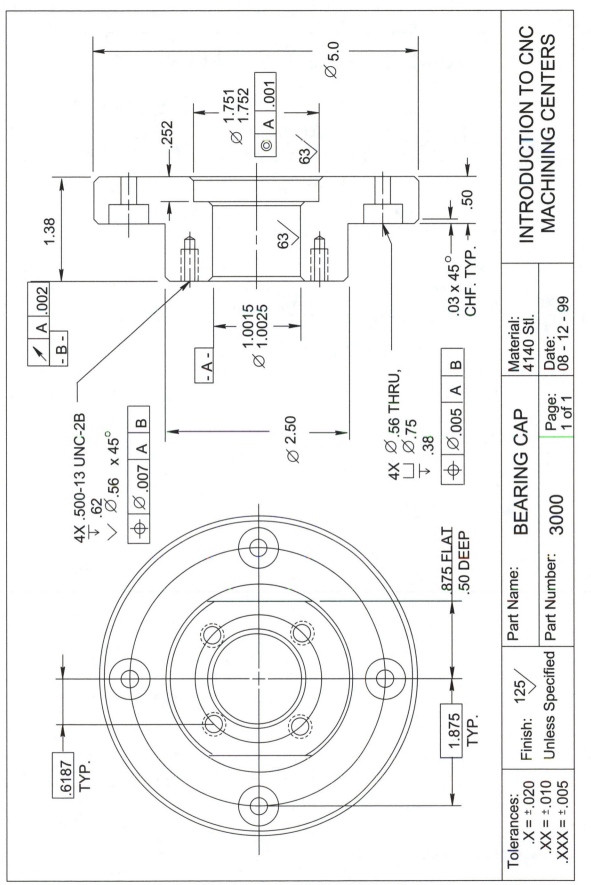

Figure 8-7 *Engineering drawing for CNC program 3000*

.6187 TYP.

1.875 TYP.

.875 FLAT
.50 DEEP

4X .500-13 UNC-2B
⤸ .62
∨ ⌀.56 × 45°
⌖ ⌀.007 A B

- A -

⌀ 1.0015
⌀ 1.0025

⌀ 2.50

4X ⌀.56 THRU,
⌀.75
⌴ .38
⌖ ⌀.005 A B

1.38

.252

⌀ 1.751
⌀ 1.752
◎ A .001

63

63

⌀ 5.0

.50

.03 × 45°
CHF. TYP.

⌗ A .002
- B -

Part Name:	BEARING CAP	Material: 4140 Stl.
Part Number:	3000	Page: 1 of 1	Date: 08 - 12 - 99

Tolerances:
.X = ±.020
.XX = ±.010
.XXX = ±.005

Finish: 125

Unless Specified

INTRODUCTION TO CNC
MACHINING CENTERS

CNC MACHINE SETUP PLAN

Part No.:	3000	Part Name: Bearing Cap	Date: 10/15/99
Oper. No.:	10	Holding Device:	Introduction To CNC Machining Centers
Program No.:	O3000	3-jaw Chuck & Soft Jaws	

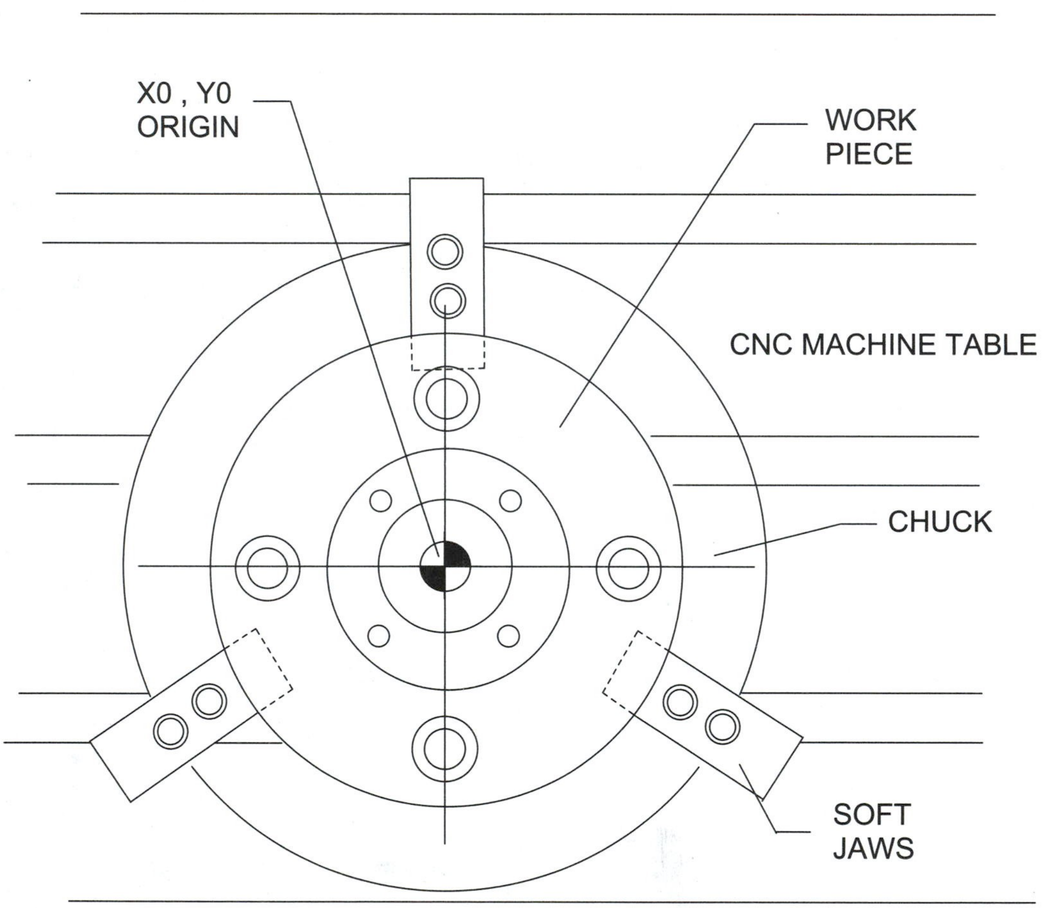

X0 , Y0
ORIGIN

WORK
PIECE

CNC MACHINE TABLE

CHUCK

SOFT
JAWS

INSTRUCTIONS:

NOTE: Indicate the 1.0015 dia. bore for X-Y zero.

Z-zero is at the top of the part.

Figure 8-8 Setup plan for CNC program 3000

CNC TOOL LIST AND OPERATION

Part No.: 3000		Prog. No.: O3000	Oper. No.: 10
Part Name: Bearing Cap		Material.: C.R. 1020	Page 1 of 1

Pocket No.	Length Offset	Radius Offset	Tool No.	Tool Dia.	Tool Description and Operation
T01	H01	D	8001	2.0	2.00 Dia. End Mill Mill (2) Flats
T02	H02	D	3002	1.0	90 Degree Spot Drill spot (8) holes
T03	H03	D	1015	.562	.56" Dia Drill Drill (4) holes
T04	H04	D	1007	.437	.437 Dia. Drill Drill (4) holes
T05	H05	D	5021	.500	.500-13 UNC Tap Tap (4) holes
T06	H06	D26	7012	.750	.75 Dia. End Mill Mill (4) C'bores

Figure 8-9 Tool list for CNC program 3000

CNC PROGRAM EXAMPLE 3000

```
%
O3000
(BEARING CAP - PROGRAM 3000)
(DATE, 08/12/99)
(TOOL - 01 2.00 DIA. - END MILL)
(TOOL - 02 1.000 DIA. - SPOTDRILL)
(TOOL - 03  .562 DIA. - DRILL)
(TOOL - 04  .437 DIA. - DRILL)
(TOOL - 05  .500-13 UNC - TAP)
(TOOL - 06  .750 DIA. - END MILL)

N010 G00 G40 G49 G80 G90
N015 G91 G28 Z0.
N020 G28 X0. Y0.

(TOOL - 01 2.00 DIA. - END MILL)
N025 T1 M06
N030 G0 G90 X1.975 Y-1.3621 S0200 M3
N035 G43 H1 Z1. M08
N040 G1 Z-.5 F50.
N045 G42 X.885 D26
N050 Y1.3349 F25.
N055 G0 G40 Z.2
N060 X1.975 Y-1.3621
N065 G1 Z-.5 F50.
N070 G42 X.875
N075 Y1.3349 F25.
N080 G0 G40 Z.2
N085 X-1.975 Y-1.3621
N090 G1 Z-.5 F50.
N095 G41 X-.885
N100 Y1.3349 F25.
N105 G0 G40 Z.2
N110 X-1.975 Y-1.3621
N115 G1 Z-.5 F50.
N120 G41 X-.875
N125 Y1.3349 F25.
N130 G0 G40 Z1. M09
N135 G91 G49 G28 Z0. M05

(TOOL - 02 1.000 DIA. - SPOTDRILL)
N140 G00 G40 G49 G80 G90
N145 T2 M06
N150 M01
N155 G0 G90 X-1.875 Y1.875 S0500 M3
```

■ ■ ■ ■

CNC PROGRAM EXAMPLE 3000 (*continued*)

N160 G43 H2 Z1. M08
N165 G98 G81 Z-1.27 R-.78 F12.
N170 X1.875
N175 Y-1.875
N180 X-1.875
N185 G80
N190 G99 G81 X-.6187 Y-.6187 Z-.28 R.1
N195 Y.6187
N200 X.6187
N205 Y-.6187
N210 G80 M09
N215 G91 G49 G28 Z0. M05

(TOOL - 03 .562 DIA. - DRILL)
N220 G00 G40 G49 G80 G90
N225 T3 M06
N230 M01
N235 G0 G90 X-1.875 Y1.875 S0950 M3
N240 G43 H3 Z1. M08
N245 G98 G81 Z-1.5988 R-.78 F12.
N250 X1.875
N255 Y-1.875
N260 X-1.875
N265 G80 M09
N270 G91 G49 G28 Z0. M05

(TOOL - 04 .437 DIA. - DRILL)
N275 G00 G40 G49 G80 G90
N280 T4 M06
N285 M01
N290 G0 G90 X-.6187 Y.6187 S1050 M3
N295 G43 H4 Z1. M08
N300 G98 G81 Z-1.0013 R.1 F15.
N305 X.6187
N310 Y-.6187
N315 X-.6187
N320 G80 M09
N325 G91 G49 G28 Z0. M05

(TOOL - 05 .500-13 UNC - TAP)
N330 G00 G40 G49 G80 G90
N335 T5 M06
N340 M01
N345 G0 G90 X-.6187 Y.6187 S0400 M3
N350 G43 H5 Z1. M08

■ ■ ■ ■

CNC PROGRAM EXAMPLE 3000 (*continued*)

```
N355 G99 G84 Z-.62 R.2 F30.77
N360 X.6187
N365 Y-.6187
N370 X-.6187
N375 G80 M09
N380 G91 G49 G28 Z0. M05

(TOOL - 06 .750 DIA. - END MILL)
N385 G00 G40 G49 G80 G90
N390 T6 M06
N395 M01
N400 G0 G90 X-1.875 Y1.875 S0535 M3
N405 G43 H6 Z1. M08
N410 G98 G81 Z-1.26 R-.78 F6.5
N415 X1.875
N420 Y-1.875
N425 X-1.875
N430 G80
N435 G0 Z.1M09
N440 G91 G49 G28 Z0. M05
N445 G90
N450 M30
%
```

■ ■

CHAPTER SUMMARY

The key concepts presented in this chapter include the following:

- Cutter diameter compensation description.

- The G41 cutter diameter compensation left code description.

- The G42 cutter diameter compensation right code description.

- The G40 cutter diameter compensation cancel code description.

- The G43 tool length compensation right code description.

- The G49 tool length compensation cancel code description.

- The G28 machine home return code description.

- Calculations for CNC machining center coordinates.

- The format description of a CNC program for CNC machining centers.

- The basic codes required of a CNC program for CNC machining centers.

■ ■ REVIEW QUESTIONS

1. Which type of tool uses cutter diameter compensation (G41 or G42)?

2. Which tools require tool length compensation (G43)?

3. When the G41 or G42 codes are programmed, what are the responsibilities of the CNC operator or setup person?

4. When the G43 code is programmed, what are the responsibilities of the CNC operator or setup person?

5. What control features can be used to avoid machine collisions?

6. Which letter code in a CNC program identifies a tool length offset number?

7. Which letter code in a CNC program identifies a tool radius offset number?

8. How is the depth of a machined feature adjusted to size on a VMC or HMC?

9. When adjusting a tool length offset, what amount should not be exceeded?

10. Discuss and identify the CNC program format structure and codes covered in this chapter. The items that should be identified include CNC program number, tool information notes, tool change blocks, default code setting blocks, spindle starts, coolant on and off, machine home command blocks, length offset activation blocks, diameter compensation activate/cancel blocks, and end of program.

■ ■ CNC PROGRAM EXERCISE FOR VMC

Using the sketch in Figure 8-10, calculate the G02 and G03 circular feed moves for the X- and Y-axes using G41/G42 and G40 cutter diameter compensation codes. Enter the codes and coordinates in the CNC program provided on the following page.

■ ■ CNC PROGRAM EXERCISE 3001

Fill in the missing program information of CNC program 3001 as indicated below:

Block N180: enter the G code, the X-Y coordinate value, and IPM value.

Block N190: enter the X-coordinate value.

Block N200: enter the G code, the X-Y coordinate value, and R value.

Block N210: enter the G code, the X-Y coordinate value, and R value.

Block N220: enter the G code, and the X-coordinate value.

Block N230: enter the G code, the X-Y coordinate value, and R value.

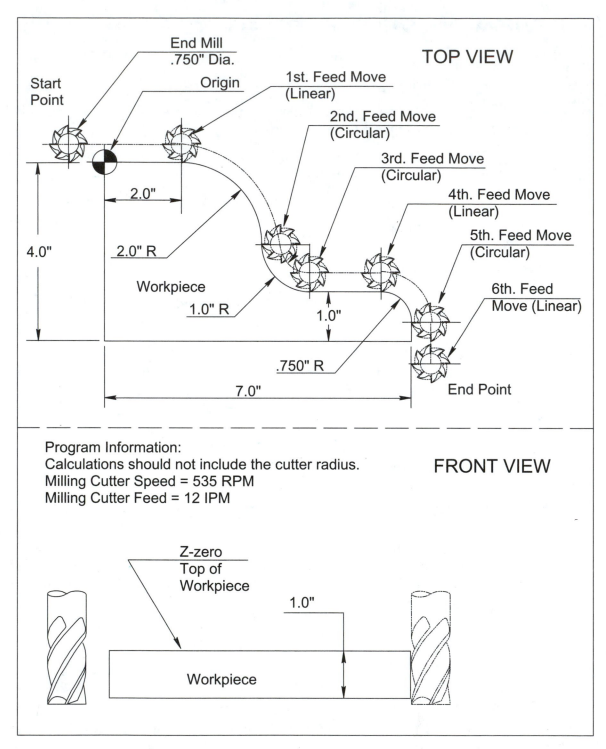

Program Information:
Calculations should not include the cutter radius.
Milling Cutter Speed = 535 RPM
Milling Cutter Feed = 12 IPM

Figure 8-10 **Program exercise**

Block N240: enter the G code, and the Y-coordinate value.

Block N250: enter the G code.

```
%
O3001
(CAM - PROGRAM NO. 3001)
(DATE, 06/22/03)
(TOOL - 01 .750 DIA. - END MILL)

N100 G00 G17 G40 G49 G80 G90
N110 G91 G28 Z0.
N120 G28 X0. Y0.

(TOOL - 01 .750 DIA. - END MILL)
N130 T1 M06
N140 M01
N150 G0 X-1.0 Y1.25 S535 M3
N160 G43 H1 Z.3 M08
N170 G1 Z-1.1 F25.

N180 G_____ X _____ Y _____ F _____
N190 X_____
N200 G_____ X _____ Y _____ R _____
N210 G_____ X _____ Y _____ R _____
N220 G_____ X _____
N230 G_____ X _____ Y _____ R _____
N240 G_____ Y _____
N250 G_____

N260 M05
N270 G91 G49 G28 Z0. M09
N280 M30
%
```

chapter 9

CNC DRILLING CANNED CYCLES

Chapter Objectives

After studying and completing this chapter, the student should have knowledge of the following:

- **G81 drilling canned cycle**
- **G82 counterboring canned cycle**
- **G83 peck drilling canned cycle**
- **G80 canned cycle cancel code**
- **G99 return to reference level code**
- **G98 return to initial level code**
- **Calculating X-Y coordinates**
- **CNC program format descriptions**

■ ■ ■ ■

CANNED CYCLE CODES

Canned cycle codes were developed to increase programming efficiency of the CNC programs for CNC machines. The CNC program example in this chapter uses some of these canned cycles to illustrate their application in machining a workpiece. Some common canned cycle codes are described in Table 9-1.

A *canned cycle* code is basically designed to execute a combination of axis moves and machine functions automatically. The axis moves usually include

■____*Table 9-1 Canned cycle descriptions*

CANNED CYCLE CODE	CNC MACHINING APPLICATION	SPINDLE OPERATION DESCRIPTION	Z-AXIS FEED OPERATION DESCRIPTION	Z-AXIS RETRACT OPERATION DESCRIPTION
G73	Drilling of deep holes	Rotates continually	Interrupted feed to Z-depth	Rapids out to R value
G74	Tapping Left-handed threads	Rotates CCW to Z-depth, then CW to R value	Feed to Z-depth, then spindle reverses	Feeds out to R value
G76	Finish boring without tool marks	Rotates to Z-depth, then shifts X-axis, stopped to R value	Feed to Z-depth, spindle stops and then orients	Rapids out to R value
G81	Drilling holes of average depth	Rotates continually	Feed to Z-depth	Rapids out to R value
G82	Counterbore holes	Rotates continually	Feed to Z-depth, then dwells	Rapids out to R value
G83	Drilling of deep holes	Rotates continually	Interrupted feed to Z-depth	Rapids out to R value
G84	Tapping right-handed threads	Rotates CW to Z-depth, then CCW to R value	Feed to Z-depth, then spindle reverses	Feeds out to R value
G85	Finish boring without tool marks	Rotates continually	Feed to Z-depth	Feeds out to R value
G86	Rough boring	Rotates to Z-depth, stopped to R value	Feed to Z-depth, then spindle stops	Rapids out to R value
G87	Back-boring	Rotates to Z-depth, stopped to R value	Feed to Z-depth, then spindle stops	Manual/rapid out to R value
G88	Boring	Rotates to Z-depth, stopped to R value	Feed to Z-depth, then dwells	Manual/rapid out to R value
G89	Boring	Rotates continually	Feed to Z-depth, then dwells	Feeds out to R value

the X-, Y-, and Z-axes in either feed mode or rapid mode. The machine functions can include any of the following:

- Spindle rotation CW

- Spindle rotation CCW

- Spindle stop

- Cycle dwell

The combination of moves and functions depends on the type of machining operation to be completed.

The CNC operator and CNC setup personnel are responsible for verifying that every canned cycle tool path move (feed and rapid) in the CNC program is safe and correct. When the control reads (Cycle Start), a canned cycle code, it becomes modal and any axis letters (X, Y, or Z) that follow in the CNC program will execute the canned cycle until the canned cycle is canceled.

These canned cycle codes are typically available on the CNC machines used throughout manufacturing plants. It should be noted that each CNC machine includes a CNC manual that lists the canned cycles available. The CNC manual typically lists and describes all the CNC programming codes that can be programmed.

The CNC programmer, CNC setup person and CNC operator should become familiar with each canned cycle code available. Additionally, it is important to note that, because these canned cycle codes are intended for cutting applications, the tool is also programmed to rotate (RPM) and move at a feedrate (IPM or IPR) value.

Consequently, it is a recommended practice to Dry Run or proveout the CNC program to ensure that collisions will not occur. In cases where the canned cycle codes have not been proved out, the CNC operator can use the control override features to slow the axis movement or completely stop the axis movement. This will allow the CNC operator time to react to a possible collision and stop the movement before any damage occurs. To verify that the coordinates are correct, the CNC operator or setup person can compare the CNC program coordinates on the CRT readout to the dimensions on the engineering drawing. This is required for all canned cycles and moves in the CNC program.

CNC DRILLING CANNED CYCLES

This chapter introduces the drilling canned cycle codes that are more commonly used on CNC machining centers. The drilling canned cycles described in this chapter include the G81, the G82, and the G83. Each code is designed with a specific application in mind that determines the characteristics of each canned cycle.

Additionally, this chapter includes a coordinate calculation worksheet designed to illustrate how the drilling coordinates are determined for the

CNC program. Also, the sample CNC program for this chapter includes the G81 drilling canned cycle to describe each feature of the canned cycle and how it appears on the CNC program.

G81 Drilling Canned Cycle

The G81 drilling canned cycle code is used for CNC drilling applications where the drilling depth range of the hole is 0.100 inch to approximately 3.000 inches deep. The drilling tool can vary depending on the type of operation required, the type of material to be drilled, or the hole diameter size. The typical tool used is the HSS 118° point twist drill. The speed (RPM) and feed (IPR or IPM) will also vary depending on the diameter, tool material, part material, and so on.

This canned cycle basically positions the tool to an X-Y coordinate, rapids to a Z-position, feeds to Z-depth, and then rapids out to the first Z-position. For this code, the spindle rotates continually throughout the cycle (see Figures 9-1 and 9-2). After the G81 drilling canned cycle is read, it becomes modal. Subsequently, the only requirement for the next hole or holes to be drilled is the X-Y coordinate for each hole. This is illustrated in CNC program 4000 of this chapter.

G82 Counterboring Canned Cycle

The G82 counterboring canned cycle code is used for CNC hole applications where the drilled hole requires a counterbore or a chamfer. The counterbore typically has a range from 0.100 inch to approximately 1.000 inch in depth depending on the workpiece design requirements. The counterbore tool can vary depending on the type of operation required. It can include a counterbore tool with a pilot, an end mill, or a chamfer tool. The speed (RPM) and feed (IPR or IPM) will also vary depending on the hole diameter size, the workpiece material, and so on.

This canned cycle basically positions the tool to an X-Y coordinate, rapids to a Z-position, feeds to Z-depth, dwells to cut completely around the surface, and then rapids out to the first Z-position (see Figure 9-3). After the G82 drilling canned cycle is read, it becomes modal. Subsequently, the only requirement for the next counterbore or counterbores to be machined is the X-Y coordinate for each counterbore.

G83 Peck Drilling Canned Cycle

The G83 *peck drilling* canned cycle code is used for CNC drilling applications where the drilling depth of the hole is greater than approximately 3.500 inches, depending also on the diameter of the drill. This canned cycle, as the name implies, pecks in and out until the required depth is reached. The purpose of this pecking feature is to break and eject the chips from the hole as the tool is retracted.

This canned cycle basically positions the tool to an X-Y coordinate, rapids to a Z-position, feeds to an increment of the Z-depth, partially retracts,

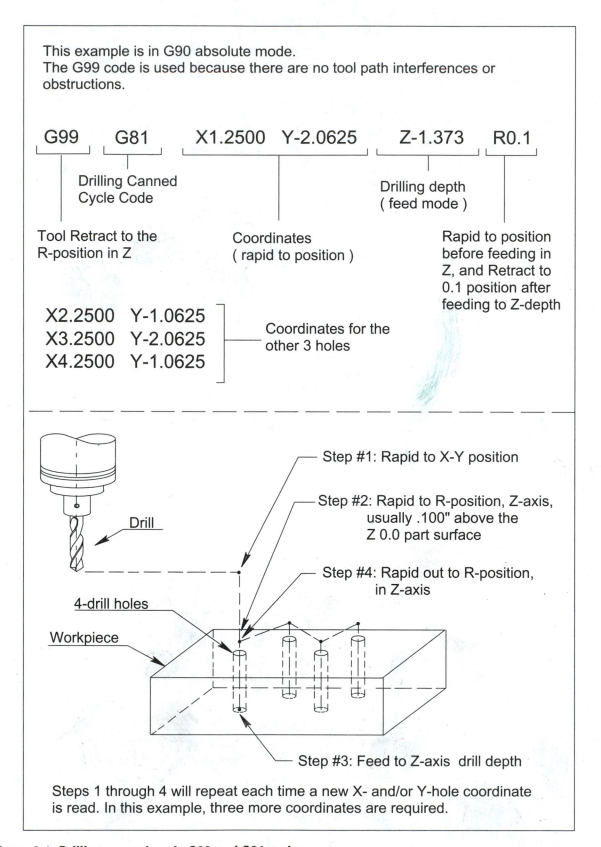

This example is in G90 absolute mode.
The G99 code is used because there are no tool path interferences or obstructions.

| G99 | G81 | X1.2500   Y-2.0625 | Z-1.373 | R0.1 |

Drilling Canned Cycle Code

Drilling depth (feed mode)

Tool Retract to the R-position in Z

Coordinates (rapid to position)

Rapid to position before feeding in Z, and Retract to 0.1 position after feeding to Z-depth

X2.2500   Y-1.0625
X3.2500   Y-2.0625
X4.2500   Y-1.0625

Coordinates for the other 3 holes

Drill

4-drill holes

Workpiece

Step #1: Rapid to X-Y position

Step #2: Rapid to R-position, Z-axis, usually .100" above the Z 0.0 part surface

Step #4: Rapid out to R-position, in Z-axis

Step #3: Feed to Z-axis drill depth

Steps 1 through 4 will repeat each time a new X- and/or Y-hole coordinate is read. In this example, three more coordinates are required.

Figure 9-1 Drilling canned cycle G99 and G81 codes

The example is in G90 absolute mode.
The G98 code is used because there is a tool path interference from the workpiece.

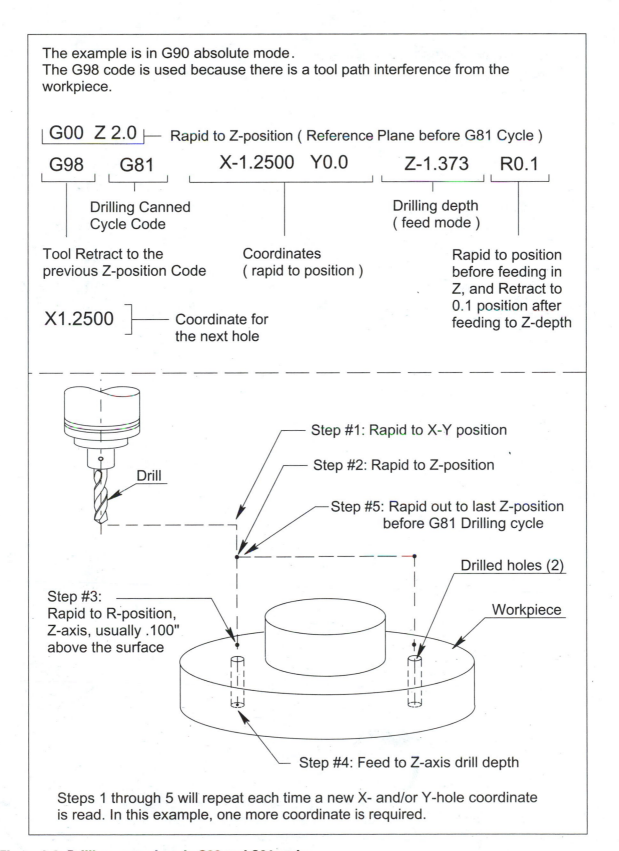

G00 Z 2.0 ⊢— Rapid to Z-position (Reference Plane before G81 Cycle)

| G98 | G81 | X-1.2500 Y0.0 | Z-1.373 | R0.1 |

Drilling Canned
Cycle Code

Drilling depth
(feed mode)

Tool Retract to the
previous Z-position Code

Coordinates
(rapid to position)

Rapid to position
before feeding in
Z, and Retract to
0.1 position after
feeding to Z-depth

X1.2500 ⊢— Coordinate for
the next hole

Drill

Step #1: Rapid to X-Y position

Step #2: Rapid to Z-position

Step #5: Rapid out to last Z-position
before G81 Drilling cycle

Drilled holes (2)

Step #3:
Rapid to R-position,
Z-axis, usually .100"
above the surface

Workpiece

Step #4: Feed to Z-axis drill depth

Steps 1 through 5 will repeat each time a new X- and/or Y-hole coordinate
is read. In this example, one more coordinate is required.

Figure 9-2 *Drilling canned cycle G98 and G81 codes*

This example is in G90 absolute mode.
The G98 code is used because there is tool path interference from the workpiece.

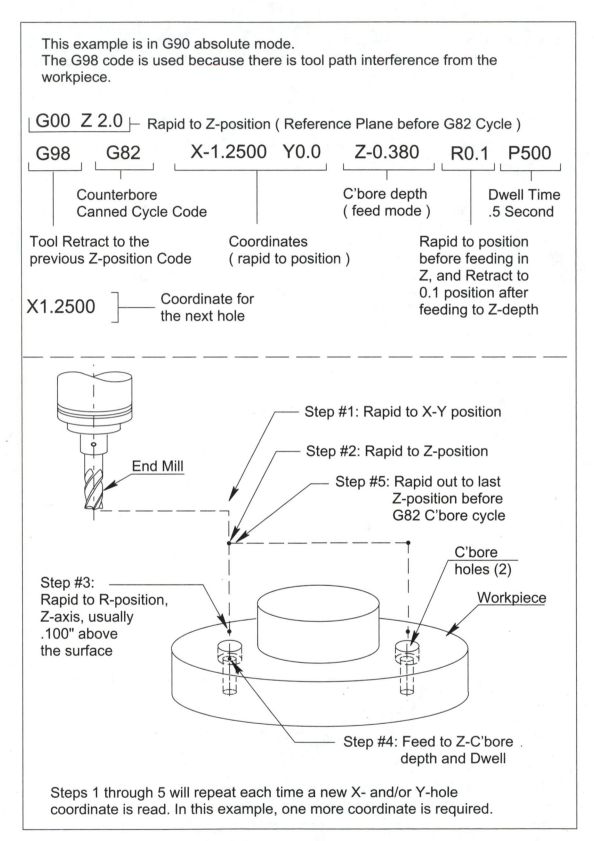

G00 Z 2.0 — Rapid to Z-position (Reference Plane before G82 Cycle)

| G98 | G82 | X-1.2500 Y0.0 | Z-0.380 | R0.1 | P500 |

Counterbore
Canned Cycle Code

C'bore depth
(feed mode)

Dwell Time
.5 Second

Tool Retract to the
previous Z-position Code

Coordinates
(rapid to position)

Rapid to position
before feeding in
Z, and Retract to
0.1 position after
feeding to Z-depth

X1.2500 — Coordinate for
the next hole

End Mill

Step #1: Rapid to X-Y position

Step #2: Rapid to Z-position

Step #5: Rapid out to last
Z-position before
G82 C'bore cycle

C'bore
holes (2)

Workpiece

Step #3:
Rapid to R-position,
Z-axis, usually
.100" above
the surface

Step #4: Feed to Z-C'bore
depth and Dwell

Steps 1 through 5 will repeat each time a new X- and/or Y-hole coordinate is read. In this example, one more coordinate is required.

*Figure 9-3 **Counterboring canned cycle G98 and G82 codes***

rapids back into the hole, repeats this until the Z-depth is reached, and then rapids out to the first Z-position (see Figure 9-4). After the G83 peck drilling canned cycle is read, it becomes modal. Subsequently, the only requirement for the next hole or holes to be peck drilled is the X-Y coordinate for each hole.

The drilling tool will vary depending on the depth of the hole operation required. The types of tools typically used for peck drilling include taper-length drills or extra-long drills. The spindle speed (RPM) and feedrate (IPR or IPM) will also vary depending on the tool diameter, tool material, part material, and so on. It is important to note that, with holes considered to be deeper than average, special care should be taken to compensate for this condition. For example, to compensate for the decrease of rigidity of the tool, the speed and feed can be reduced or the tool can be supported with a bushing and so on.

In almost all cases when drilling, coolant is applied to reduce wear and increase productivity. However, when deep-hole drilling the use of coolant becomes a major factor that affects the condition and quality of the holes produced. Therefore, when deep-hole peck drilling it is recommended that the CNC machine be equipped with high-pressure coolant and coolant through-the-tool drills.

G80 Canned Cycle Cancel

The canned cycle codes are all modal and each will remain active for any subsequent line until the code is cancelled by a G80 code. Therefore, any axis movement (X, Y, or Z) that is present in a subsequent line of the CNC program will execute the active canned cycle whether or not the canned cycle code is present in the subsequent line.

The canned cycle operations typically involve depth, either through the workpiece or to a specified amount, which is specified by the engineering drawing. In either situation, the CNC operator or CNC setup person must first measure the depth of the hole after it is drilled, determine if an adjustment is required, and then adjust the depth by adding or subtracting the required amount from the tool length offset. As stated previously, if the depth adjustment amount exceeds 0.100 inch, the CNC program, tool and toolholder, and tool length offset value should be checked for a possible programming or setup error.

These CNC canned cycle codes discussed are common in format and application to most CNC programs used in industrial applications. The CNC programmer can vary the format order of each canned cycle. However, in order to increase the efficiency of all related CNC operations, the canned cycle codes should be as consistent as possible.

Z-Return Codes

As described previously, the canned cycle codes will automatically return to the "R" retract position in the Z-axis. In most situations there is clearance

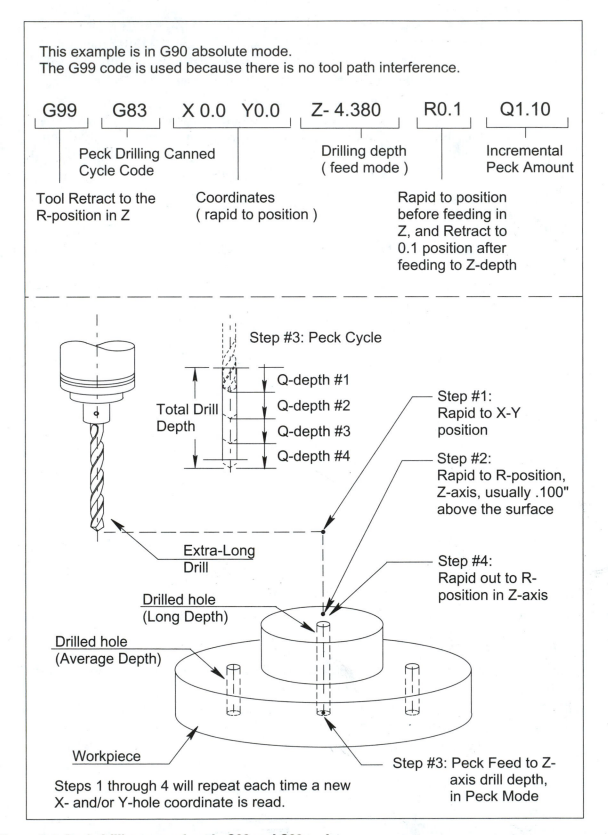

This example is in G90 absolute mode.
The G99 code is used because there is no tool path interference.

| G99 | G83 | X 0.0 Y0.0 | Z- 4.380 | R0.1 | Q1.10 |

Peck Drilling Canned
Cycle Code

Drilling depth
(feed mode)

Incremental
Peck Amount

Tool Retract to the
R-position in Z

Coordinates
(rapid to position)

Rapid to position
before feeding in
Z, and Retract to
0.1 position after
feeding to Z-depth

Step #3: Peck Cycle

Q-depth #1
Q-depth #2
Q-depth #3
Q-depth #4

Total Drill
Depth

Step #1:
Rapid to X-Y
position

Step #2:
Rapid to R-position,
Z-axis, usually .100"
above the surface

Extra-Long
Drill

Step #4:
Rapid out to R-
position in Z-axis

Drilled hole
(Long Depth)

Drilled hole
(Average Depth)

Workpiece

Step #3: Peck Feed to Z-
axis drill depth,
in Peck Mode

Steps 1 through 4 will repeat each time a new
X- and/or Y-hole coordinate is read.

Figure 9-4 Peck drilling canned cycle G98 and G83 codes

between the holes to be drilled. However, in some situations there may be interference between the hole locations. Consequently, there are two codes designed to accommodate either situation. These two codes, G99 or G98, can be used to specify the retract method for all the canned cycles.

G99 Return to Reference (R) Level Code

The G99 code (see Figure 9-1) retracts the tool to a Z-axis (R code) position that is minimal yet sufficient to clear the part and workholding when moving to the next coordinate. The G99 typically is used when a clearance of 0.100 inch above the machined surface is sufficient.

G98 Return to Initial Level Code

The G98 code (see Figure 9-2) is used in situations where more clearance is required due to an obstruction in the path of the tool as it rapid traverses from one coordinate position to the next coordinate position. The obstruction may include such items as clamps, studs, fixture components, or features of the workpiece itself. In these situations, the CNC programmer is responsible for programming the appropriate return code (G99 or G98). These G99 and G98 return codes are both modal and are used for all the other canned cycle codes available.

INSTRUCTIONS FOR CNC PROGRAM EXAMPLE 4000

This section includes a coordinate worksheet, CNC documentation, and a CNC program for a CNC vertical machining center. As explained previously, each of these items is designed to illustrate the CNC setup and operation process. Therefore, it is important to review and understand the purpose and role of each CNC item. These items are listed below:

- Coordinate worksheet (for instructional purposes)

- Engineering drawing

- CNC machine setup plan

- CNC tool list

This program is similar in format to the other sample programs. The similarities that should be noted include

- Program number

- Information notes

- Tool changes

- Default code setting

- Spindle starts

- Coolant on

- Machine home command

- Length offset activation

- End of program

Instructions for CNC program example 4000

For this CNC program, the student is required to identify and note the main CNC program items that are listed below:

1. Program number

2. Information notes

3. Tool changes

4. Default code setting blocks

5. Rapid moves

6. Feed moves

7. Radius feed moves

8. Canned cycle codes

9. X- and Y-coordinate positions

10. End of program

The example below illustrates how to identify and note the CNC program items above.

```
N025 T1 M06
N030 G0 G90 X1.975 Y-1.3621 S0200 M3
N035 G43 H1 Z1. M08 --------------- Activate Length Offset 1
N040 G1 Z-.5 F50. --------------------- Feed Move Z-.5, @50 IPM
N045 G42 X.885 D26 ---------------- Activate Rad. Offset 26, Feed Move
                                      X.885
N050 Y1.3349 F25. -------------------- Feed Move Y1.3349 @25 IPM
```

■ ■ ■ ■

COORDINATE WORKSHEET 4000

Instructions: Mark the X-Y zero origin, then calculate the coordinates for each point. Record the calculations in the chart. For the dimensions, use part drawing 4000 illustrated in Figure 9-5.

Use the standard origin symbol at right to mark the X-Y zero.

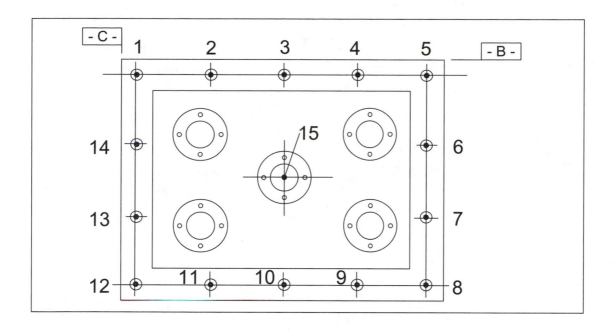

PART NO.	X-COORDINATE	Y-COORDINATE
1		
2		
3		
4		
5		
6		
7		
8		
9		
10		
11		
12		
13		
14		
15		

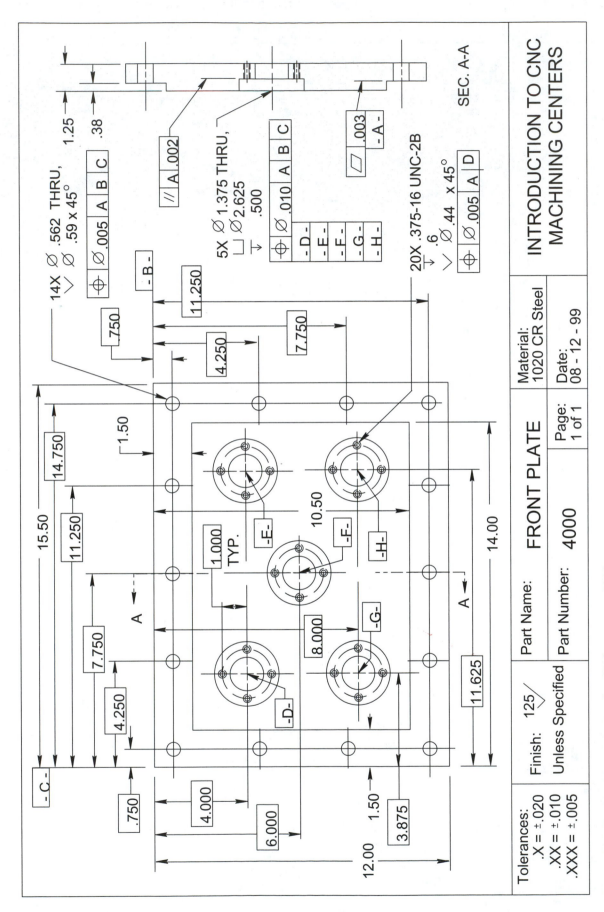

Figure 9-5 Engineering drawing for CNC program 4000

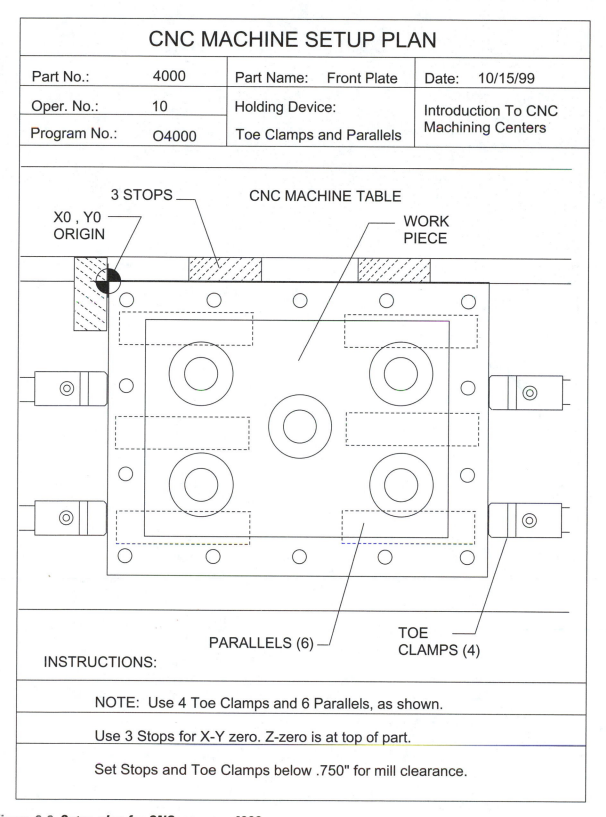

CNC MACHINE SETUP PLAN

Part No.:	4000	Part Name:	Front Plate	Date:	10/15/99
Oper. No.:	10	Holding Device:		Introduction To CNC Machining Centers	
Program No.:	O4000	Toe Clamps and Parallels			

3 STOPS

CNC MACHINE TABLE

X0 , Y0 ORIGIN

WORK PIECE

PARALLELS (6)

TOE CLAMPS (4)

INSTRUCTIONS:

NOTE: Use 4 Toe Clamps and 6 Parallels, as shown.

Use 3 Stops for X-Y zero. Z-zero is at top of part.

Set Stops and Toe Clamps below .750" for mill clearance.

Figure 9-6 Setup plan for CNC program 4000

CNC TOOL LIST AND OPERATION

Part No.: 4000			Prog. No.: O4000		Oper. No.: 10
Part Name: Front Plate			Date: 03/11/00		Page 1 of 1

Pocket No.	Length Offset	Radius Offset	Tool No.	Tool Dia.	Tool Description and Operation
T01	H01	D	8001	2.0	2.00 Dia. End Mill Mill 1.5 Wide (4) Sides
T02	H02	D	1015	1.250	1.250" Dia Drill Drill (5) holes
T03	H03	D23	7012	.750	.75 Dia. End Mill Mill (5) C'bores
T04	H04	D	3005	.750	90 Degree Spot Drill spot (34) holes
T05	H05	D	1015	.562	.56" Dia Drill Drill (14) holes
T06	H06	D	1004	.312	.312 Dia. Drill Drill (20) holes
T07	H07	D	5028	.375	.375-16 UNC Tap Tap (20) holes
T08	H08	D	1015	1.375	1.375" Dia Boring Bar Bore (5) holes

Figure 9-7 Tool list for CNC program 4000

■ ■ ■ ■
CNC PROGRAM EXAMPLE 4000

```
%
O4000
(FRONT PLATE - PROGRAM 4000)
(DATE, 08/12/99 )
(TOOL - 01 2.000 DIA. - CARBIDE END MILL)
(TOOL - 02 1.250 DIA. - DRILL)
(TOOL - 03  .750 DIA. - HSS END MILL)
(TOOL - 04  .750 DIA. - SPOT DRILL)
(TOOL - 05  .562 DIA. - DRILL)
(TOOL - 06  .312 DIA. - DRILL)
(TOOL - 07  .375-16 UNC - TAP)
(TOOL - 08 1.375 DIA. - BORING BAR)
N010 G00 G40 G49 G80 G90
N015 G91 G28 Z0.
N020 G28 X0. Y0.

(TOOL - 01 2.00 DIA. - CARBIDE END MILL)
N025 T1 M06
N030 G0 G90 X16.8 Y-.5 S1000 M3
N035 G43 H1 Z1. M08
N040 G1 Z-.38 F40.
N045 X.5 F25.
N050 Y-11.5
N055 X15.
N060 Y1.25
N070 G0 Z1. M09
N075 G91 G49 G28 Z0. M05

(TOOL - 02 1.250 DIA. - DRILL)
N100 G00 G40 G49 G80 G90
N105 T2 M06
N110 M01
N115 G0 G90 X3.875 Y-4. S0320 M3
N120 G43 H2 Z1. M08
N125 G99 G81 Z-1.6755 R.1 F14.
N130 X7.75 Y-6.
N135 X11.625 Y-4.
N140 Y-8.
N145 X3.875
N150 G80 M09
N155 G91 G49 G28 Z0. M05

(TOOL - 03 .750 DIA. - HSS END MILL)
N200 G00 G40 G49 G80 G90
N205 T3 M06
```

■ ■ ■ ■

CNC PROGRAM EXAMPLE 4000 (*continued*)

N210 M01
N215 G0 G90 X3.875 Y-4. S0535 M3
N220 G43 H3 Z1. M08
N225 G1 Z.2 F40.
N230 G1 Z-.5 F5.
N235 G41 D23 G1 X5.1875
N240 G3 X5.1875 Y-4. I-1.3125 J0.
N245 G40 G1 X3.875
N250 G0 Z.2
N255 X7.75 Y-6.
N260 G1 Z-.5
N265 G41 X9.0625
N270 G3 X9.0625 Y-6. I-1.3125 J0.
N275 G40 G1 X7.75
N280 G0 Z.2
N285 X11.625 Y-4.
N290 G1 Z-.5
N295 G41 X12.9375
N300 G3 X12.9375 I-1.3125 J0.
N305 G40 G1 X11.625
N310 G0 Z.2
N315 Y-8.
N320 G1 Z-.5
N325 G41 X12.9375
N330 G3 X12.9375 Y-8. I-1.3125 J0.
N335 G40 G1 X11.625
N340 G0 Z.2
N345 X3.875
N350 G1 Z-.5
N355 G41 X5.1875
N360 G3 X5.1875 Y-8. I-1.3125 J0.
N365 G40 G1 X3.875
N370 G0 Z1. M09
N375 G91 G49 G28 Z0. M05

(TOOL - 04 .750 DIA. - SPOT DRILL)
N400 G00 G40 G49 G80 G90
N405 T4 M06
N410 M01
N415 G0 G90 X.75 Y-.75 S0600 M3
N420 G43 H4 Z1. M08
N425 G98 G81 Z-.6765 R-.28 F5.
N430 X4.25
N435 X7.75

■ ■ ■ ■

CNC PROGRAM EXAMPLE 4000 (*continued*)

```
N440 X11.25
N445 X14.75
N450 Y-4.25
N455 Y-7.75
N460 Y-11.25
N465 X11.25
N470 X7.75
N475 X4.25
N480 X.75
N485 Y-7.75
N490 Y-4.25
N495G80 G0 Z1.
N500 G98 G81 X2.875 Y-4. Z-.7185 R-.4 F5.
N505 X3.875 Y-3.
N510 X4.875 Y-4.
N515 X3.875 Y-5.
N520 X7.75
N525 X8.75 Y-6.
N530 X7.75 Y-7.
N535 X6.75 Y-6.
N540 X10.625 Y-4.
N545 X11.625 Y-3.
N550 X12.625 Y-4.
N555 X11.625 Y-5.
N560 Y-7.
N565 X12.625 Y-8.
N570 X11.625 Y-9.
N575 X10.625 Y-8.
N580 X4.875
N585 X3.875 Y-7.
N590 X2.875 Y-8.
N595 X3.875 Y-9.
N600 G80 M09
N605 G91 G49 G28 Z0. M05

(TOOL - 05 .562 DIA. - DRILL)
N610 G00 G40 G49 G80 G90
N605 T5 M06
N610 M01
N615 G0 G90 X.75 Y-.75 S0715 M3
N620 G43 H5 Z1. M08
N625 G98 G81 Z-1.4688 R-.28 F13.
N630 X4.25
N635 X7.75
```

■ ■ ■ ■
CNC PROGRAM EXAMPLE 4000 (*continued*)

N640 X11.25
N645 X14.75
N650 Y-4.25
N655 Y-7.75
N660 Y-11.25
N665 X11.25
N670 X7.75
N675 X4.25
N680 X.75
N685 Y-7.75
N690 Y-4.25
N695 G80 M09
N700 G91 G49 G28 Z0. M05

(TOOL - 06 .312 DIA. - DRILL)
N705 G00 G40 G49 G80 G90
N710 T6 M06
N715 M01
N720 G0 G90 X2.875 Y-4. S1285 M3
N725 G43 H6 Z1. M08
N730 G98 G81 Z-1.3736 R-.4 F10.
N735 X3.875 Y-3.
N740 X4.875 Y-4.
N745 X3.875 Y-5.
N750 X7.75
N755 X8.75 Y-6.
N760 X7.75 Y-7.
N765 X6.75 Y-6.
N770 X10.625 Y-4.
N775 X11.625 Y-3.
N780 X12.625 Y-4.
N785 X11.625 Y-5.
N790 Y-7.
N795 X12.625 Y-8.
N800 X11.625 Y-9.
N805 X10.625 Y-8.
N810 X4.875
N815 X3.875 Y-7.
N820 X2.875 Y-8.
N825 X3.875 Y-9.
N830 G80 M09
N835 G91 G49 G28 Z0. M05

(TOOL - 07 .375-16 UNC - TAP)
N840 G00 G40 G49 G80 G90

CNC PROGRAM EXAMPLE 4000 (*continued*)

```
N845 T7 M06
N850 M01
N855 G0 G90 X2.875 Y-4. S0530 M3
N860 G43 H7 Z1. M08
N865 G98 G84 Z-1.35 R-.3 F33.13
N870 X3.875 Y-3.
N875 X4.875 Y-4.
N880 X3.875 Y-5.
N885 X7.75
N890 X8.75 Y-6.
N895 X7.75 Y-7.
N900 X6.75 Y-6.
N905 X10.625 Y-4.
N910 X11.625 Y-3.
N915 X12.625 Y-4.
N920 X11.625 Y-5.
N925 Y-7.
N930 X12.625 Y-8.
N935 X11.625 Y-9.
N940 X10.625 Y-8.
N945 X4.875
N950 X3.875 Y-7.
N955 X2.875 Y-8.
N960 X3.875 Y-9.
N965 G80 M09
N970 G91 G49 G28 Z0. M05

(TOOL - 08 1.375 DIA. - BORING BAR)
N975 G00 G40 G49 G80 G90
N980 T8 M06
N985 M01
N990 G0 G90 X3.875 Y-4. S1500 M3
N995 G43 H8 Z1. M08
N1000 G98 G86 Z-1.28 R-.4 F8.
N1005 X7.75 Y-6.
N1010 X11.625 Y-4.
N1015 Y-8.
N1020 X3.875
N1025 G80
N1030 G0 Z.1M09
N1035 G91 G49 G28 Z0. M05
N1040 G90
N1045 M30
%
```

■ ■

CHAPTER SUMMARY

The key concepts presented in this chapter include the following:

- Canned cycle code types and descriptions.
- The G81 code and machine movement descriptions.
- The G82 code and machine movement descriptions.
- The G83 code and machine movement descriptions.
- The G80 canned cycle cancel code description.
- The G99 return to reference level code description.
- The G98 return to initial level code description.
- Calculations for CNC machining center coordinates.
- The format description of a CNC program for CNC machining centers.
- The basic codes required of a CNC program for CNC machining centers.

■ ■ ## REVIEW QUESTIONS

1. List the types of tools used with the G81 canned cycle
2. List the types of tools used with the G82 canned cycle
3. List the types of tools used with the G83 canned cycle
4. What should the CNC operator or setup person be aware of when a canned cycle code is proved out?
5. Which canned cycle code is used for right-handed tapping operations?
6. Which canned cycle code is used for finish boring without tool marks?
7. What is the difference between the G99 and the G98 codes?
8. What are the "P" and "Q" codes used for in the canned cycle?
9. Why is it necessary to program a G80 code?
10. When drilling with a canned cycle, how is the depth controlled?
11. Discuss and identify the CNC program format structure and codes covered in this chapter. The items that should be identified include CNC program number, tool information notes, tool change blocks, default code setting blocks, canned cycle codes, coolant on and off, machine home command blocks, diameter compensation activate/cancel blocks, length offset activation blocks, and end of program block.

■ ■ ## CNC PROGRAM EXERCISE FOR VMC

Using the sketch in Figure 9-8, calculate the drilling canned cycle values for the X-, Y-, and Z-axes and R-plane. Also determine and enter the drilling canned cycle G codes. Enter the codes and coordinates in the CNC program provided on the following page.

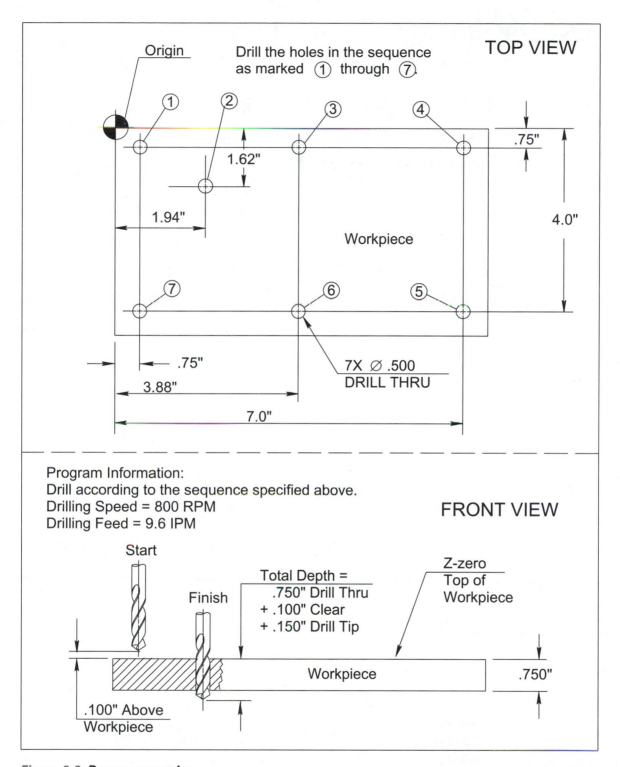

TOP VIEW

Origin

Drill the holes in the sequence as marked ① through ⑦.

① ② ③ ④

.75"

1.62"

1.94"

4.0"

Workpiece

⑦ ⑥ ⑤

.75"

3.88"

7X ⌀ .500
DRILL THRU

7.0"

Program Information:
Drill according to the sequence specified above.
Drilling Speed = 800 RPM
Drilling Feed = 9.6 IPM

FRONT VIEW

Start

Finish

Total Depth =
.750" Drill Thru
+ .100" Clear
+ .150" Drill Tip

Z-zero
Top of
Workpiece

Workpiece

.750"

.100" Above
Workpiece

Figure 9-8 **Program exercise**

▪ ▪ CNC PROGRAM EXERCISE 4001

Fill in the missing program information of CNC program 4001 as indicated below:

Block N170: enter the G codes, X-Y-Z-R coordinate values, and IPM value.

Block N180: enter the X-Y coordinate values.

Block N190: enter the X-Y coordinate values.

Block N200: enter the X-coordinate value.

Block N210: enter the Y-coordinate value.

Block N220: enter the X-coordinate value.

Block N230: enter the X-coordinate value.

Block N240: enter the G code.

```
%
O4001
(COVER - PROGRAM NO. 4001)
(DATE, 07/15/03)
(TOOL - 01 .500 DIA. - DRILL)

N100 G00 G17 G40 G49 G80 G90
N110 G91 G28 Z0.
N120 G28 X0. Y0.

(TOOL - 01 .500 DIA. - DRILL)
N130 T1 M06
N140 M01
N150 G0 X0 Y0 S800 M3
N160 G43 H1 Z1.0 M08

N170 G_____ G_____ X _____ Y _____ Z _____
        R _____ F _____
N180 X _____ Y _____
N190 X _____ Y _____
N200 X _____
N210 Y _____
N220 X _____
N230 X _____
N240 G_____
N250 M05
N260 G91 G49 G28 Z0. M09
N270 M30
%
```

chapter 10

CNC CANNED CYCLES G84, G86, AND G76

Chapter Objectives

After studying and completing this chapter, the student should have knowledge of the following:

- *G84 tapping canned cycle*

- *G86 boring canned cyles*

- *G76 fine boring canned cycle*

- *G20 inch format code*

- *G21 metric format code*

- *G54–G59 work coordinate system codes*

- *Calculating X-Y coordinates*

- *HMC table rotation B-axis*

- *HMC CNC program format description*

■ ■ ■ ■

CNC CANNED CYCLES G84, G86, AND G76

This chapter introduces other canned cycle codes commonly used on CNC machining centers. The canned cycle codes described in this chapter—G84 tapping canned cycle, G86 boring canned cycle, and G76 fine boring canned cycle—have been previously listed in Table 9-1. Additionally, this chapter describes the G20 inch format code and G21 metric format code, the G54–G59 work coordinate system codes, and the B-axis. It also includes a coordinate calculation worksheet for X-Y program coordinate analysis and introduces horizontal machining center CNC program format and applications.

The application of these canned cycle codes has increased the productivity of CNC machining for both vertical and horizontal CNC machining centers. The example CNC program in this chapter, which is for a CNC horizontal machining center, includes some of these commonly used canned cycles to illustrate their application in machining a workpiece.

As previously mentioned, these canned cycle codes are typically available on most CNC machines used in industry as well. It is also important to restate that the CNC operator and setup person are responsible for verifying that all the canned cycle tool-path moves (feed and rapid) in the CNC program are safe and correct. When the control reads "Cycle Start," a canned cycle code, it becomes modal and will execute the canned cycle automatically.

Therefore, with these canned cycle codes it is also a recommended practice to Dry Run or prove out the CNC program to ensure that collisions will not occur. In cases where the canned cycle codes have not been proved out, the CNC operator can use the override features to slow the axis movement or completely stop the axis movement. The Feed Hold button is another method of controlling the CNC machine during proveout. This should also allow the CNC operator time to react to a possible collision and stop the movement before any damage occurs. However, it is important to note that the Feed Hold button and the Feedrate Override control will not stop the G84 tapping canned cycle during the Z-axis cutting move. Consequently, care should be taken when the G84 tapping canned cycle executes the Z-axis cutting move. If a collision occurs, the Emergency Stop button should be pressed.

The set of illustrations in this chapter describes the tapping canned cycle, the rough boring canned cycle, and the fine boring canned cycle. These canned cycle codes are identical for both VMCs and HMCs. The CNC program that follows (5000) applies these codes as it machines a workpiece using the B-axis and G54–G59 work coordinate system. This CNC program example is designed to describe the differences between VMC machining programs and HMC machining programs.

To better understand and distinguish the differences between the CNC horizontal machining center and the CNC vertical machining center, it is recommended that the previous chapters be reviewed. The illustrations that describe the VMC and HMC machine components, the X-, Y-, Z-, and B-axis descriptions, and the workholding method descriptions should be reviewed. For all the canned cycles it should be noted that the CNC program is identical because the Z-axis represents the spindle on both machines.

G84 Tapping Canned Cycle

The G84 tapping canned cycle code is used for CNC right-handed tapping applications where the *thread depth* is approximately 0.100–1.000 inch. The tapping tool can vary depending on the type of operation that is required. The speed (RPM) will also vary depending on the diameter, tool material, workpiece material, and so on. The feedrate is fixed for each tap thread designation. The tapping canned cycle positions the tool to the X-Y coordinate, rapids to a Z-position, the spindle rotates CW, feeds to Z-depth, reverses the spindle CCW, and then feeds out to the first Z-position (see Figure 10-1).

G86 Boring Canned Cycle

These boring canned cycle codes are used for CNC machining applications when an existing hole requires further machining. The existing hole is typically a drilled hole or a casting cored hole. These boring canned cycle codes are applied to a wide range of CNC machining applications depending on the workpiece design requirements. These various applications are described in the canned cycle description chart, Table 9-1, from the previous chapter.

The boring tool can vary depending on the type of operation required. The tool can be a double cutter type for roughing, a single microadjustable type for finishing, or a back-boring cartridge type. The speed (RPM) and feed (IPR or IPM) will also vary depending on the bore diameter, tool material, workpiece material, and so on.

The G86 rough boring canned cycle (Figure 10-2) is primarily used in situations where a smooth surface finish is not required. This boring canned cycle positions the tool to the X-Y coordinate, rapids to a Z-position, feeds to Z-depth, stops the spindle rotation, and then rapids out to the first Z-position.

G76 Fine Boring Canned Cycle

The G76 fine boring canned cycle (Figures 10-3 and 10-4) is used for CNC boring applications where an accurate bore size or smooth surface finish is required. This fine boring canned cycle positions the tool to the X-Y coordinate, rapids to a Z-position, feeds to the Z-depth, stops and orients the spindle, moves the tool away from the bore, then rapids out to the first Z-position.

G98 and G99 Return Codes

These canned cycle return codes, which have been previously described, are also used to retract the tool in these examples. These codes are used for all canned cycles and for VMC programs as well as HMC programs.

This example is in G90 absolute mode.
The G99 code is used because there are no tool path interferences or obstructions.

G99	G84	X1.2500 Y-2.0625	Z-0.62	R0.2	P500	F.100

Tapping Canned
Cycle Code

Tapping depth
(feed mode)

Dwell
Time

Tool Retract to the
R-position in Z

Coordinates
(rapid to position)

Rapid to position
before feeding in Z,
and Feed out to R-
position after feeding
to Z

Feedrate

X2.2500 Y-1.0625

X3.2500 Y-2.0625

X4.2500 Y-1.0625

Coordinates
for the other
3 holes

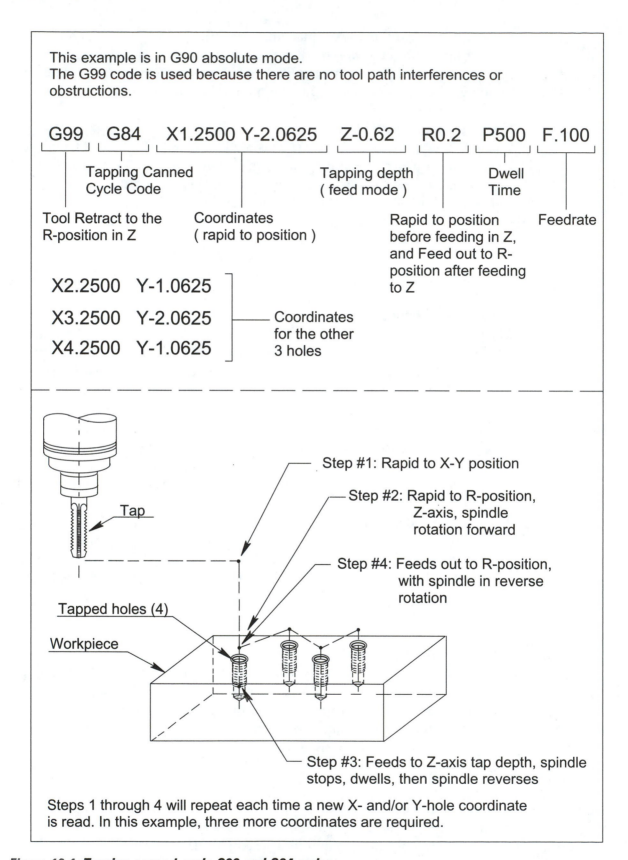

Tap

Step #1: Rapid to X-Y position

Step #2: Rapid to R-position,
Z-axis, spindle
rotation forward

Step #4: Feeds out to R-position,
with spindle in reverse
rotation

Tapped holes (4)

Workpiece

Step #3: Feeds to Z-axis tap depth, spindle
stops, dwells, then spindle reverses

Steps 1 through 4 will repeat each time a new X- and/or Y-hole coordinate
is read. In this example, three more coordinates are required.

Figure 10-1 Tapping canned cycle G99 and G84 codes

This example is in G90 absolute mode.
The G99 code is used because there are no tool path interferences or obstructions.

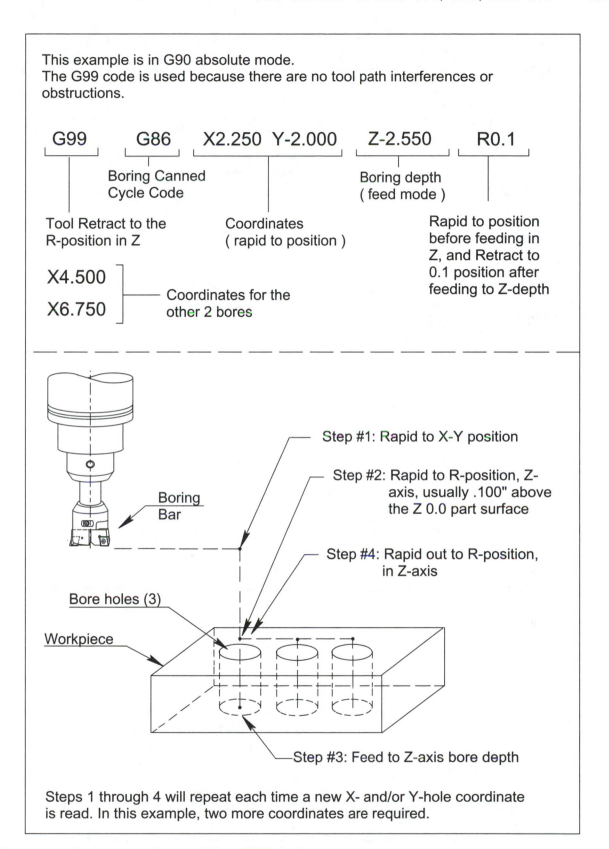

| G99 | G86 | X2.250 Y-2.000 | Z-2.550 | R0.1 |

Boring Canned Cycle Code

Boring depth (feed mode)

Tool Retract to the R-position in Z

Coordinates (rapid to position)

Rapid to position before feeding in Z, and Retract to 0.1 position after feeding to Z-depth

X4.500

X6.750

Coordinates for the other 2 bores

Boring Bar

Step #1: Rapid to X-Y position

Step #2: Rapid to R-position, Z-axis, usually .100" above the Z 0.0 part surface

Step #4: Rapid out to R-position, in Z-axis

Bore holes (3)

Workpiece

Step #3: Feed to Z-axis bore depth

Steps 1 through 4 will repeat each time a new X- and/or Y-hole coordinate is read. In this example, two more coordinates are required.

Figure 10-2 Boring canned cycle G99 and G86 codes

This example is in G90 absolute mode.
The G99 code is used because there are no tool path interferences or obstructions.

| G99 | G76 | X2.2500 Y-2.000 | Z-2.550 | R0.1 | P1000 | Q.03 |

Fine Boring
Canned
Cycle Code

Boring depth
(feed mode)

Dwell Time
No Decimal
1-second

Tool Retract to the
R-position in Z

Coordinates
(rapid to position)

Rapid to position
before feeding in
Z, Dwell, Spindle
stop, Tool Shift,
and Retract to
0.1 position

Shift
Amount

X4.500

X6.750

Coordinates for the
other 2 bores

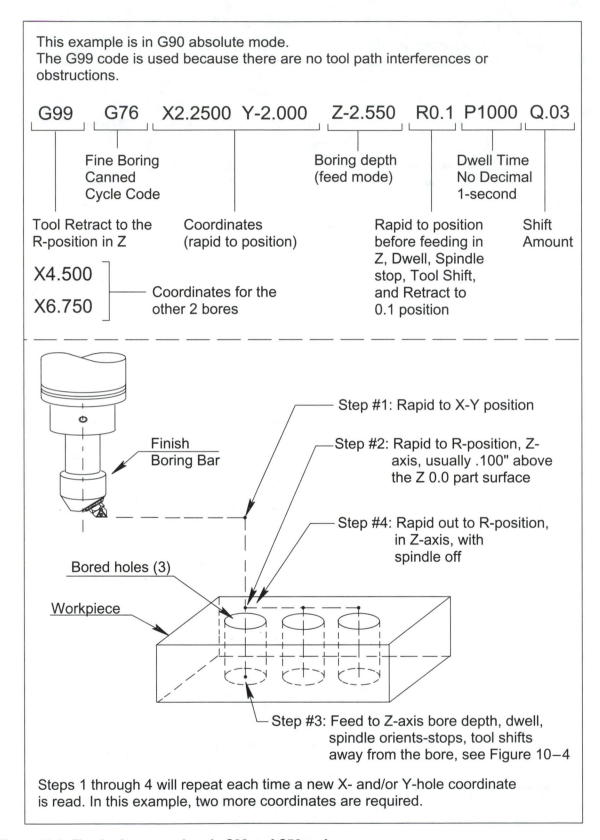

Step #1: Rapid to X-Y position

Finish
Boring Bar

Step #2: Rapid to R-position, Z-
axis, usually .100" above
the Z 0.0 part surface

Step #4: Rapid out to R-position,
in Z-axis, with
spindle off

Bored holes (3)

Workpiece

Step #3: Feed to Z-axis bore depth, dwell,
spindle orients-stops, tool shifts
away from the bore, see Figure 10–4

Steps 1 through 4 will repeat each time a new X- and/or Y-hole coordinate
is read. In this example, two more coordinates are required.

Figure 10-3 ***Fine boring canned cycle G99 and G76 codes***

This figure illustrates the boring bar shift that was described in Figure 10–3, step #3.

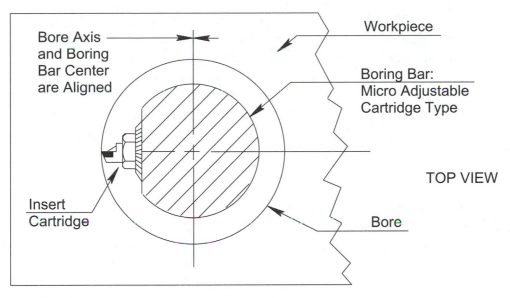

The view above illustrates the finish boring bar aligned with the center axis of the bore. This is the position of the cartridge when the spindle has been stopped and orientated.

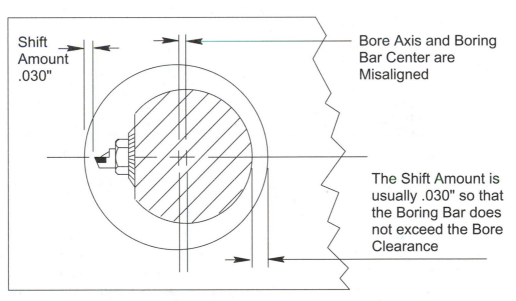

The view above illustrates the finish boring bar shifted away from the bore. The boring bar insert cartridge must be mounted in the spindle so that it shifts away from the bore when it is orientated. At this point the boring bar can retract.

Figure 10-4 **Fine boring canned cycle G76 code**

G80 Canned Cycle Cancel

As explained previously, these canned cycle codes also require the G80 *cancel code* when the operation is completed. The G80 code is used for all canned cycles and for VMC programs as well as HMC programs.

These canned cycle operations also require depth setting, either through the workpiece or to a specified depth. In either case, the CNC operator or setup person must first measure the depth, determine if an adjustment is required, then adjust the depth by adding or subtracting the required amount from the length offset. As stated previously, if the depth adjustment amount exceeds 0.100 inch, the CNC program, tool and toolholder, and tool length offset value should be checked for a possible programming or setup error.

G20 Inch and G21 Metric Format Codes

The engineering drawing, which determines the CNC machining, can be dimensioned in either inches or millimeters, and some may include both. Whichever dimensioning format is applied to the drawing, efficiency is increased if the drawing dimensions match the CNC program coordinates. For this reason, the CNC machine can be programmed to operate in either metric format or inch format.

The G20 code is used to set the CNC machine to inch format. The G21 code is used to set the CNC machine to metric format. When the control reads the G20 code, any axis (X, Y, or Z) movement command or feedrate value in the CNC program will be executed in inches. When the control reads the G21 code, any axis (X, Y, or Z) movement command or feedrate value in the CNC program will be executed in millimeters. Depending on the manufacturing company, the CNC programs may be mostly in G20 inch format or mostly in G21 metric format. It is also possible that the CNC programs may be a combination of both G20 inch and G21 metric format.

G54–G59 Work Coordinate System Codes

The work coordinate system codes (G54–G59) were developed to increase the efficiency of the CNC machine with respect to the location and adjustments of the workpiece origin. These codes enable the CNC machine to store and use six different origins in a CNC program. Additionally, the operator can adjust the workpiece origin by entering an amount into the work coordinate offset memory.

The work coordinate system codes can be used with multiple origins on a single workpiece or multiple fixture origins on the CNC machine. The HMC pallet system, which allows machining on multiple sides, uses this coordinate system. This increases CNC efficiency by using the machine's ability to automatically machine multiple parts and decrease load time. The CNC program example in this chapter, which machines three different operations in the same CNC program, demonstrates this method.

HMC B-Axis

The HMC is equipped with the ability to rotate the pallet 360° around. The letter address code B along with a numerical degree value (B180.000) rotates the pallet automatically. The pallet rotation can be CW, which is a (+) value, or CCW which is a (−) value. The pallet can be rotated manually with the machine control, by entering a B-axis command in MDI mode, or during a CNC program B-axis command cycle. This HMC pallet rotation feature is demonstrated in the example program of this chapter.

As with all CNC machine functions, it is a recommended practice to Dry Run or proveout any pallet rotation command. The CNC operator and setup person should take precautions to avoid a collision as the pallet rotates. In cases where the pallet rotations have not been proved out, the CNC operator can use the machine override features to slow the B-axis movement or completely stop it if necessary. This should allow the CNC operator time to react to a possible collision and stop the pallet rotation before any damage can occur. This is required for all B-axis commands in a CNC program or during a manual rotation.

INSTRUCTIONS FOR CNC PROGRAM EXAMPLE 5000

This section includes a coordinate worksheet, CNC documentation, and a CNC program for a HMC. As explained previously, each of these items is designed to illustrate the CNC setup and operation process. Therefore, it is important to review and understand the purpose and role of each CNC item. These items, which are required for both VMCs and HMCs, are listed below:

- Coordinate worksheet (for instructional purposes)
- Engineering drawing or process drawing
- CNC machine setup plan
- CNC tool list

This CNC program is similar in format to programs commonly used in industry. The similarities that should be noted include

- Program number
- Information notes
- Tool changes

- Default code setting
- Spindle starts
- Coolant on
- Machine home command
- Length offset activation
- End of program

Instructions for CNC Program Example 5000

For this CNC program, the student is required to identify and note the main CNC program items listed below:

1. Program number
2. Information notes
3. Tool changes
4. Default code setting blocks
5. Rapid moves
6. Feed moves
7. Radius feed moves
8. Tool radius compensation codes
9. Canned cycles
10. End of program

The example below illustrates how to identify and note the CNC program items above.

N865 G98 G84 Z-1.35 R-.3 F33.13	*Tapping Canned Cycle with Rapid above R-point*
N870 X3.875 Y-3.	*Coordinate Position 2*
N875 X4.875 Y-4.	*Coordinate Position 3*
N880 X3.875 Y-5.	*Coordinate Position 4*
N885 X7.75	*Coordinate Position 5*

■ ■ ■ ▪

COORDINATE WORKSHEET 5000

Instructions: Mark the X-Y zero origin, then calculate the coordinates for each point. Record the calculations in the chart. For the dimensions, use part drawing 5000 illustrated in Figures 10-5 and 10-6.

Use the standard origin symbol at right to mark the X-Y zero.

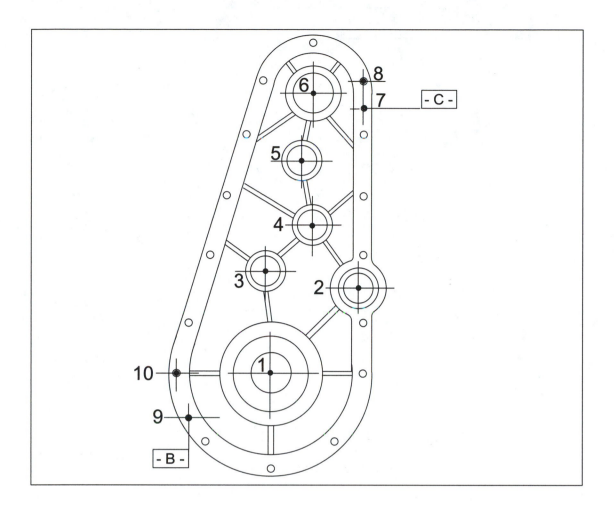

PART NO.	X-COORDINATE	Y-COORDINATE
1		
2		
3		
4		
5		
6		
7		
8		
9		
10		

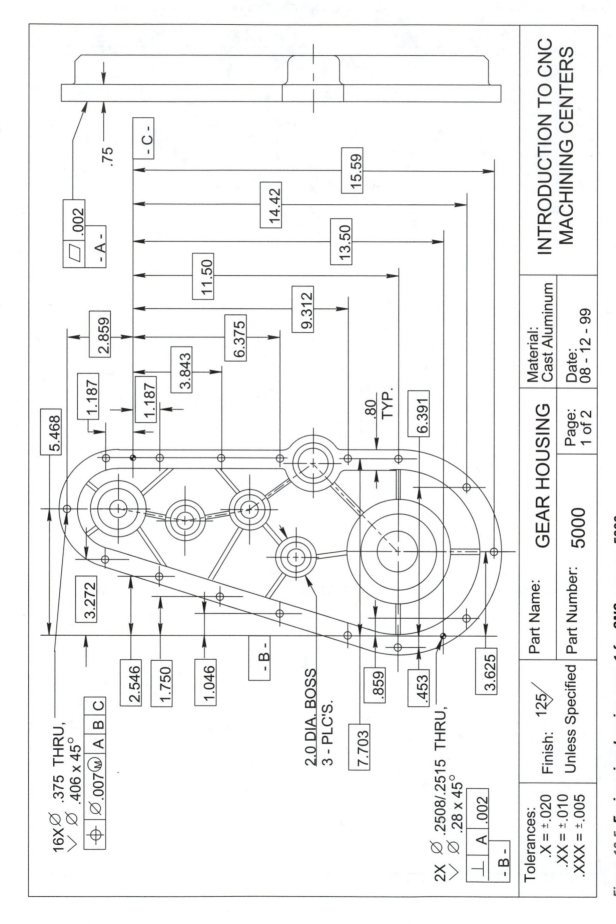

Figure 10-5 Engineering drawing page 1 for CNC program 5000

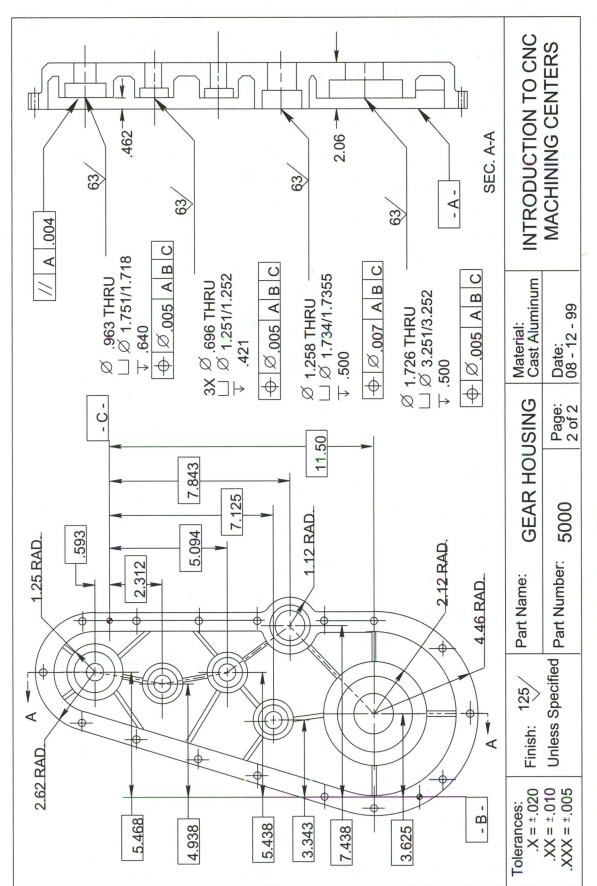

Figure 10-6 *Engineering drawing page 2 for CNC program 5000*

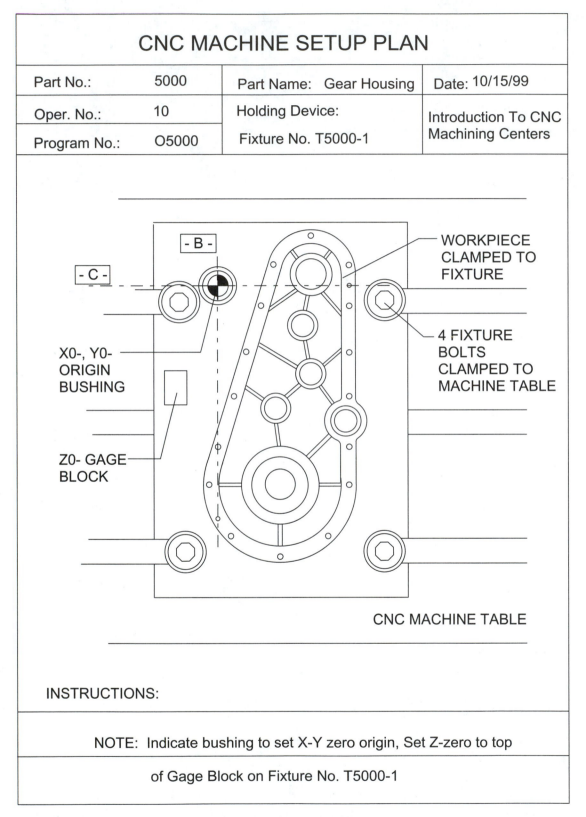

CNC MACHINE SETUP PLAN

Part No.:	5000	Part Name: Gear Housing	Date: 10/15/99
Oper. No.:	10	Holding Device:	Introduction To CNC Machining Centers
Program No.:	O5000	Fixture No. T5000-1	

- B -

- C -

WORKPIECE CLAMPED TO FIXTURE

4 FIXTURE BOLTS CLAMPED TO MACHINE TABLE

X0-, Y0- ORIGIN BUSHING

Z0- GAGE BLOCK

CNC MACHINE TABLE

INSTRUCTIONS:

NOTE: Indicate bushing to set X-Y zero origin, Set Z-zero to top

of Gage Block on Fixture No. T5000-1

Figure 10-7 Setup plan for CNC program 5000

CNC TOOL LIST AND OPERATION

Part No.: 5000		Prog. No.: O5000	Oper. No.: 10
Part Name: Gear Housing		Date: 03/11/00	Page 1 of 2

Pocket No.	Length Offset	Radius Offset	Tool No.	Tool Dia.	Tool Description and Operation
T01	H01	D	8001	2.0	2.00 Dia. Carbide End Mill Mill Flange and Bosses (5)
T02	H02	D	1012	.500	90 Degree Spot Drill spot (18) holes
T03	H03	D	1009	.375	.375" Dia. Drill Drill (16) holes
T04	H04	D	1104	6mm (.236)	6mm (.236") Dia. Drill Drill (2) holes
T05	H05	D	2025	.250	.250" Dia. Ream Ream (2) holes
T06	H06	D	4026	1.726	1.726" Dia. Boring Bar Bore (1) hole
T07	H07	D	4027	3.2365	3.2365" Dia. Boring Bar C'bore (1) hole
T08	H08	D	4028	3.2515	3.2515" Dia. Boring Bar C'bore (1) hole
T09	H09	D	4029	1.258	1.258" Dia. Boring Bar Bore (1) hole
T10	H10	D	4030	1.719	1.719" Dia. Boring Bar C'bore (1) hole

Figure 10-8 Tool list page 1 of 2 for CNC program 5000

CNC TOOL LIST AND OPERATION

Part No.: 5000		Prog. No.: O5000		Oper. No.: 10

Part Name: Gear Housing		Date: 03/11/00		Page 2 of 2

Pocket No.	Length Offset	Radius Offset	Tool No.	Tool Dia.	Tool Description and Operation
T11	H11	D	4031	1.7345	1.7345" Dia. Boring Bar C'bore (1) hole
T12	H12	D	4032	0.696	.696" Dia. Boring Bar Bore (3) holes
T13	H13	D	4033	1.2365	1.2365" Dia. Boring Bar C'bore (3) holes
T14	H14	D	4034	1.2515	1.2515" Dia. Boring Bar C'bore (3) holes
T15	H15	D	4035	0.963	.963" Dia. Boring Bar Bore (1) hole
T16	H16	D	4036	1.7364	1.7364" Dia. Boring Bar C'bore (1) hole
T17	H17	D	4037	1.7514	1.7514" Dia. Boring Bar C'bore (1) hole

*Figure 10-9 **Tool list page 2 of 2 for CNC program 5000***

CNC PROGRAM EXAMPLE 5000

```
%
O5000
(GEAR HOUSING - PROGRAM 5000)
(DATE, 08/12/99)
(TOOL - 01 2.000 DIA. - CARBIDE END MILL)
(TOOL - 02   .500 DIA. - SPOT DRILL)
(TOOL - 03   .375 DIA. - HSS DRILL)
(TOOL - 04   .236 DIA. - HSS 6MM DRILL)
(TOOL - 05   .250 DIA. - HSS REAMER)
(TOOL - 06 1.726 DIA. - CARBIDE INSERT BORING BAR)
(TOOL - 07 3.236 DIA. - CARBIDE INSERT BORING BAR)
(TOOL - 08 3.251 DIA. - CARBIDE INSERT BORING BAR)
(TOOL - 09 1.258 DIA. - CARBIDE INSERT BORING BAR)
(TOOL - 10 1.719 DIA. - CARBIDE INSERT BORING BAR)
(TOOL - 11 1.734 DIA. - CARBIDE INSERT BORING BAR)
(TOOL - 12   .696 DIA. - CARBIDE INSERT BORING BAR)
(TOOL - 13 1.236 DIA. - CARBIDE INSERT BORING BAR)
(TOOL - 14 1.251 DIA. - CARBIDE INSERT BORING BAR)
(TOOL - 15   .963 DIA. - CARBIDE INSERT BORING BAR)
(TOOL - 16 1.736 DIA. - CARBIDE INSERT BORING BAR)
(TOOL - 17 1.751 DIA. - CARBIDE INSERT BORING BAR)

N010 G0 G17 G20 G40 G49 G80 G90
N015 G91 G28 Z0.
N020 G28 X0. Y0.

(TOOL - 01 2.00 DIA. – CARBIDE END MILL)
N025 T1 M06
N030 G0 G90 G54 X10.0859 Y-7.843 B0. S1909 M3
N035 G43 H1 Z1. M08
N045 G1 Z0. F9.55
N050 X7.2859 F19.09
N055 X7.288 Y.5925
N60 G3 X3.7305 Y1.1347 I-1.82 J.0005
N065 G1 X.1309 Y-10.4106
N070 G3 X7.285 Y-11.5009 I3.4941 J-1.0894
N075 G1 X7.2858 Y-8.4502
N080 G0 Z1.
N085 X10.0859 Y-7.843
N090 G1 Z0. F9.55
N095 X7.2859 F19.09
N100 X7.288 Y.5925
N105 G3 X3.7305 Y1.1347 I-1.82 J.0005
N110 G1 X.1309 Y-10.4106
```

■ ■ ■ ■

CNC PROGRAM EXAMPLE 5000 (*continued*)

N115 G3 X7.285 Y-11.5009 I3.4941 J-1.0894
N120 G1 X7.2858 Y-8.4502
N125 G0 Z1.
N130 X3.625 Y-11.5
N135 G1 Z-.462 F9.55
N140 Y-9.8743 F19.09
N145 G3 X5.2508 Y-11.5 I0. J-1.6258
N150 X3.7369 Y-9.8781 I-1.6258 J0.
N155 G1 X3.343 Y-8.125
N160 Y-7.125
N165 X5.438 Y-5.094
N170 Y-4.094
N175 X4.938 Y-3.312
N180 Y-1.312
N185 X5.468 Y-.657
N190 Y-.032
N195 G2 X6.093 Y.593 I0. J.625
N200 X5.488 Y-.0317 I-.625 J0.
N205 G0 Z1.
N210 X3.625 Y-11.5
N215 G1 Z-.462 F9.55
N220 Y-9.8743 F19.09
N225 G3 X5.2508 Y-11.5 I0. J-1.6258
N230 X3.7369 Y-9.8781 I-1.6258 J0.
N235 G1 X3.343 Y-8.125
N240 Y-7.125
N245 X5.438 Y-5.094
N250 Y-4.094
N255 X4.938 Y-3.312
N260 Y-1.312
N265 X5.468 Y-.657
N270 Y-.032
N275 G2 X6.093 Y.593 I0. J.625
N280 X5.488 Y-.0317 I-.625 J0.
N285 G0 Z1.
N290 G0 Z1. M09
N295 G91 G49 G28 Z0. M05

(TOOL - 02 .500 DIA. - SPOT DRILL)
N300 G0 G17 G20 G40 G49 G80 G90
N305 T2 M06
N310 M01
N315 G0 G90 G55 X7.703 Y0 B0. S1000 M3

■ ■ ■ ▪

CNC PROGRAM EXAMPLE 5000 (*continued*)

```
N320 G43 H2 Z1. M08
N325 G99 G81 Z-.138 R.1 F5.
N330 X0. Y-13.5
N335 G80
N340 G0
N345 G99 G81 X-.453 Y-11.5 Z-.203 R.1 F5.
N350 X0. Y-9.312
N355 X1.046 Y-6.375
N360 X1.75 Y-3.843
N365 X2.546 Y-1.187
N370 X3.272 Y1.187
N375 X5.468 Y2.859
N380 X7.703 Y1.187
N385 Y-1.187
N390 Y-3.843
N395 Y-6.375
N400 Y-9.312
N405 Y-11.5
N410 X6.391 Y-14.42
N415 X3.625 Y-15.59
N420 X.859 Y-14.42
N425 G80 M09
N430 G91 G49 G28 Z0. M05

(TOOL - 03  .375 DIA. - DRILL)
N500 G0 G17 G20 G40 G49 G80 G90
N505 T3 M06
N510 M01
N515 G0 G90 G54 X-.453 Y-11.5 B0. S2100 M3
N520 G43 H3 Z1. M08
N525 G99 G81 Z-.9127 R.1 F16.5
N530 X0. Y-9.312
N535 X1.046 Y-6.375
N540 X1.75 Y-3.843
N545 X2.546 Y-1.187
N550 X3.272 Y1.187
N555 X5.468 Y2.859
N560 X7.703 Y1.187
N565 Y-1.187
N570 Y-3.843
N575 Y-6.375
N580 Y-9.312
N585 Y-11.5
```

■ ■ ■ ■

CNC PROGRAM EXAMPLE 5000 (*continued*)

N590 X6.391 Y-14.42
N595 X3.625 Y-15.59
N600 X.859 Y-14.42
N605 G0 Z1. M09
N610 G91 G49 G28 Z0. M05

(TOOL - 04 .236 DIA. – 6MM DRILL)
N615 G0 G17 G20 G40 G49 G80 G90
N620 T4 M06
N625 M01
N630 G0 G90 G54 X0. Y-13.5 B0. S1000 M3
N635 G43 H4 Z1. M08
N640 G99 G81 Z-.8709 R.1 F23.
N645 X7.703 Y0.
N650 G80 M09
N655 G91 G49 G28 Z0. M05

(TOOL - 05 .250 DIA. - REAMER)
N660 G0 G17 G20 G40 G49 G80 G90
N665 T5 M06
N670 M01
N675 G0 G90 G54 X0. Y-13.5 B0. S3000 M3
N680 G43 H5 Z1. M08
N685 G99 G81 Z-.8751 R.1 F18.
N690 X7.703 Y0.
N695 G80 M09
N700 G91 G49 G28 Z0. M05

(TOOL - 06 1.726 DIA. - BORING BAR)
N705 G0 G17 G20 G40 G49 G80 G90
N710 T6 M06
N715 M01
N720 G0 G90 G54 X3.625 Y-11.5 B0. S1163 M3
N725 G43 H6 Z1. M08
N730 G98 G86 Z-2.12 R-.36 F6.98
N735 G80 M09
N740 G91 G49 G28 Z0. M05

(TOOL - 07 3.2365 DIA. - BORING BAR)
N745 G0 G17 G20 G40 G49 G80 G90
N750 T7 M06
N755 M01
N760 G0 G90 G54 X3.625 Y-11.5 B0. S617 M3
N765 G43 H7 Z1. M08
N770 G98 G86 Z-.962 R-.36 F6.79

■ ■ ■ ■

CNC PROGRAM EXAMPLE 5000 (*continued*)

N775 G80 M09
N780 G91 G49 G28 Z0. M05

(TOOL - 08 3.2515 DIA. - BORING BAR)
N785 G0 G17 G20 G40 G49 G80 G90
N790 T8 M06
N795 M01
N800 G0 G90 G54 X3.625 Y-11.5 B0. S738 M3
N805 G43 H8 Z1. M08
N820 G98 G86 Z-.962 R-.36 F5.98
N825 G80 M09
N830 G91 G49 G28 Z0. M05

(TOOL - 09 1.258 DIA. - BORING BAR)
N835 G0 G17 G20 G40 G49 G80 G90
N840 T9 M06
N845 M01
N850 G0 G90 G54 X7.438 Y-7.843 B0. S1590 M3
N855 G43 H9 Z1. M08
N860 G99 G86 Z-2.12 R.1 F12.5
N865 G80 M09
N870 G91 G49 G28 Z0. M05

(TOOL - 10 1.719 DIA. - BORING BAR)
N875 G0 G17 G20 G40 G49 G80 G90
N880 T10 M06
N885 M01
N890 G0 G90 G54 X7.438 Y-7.843 B0. S1163 M3
N895 G43 H10 Z1. M08
N900 G99 G86 Z-.85 R.1 F6.98
N905 G80 M09
N910 G91 G49 G28 Z0. M05

(TOOL - 11 1.7345 DIA. - BORING BAR)
N915 G0 G17 G20 G40 G49 G80 G90
N920 T11 M06
N925 M01
N930 G0 G90 G54 X7.438 Y-7.843 B0. S1384 M3
N935 G43 H11 Z1. M08
N940 G99 G86 Z-.85 R.1 F11.07
N945 G80 M09
N950 G91 G49 G28 Z0. M05

(TOOL - 12 0.696 DIA. - BORING BAR)
N955 G0 G17 G20 G40 G49 G80 G90
N960 T12 M06

CNC PROGRAM EXAMPLE 5000 (*continued*)

N965 M01
N970 G0 G90 G54 X3.343 Y-7.125 B0. S2873 M3
N975 G43 H12 Z1. M08
N980 G98 G86 Z-2.12 R-.36 F22.98
N985 X5.438 Y-5.094
N990 X4.938 Y-2.312
N995 G80 M09
N1000 G91 G49 G28 Z0. M05

(TOOL - 13 1.2365 DIA. - BORING BAR)
N1005 G0 G17 G20 G40 G49 G80 G90
N1010 T13 M06
N1015 M01
N1020 G0 G90 G54 X3.343 Y-7.125 B0. S1612 M3
N1025 G43 H13 Z1. M08
N1030 G98 G86 Z-.883 R-.36 F12.9
N1035 X5.438 Y-5.094
N1040 X4.938 Y-2.312
N1045 G80 M09
N1050 G91 G49 G28 Z0. M05

(TOOL - 14 1.2515 DIA. - BORING BAR)
N1055 G0 G17 G20 G40 G49 G80 G90
N1060 T14 M06
N1065 M01
N1070 G0 G90 G54 X3.343 Y-7.125 B0. S1920 M3
N1075 G43 H14 Z1. M08
N1080 G98 G86 Z-.883 R-.36 F15.36
N1085 X5.438 Y-5.094
N1090 X4.938 Y-2.312
N1095 G80 M09
N1100 G91 G49 G28 Z0. M05

(TOOL - 15 0.963 DIA. - BORING BAR)
N1105 G0 G17 G20 G40 G49 G80 G90
N1110 T15 M06
N1115 M01
N1120 G0 G90 G54 X5.468 Y.593 B0. S2077 M3
N1125 G43 H15 Z1. M08
N1130 G98 G86 Z-2.12 R-.36 F16.62
N1135 G80 M09
N1140 G91 G49 G28 Z0. M05

(TOOL - 16 1.7364 DIA. - BORING BAR)
N1145 G0 G17 G20 G40 G49 G80 G90

CNC PROGRAM EXAMPLE 5000 (*continued*)

```
N1150 T16 M06
N1155 M01
N1160 G0 G90 G54 X5.468 Y.593 B0. S1149 M3
N1165 G43 H16 Z1. M08
N1170 G98 G86 Z-1.102 R-.36 F9.2
N1175 G80 M09
N1180 G91 G49 G28 Z0. M05

(TOOL - 17  1.7514 DIA. - BORING BAR)
N1185 G0 G17 G20 G40 G49 G80 G90
N1190 T17 M06
N1195 M01
N1200 G0 G90 G54 X5.468 Y.593 B0. S1371 M3
N1205 G43 H17 Z1. M08
N1210 G98 G86 Z-1.102 R-.36 F10.97
N1215 G80
N1220 G0 Z.1M09
N1225 G91 G49 G28 Z0. M05
N1230 G90
N1235 M30
%
```

CHAPTER SUMMARY

The key concepts presented in this chapter include the following:

- Canned cycle code descriptions.
- The G84 code and machine movement descriptions.
- The G86 code and machine movement descriptions.
- The G76 code and machine movement descriptions.
- The G20 inch format code description.
- The G21 metric format code description.
- The G54–G59 work coordinate system code description.
- Calculations for CNC horizontal machining center coordinates.
- The format description of a CNC program for CNC horizontal machining centers.
- The basic codes required of a CNC program for CNC horizontal machining centers.

■ ■ REVIEW QUESTIONS

1. Which types of tools are used with the G84 canned cycle?

2. Which types of tools are used with the G86 canned cycle?

3. Which types of tools are used with the G76 canned cycle?

4. Which three G codes are common to all canned cycle codes?

5. How is the Z-axis stopped during a G86 boring cycle?

6. How is the Z-axis stopped during a G84 tapping cycle?

7. What are the functions of the G54–G59 codes?

8. What are the differences between the G86 and the G76 codes?

9. What is the function of the B-axis on a horizontal machining center?

10. How is the HMC pallet rotated in a CW or CCW direction?

11. What precautions can the CNC operator or setup person take to avoid collisions during proveout of CNC canned cycles?

12. Discuss and identify the CNC program format structure and codes covered in this chapter. The items that should be identified include CNC program number, work coordinate number blocks, tool information notes, tool change blocks, default code setting blocks, spindle starts, canned cycle codes, coolant on and off, machine home command blocks, length offset activation blocks, circular interpolation blocks, table rotation blocks, and end of program block.

■ ■ CNC PROGRAM EXERCISE FOR VMC

Using the sketch in Figure 10-10, calculate the boring canned cycle values for the X-, Y-, and Z-axes and R-plane. Also determine and enter the boring canned cycle G codes. Enter the codes and coordinates in the CNC program provided on the following page.

■ ■ CNC PROGRAM EXERCISE 5001

Fill in the missing program information of CNC program 5001 as indicated below:

Block N170: enter the G codes, X-Y-Z-R coordinate values, and IPM value.

Block N180: enter the X-coordinate value.

Block N190: enter the G code.

Block N250: enter the G codes, X-Y-Z-R-P-Q values, and IPM value.

Block N260: enter the X-coordinate value.

Block N270: enter the G code.

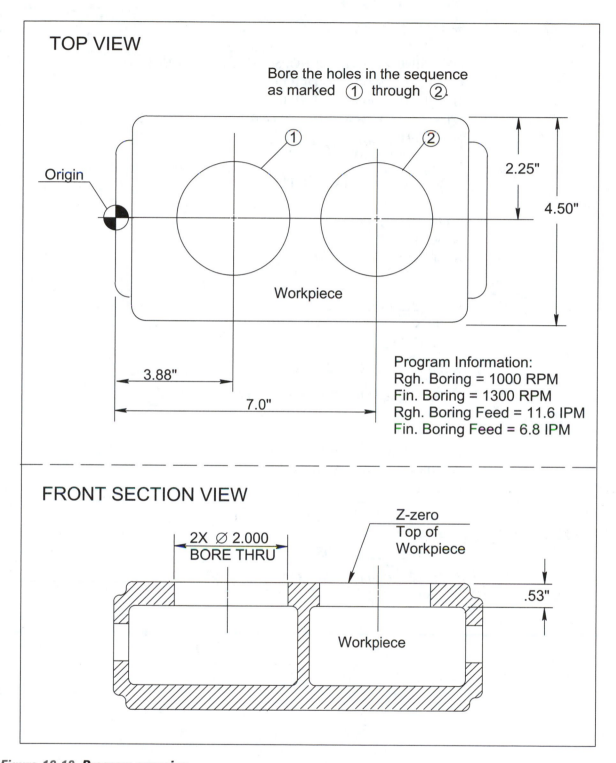

TOP VIEW

Bore the holes in the sequence
as marked ① through ②.

Origin

① ②

2.25"

4.50"

Workpiece

3.88"

7.0"

Program Information:
Rgh. Boring = 1000 RPM
Fin. Boring = 1300 RPM
Rgh. Boring Feed = 11.6 IPM
Fin. Boring Feed = 6.8 IPM

FRONT SECTION VIEW

2X ⌀ 2.000
BORE THRU

Z-zero
Top of
Workpiece

.53"

Workpiece

Figure 10-10 **Program exercise**

```
%
O5001
(HOUSING - PROGRAM NO. 5001 DATE, 07/12/03)
(TOOL - 01 .500 DIA. - DRILL)
(TOOL - 02 .500 DIA. - DRILL)
N100 G00 G17 G40 G49 G80 G90
N110 G91 G28 Z0.
N120 G28 X0. Y0.

(TOOL - 01 .500 DIA. - DRILL)
N130 T1 M06
N140 M01
N150 G0 X0 Y0 S800 M3
N160 G43 H1 Z1.0 M08

N170 G _____ G _____ X _____ Y _____ Z _____
R _____ F _____
N180 X _____
N190 G _____ M05

N200 G91 G49 G28 Z0. M09

(TOOL - 01 .500 DIA. - DRILL)
N210 T2 M06
N220 M01
N230 G0 X0 Y0 S800 M3
N240 G43 H1 Z1.0 M08

N250 G _____ G _____ X _____ Y _____ Z _____
R _____ P _____ Q _____ F _____
N260 X _____
N270 G _____ M05

N280 G91 G49 G28 Z0. M09
N290 M30
%
```

chapter 11

INTRODUCTION TO CNC TURNING

Chapter Objectives

After studying and completing this chapter, the student should have knowledge of the following:

- *CNC turning machines*
- *CNC lathes*
- *CNC vertical turret lathes*
- *CNC dual spindle turning centers*
- *Live tooling for CNC turning machines*
- *CNC automatic screw machines*
- *Safety rules for CNC turning machines*
- *CNC lathe processing*
- *CNC lathe documentation*

The remaining chapters (11–20) focus on CNC turning machines (lathes) that are used to perform the same lathe cutting functions that conventional lathes have performed for many decades. The major difference and the principal advantage of CNC turning machines is the increased control and repeatability of the cutting tool. CNC machining allows for the manufacture of parts that would otherwise be difficult or impossible to produce with conventional lathes and tooling methods.

■ ■ ■ ■

CNC TURNING MACHINES

CNC turning machines (lathes) have been continually improved to increase their accuracy, efficiency, and productivity. Most modern CNC lathes are accurate to within 0.0002 inch *positioning* and repeatability. They are also capable of positioning the cutter (toolholder) at a rate of 700 inches per minute or greater. Additionally, CNC lathes can be designed with various options. For example, CNC lathes can be equipped with tool sensing sytems, with two or more spindles, additional turrets, and "live tooling." The tool sensing system is designed to quickly and accurately find and store each tool tip offset value. The dual spindle option allows turning of two parts at the same time. By adding another tool *turret* it provides greater workpiece machining flexibility and reduces setup requirements. The live tooling option allows secondary machining such as milling and drilling operations to be performed in the same turning operation.

The following is a list of advantages that can be derived from CNC lathes:

- CNC lathes reduce setup time.

- CNC lathes reduce machining cycle time.

- CNC lathes output maximum part accuracy.

- CNC lathes reduce scrap and rework.

- CNC lathes require less inspection time.

- CNC lathes reduce handling, which improves quality and production.

- CNC lathes can minimize tooling costs.

- CNC lathes can quickly and efficiently produce complex parts.

- CNC lathes run automatically without operator intervention.

CNC turning machines use CNC-coded programs that provide the information used by the MCU to control and position the cutting tools. The MCU is the brain of the CNC lathe. It reads, interprets, and converts programmed input into appropriate movement. It also controls various functions such as spindle rotation, axis motion, coolant operation, automatic tool indexing, and program proveout graphics. The MCU converts the CNC-coded part program information into voltage or current pulses of varying frequency or magnitude. The converted pulses are used to position the axis slides and control the various functions required to operation of the CNC lathe.

There are various types of CNC turning machine designs that can vary in type and or size. However, they all basically operate by the same basic concept, which involves a spindle and chuck to hold and rotate the workpiece and a turret that clamps the cutting tools to cut the workpiece. There are some similarities between all the various types of CNC turning machines.

For example, the turret travels in X- and Z-axes on all types of turning machines. Common to most turning machines are the interchangeable cutting toolholders (see Figure 11-1). CNC lathes typically vary in size, number of turrets, and orientation (horizontal or vertical) of the spindle and the turret. For example, CNC turning machines usually have a horizontal spindle, but some larger CNC turning machines are designed with the spindle in the vertical position. The CNC vertical turret lathe (CNC VTL) is a machine designed with the spindle in the vertical position (see Figure 11-2).

CNC Turning Machine Options

The standard CNC lathe is typically designed with one spindle and one tool post block or tool turret. However, due to the various manufacturing and production requirements, CNC turning machines can be designed with various options that include

- Additional tool turret/s

- Quick tool change system

- Automatic tool offset setter

- Live tooling system

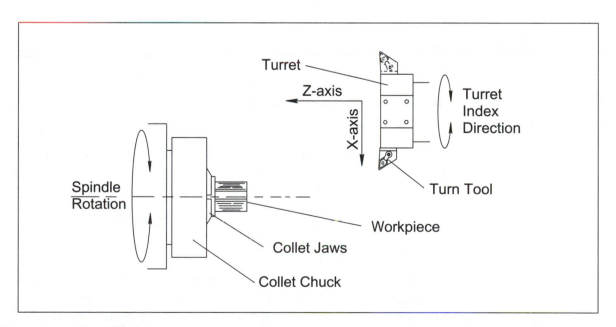

*Figure 11-1 **The CNC lathe***

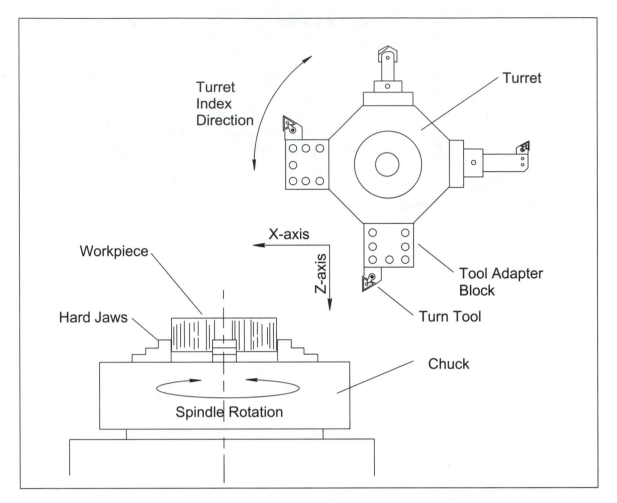

Figure 11-2 ***The CNC vertical turret lathe***

- Bar feeding system

- Dual spindle

- Higher speed (RPM) spindle

- Graphics for CNC program proveout

- Conversational programming

The CNC lathe options listed above are basically designed to reduce the items listed below:

- Part handling

- Machine setup time

- Part cycle time

Typically, the dual turret CNC lathe is designed with one turret on top and the other turret below. The dual spindle and dual turret CNC lathe design (Figure 11-3) is typically designed with a spindle at each end and with one turret on top and the other turret below.

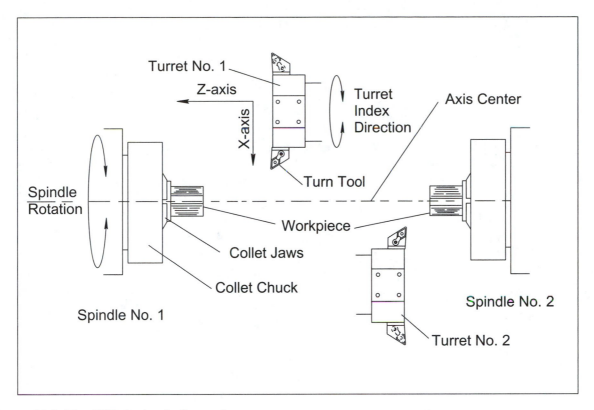

Figure 11-3 **The CNC dual spindle turning center**

The combination turning and milling center was developed to perform additional operations such as milling, drilling, and tapping on machined round parts. This is possible because this type of machine is designed with a special tool turret containing pockets with drive mechanisms that rotate the mill-type cutting tools (see Figure 11-4). Previous to this development, a round lathe part was machined in a turning or chucking machine, and then moved to a milling or drilling-type machine to complete the additional operations.

However, there are also some limitations and disadvantages. For example, the live tooling usually is powered by a low power motor of some type. This limits the size of secondary milling and drilling that can be performed. Another consideration is that, usually, only a few parts are suitable for machining on these higher cost lathes. Therefore, these machines sometimes sit idle due to lack of appropriate production parts.

In order to perform operations such as milled slots, milled flats, drilling, and tapping, the machine design must include a contouring spindle that can be indexed and locked at exact locations around the work circumference. For example, if four holes were required at 90° intervals around the part circumference, the CNC program procedure would be as follows:

- The spindle is located at a 90° position to the tool (drill) position.

- The revolving drill is moved to the length coordinate along the Z-axis.

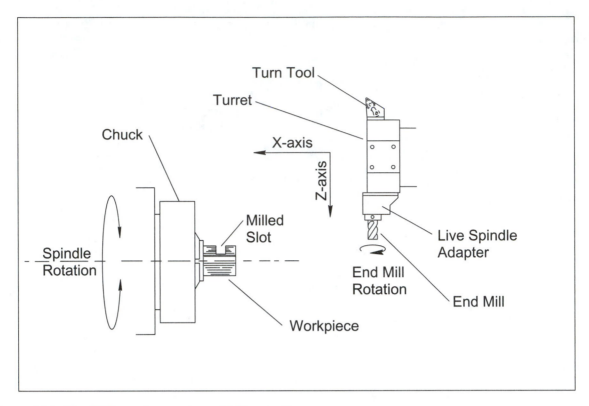

Figure 11-4 **The CNC turning center with live tooling**

- The hole is drilled to depth along the X-axis and then retracted from the hole.

- The spindle then indexes 90° and the cycle is repeated until all four holes are drilled.

CNC Automatic Screw Machine

Another type of CNC turning machine is the CNC automatic screw machine, which is a specialized CNC lathe. The CNC screw machine is designed to mainly run high-production-type parts. The parts that are ideal for CNC screw machines are 1.500 inches in diameter or smaller. The CNC screw machine is typically equipped with a bar feeder, a parts catcher, a main turret, a subturret, and a collet chuck (see Figure 11-5).

The function of the bar feeder is to quickly load stock automatically. This automatic method of loading the stock is designed to reduce the part load and handling time.

The function of the parts catcher is to catch the part immediately after it is cut off and avoid damage to the part, tools, and/or the machine. It also provides a safer part retrieval method for the operator.

The main turret and the subturret are designed to work independent of each other. This design of the turrets allows simultaneous cutting with two tools to reduce the cycle time. The main turret, which is a two-axis design, is

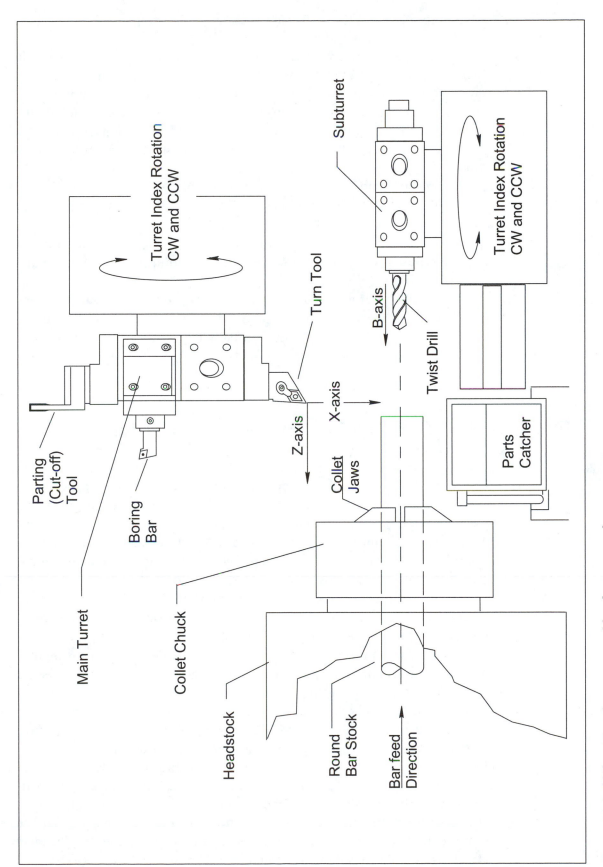

Figure 11-5 CNC automatic screw machine features and components

assigned the X- and Z-axes. The subturret, which is a single-axis design, is assigned the B-axis (longitudinal) travel.

The dual tool turret system is another method of improving machining efficiency and method flexibility. Most CNC automatic screw machines are equipped with a main turret designed for turning, facing, chamfering, drilling, and boring-type operations. The second, or subturret, is designed for center drilling, drilling, tapping, reaming, and broaching-type operations. Either turret can be used for controlling the stock feed out length position.

The most common workholding method used on CNC screw machines is the collet chuck. The collet chuck is ideal for small parts where accuracy and quickness is a factor. Collets are available for holding hexagonal, square, round, or custom shapes. Due to their efficiency and flexibility, they are typically used on CNC screw machines equipped with bar-feeding systems.

These added features are mainly designed to reduce the cycle time and thereby increase the CNC machine productivity. Due to the added features of the CNC screw machine, more time is usually required for setup, timing of the turrets and tools, and size adjustments.

SAFETY RULES FOR CNC TURNING MACHINES

CNC lathes are extremely fast and powerful machine tools; therefore, it is imperative that all safety rules and operating instructions are adhered to in every situation. Most machine shop safety rules and safeguards for operating CNC lathes are specified and enforced by OSHA and ANSI guidelines are applied to safety equipment. Therefore, most CNC lathes also have an operator's manual, which includes safe operating instructions and warning labels to alert the operator of possible dangers that may occur.

Due to the higher cutting speeds that are typical of CNC lathes, most machine tool manufacturers have designed standard built-in safety features into the CNC lathe. Some built-in safety features for CNC lathes includes a full enclosure that prevents metal chips from being ejected and causing injuries to the operator or other personnel. The enclosure usually includes sliding doors and operator viewing panels. Additionally, the doors are usually hardwired with switches, which will prevent the CNC lathe spindle from operating whenever the doors are open. Another built-in safety feature prevents the CNC lathe spindle from operating whenever the chuck jaws are unclamped. An important safety feature that is built-in for CNC lathes and most CNC machines alike is the "way-lube-fault" warning. This safety feature will also prevent the spindle from operating whenever the lubricant in the way lube reservoir diminishes to a low level. The following is a categorized list of recommended safety rules that the reader is strongly advised to fully read and understand before attempting to operate CNC lathes.

Personal Safeguards

- Wear ANSI-approved safety glasses with side shields at all times in the shop areas.

- Wear safety shoes when working with heavy tools and equipment.

- Wear hearing protection for noise levels that exceed OSHA specifications.

- Wear an approved face mask for dust levels that exceed OSHA specifications.

- Keep long hair covered when operating or standing near lathes.

- Do not wear jewelry or loose clothing while operating or standing near lathes.

- When the spindle is rotating, avoid standing directly in front of the chuck.

- Lift with the legs, not the back.

- Avoid skin contact with any cutting fluids or oils.

- Do not operate any lathes or equipment while under the influence of drugs (prescribed or otherwise).

- Always report any injury and apply first-aid treatment.

Shop Environment Safeguards

- Remove chips from the floor.

- Clean all liquid, oil, and grease spills immediately.

- Keep floors and walk aisles clean.

- Report any fumes or odors.

Lathe and Tool Handling Safeguards

- Always remove hand wrenches after tightening or untightening.

- Always remove chuck wrenches after tightening or untightening.

- Store tools in their appropriate tool trays and racks.

- Keep tools sharp and in good working condition.

- Check that all safety guards and devices are in place and working before operating any lathe.

- Keep all electrical and mechanical panels secured in place.

- Do not handle loose wires or electrical components.

- Check that all oil levels are maintained.

- Check that all compressed air equipment is in good working condition.

- Do not use compressed air to clean lathe slides.

- Keep tools, parts, and any other items off the lathe and part.

- Use gloves when handling tools by their cutting edges.

- Never use gloves when operating a lathe.

- Check that the lathe is electrically sound and use lock-out tag-out practices.

- Check that all lights in the lathe work area are in good working condition.

- Keep clear from obstructions and sharp tools when leaning into the lathe work area.

Safe Machining Practices

- In case of any emergency while operating a lathe, immediately press the Emergency Stop button.

- Keep hands away from the spindle while it is rotating.

- Do not open the electrical panel or control doors.

- Keep hands clear of all moving lathe components.

- Check each tool for possible collisions with the part or lathe components before starting any operation.

- Do not operate controls unless you have been properly instructed about the Emergency Stop button, Feed Hold button, Spindle Stop button, and the various other operation functions.

- Use caution to avoid inadvertently bumping any CNC control buttons.

- Proveout new programs in Dry Run mode before actual cutting.

- Setup the workholding device and cutting tools as rigidly as possible.

- Do not remove chips or debris by hand or while the spindle is on or in operation.

- Check that all feeds and speeds do not exceed recommended values.

- Maintain a continuous flow of coolant to the cutting tool when it is required.

- Always consult with an authorized person if you are uncertain or unfamiliar with any operation.

These safety rules are generally applicable to most CNC machine operating situations. However, due to the many variables that exist in manufacturing there may be other safety concerns not listed above. Therefore, it is recommended that each CNC machining situation be evaluated on an individual basis to determine every safety concern and safeguard before operating the CNC lathe.

CNC LATHE PROCESSING

Planning and documenting the CNC lathe process is required to efficiently complete a job on CNC lathes. Basically, the process begins when the CNC programmer reviews the engineering drawing to determine what machining is required. This will also result in determining how many operations are required (one, two, or more), and which type of CNC lathe to use. Generally, the lathe machining determination follows the order listed below:

CNC Lathe Process Determination

- Rough facing and/or turning

- Rough drilling and/or boring

- Finish facing and/or turning

- Finish boring

- Finish grooving

- Finish tapping and/or threading

- Cut off (barstock-type work)

The CNC programmer also determines how the part will be held based on the part drawing datums and requirements. Usually, the part is held with standard hard or soft jaws. If special workholding (jaws or fixture) is required, it is immediately designed and ordered to be built. This is due to the fact that special items usually require longer than usual lead time to complete.

The CNC programmer then selects, from the tool library, the tools that will be used to machine the part. Again, standard lathe tools are capable in most situations, and if special tools are required, they must be ordered as soon as possible. Tool drawings are typically used to specify standard and nonstandard tool features (see Figure 11-6). The person responsible for assembling the tools uses these drawings as a reference. It is important to note that besides the long lead time to build special fixtures and tools, they are also very costly to design and build. Therefore, whenever possible the first choice is always to use standard workholding jaws and tools.

Now the CNC programmer can begin to create the CNC program. This can be completed manually, on a conversational control or with computer programming software. There are various types of computer software systems for creating CNC lathe programs. CAD/CAM, which is one method of CNC computer programming, has become very popular with most manufacturing plants. After the CNC program is completed each item of the process must be documented to insure efficiency and accuracy.

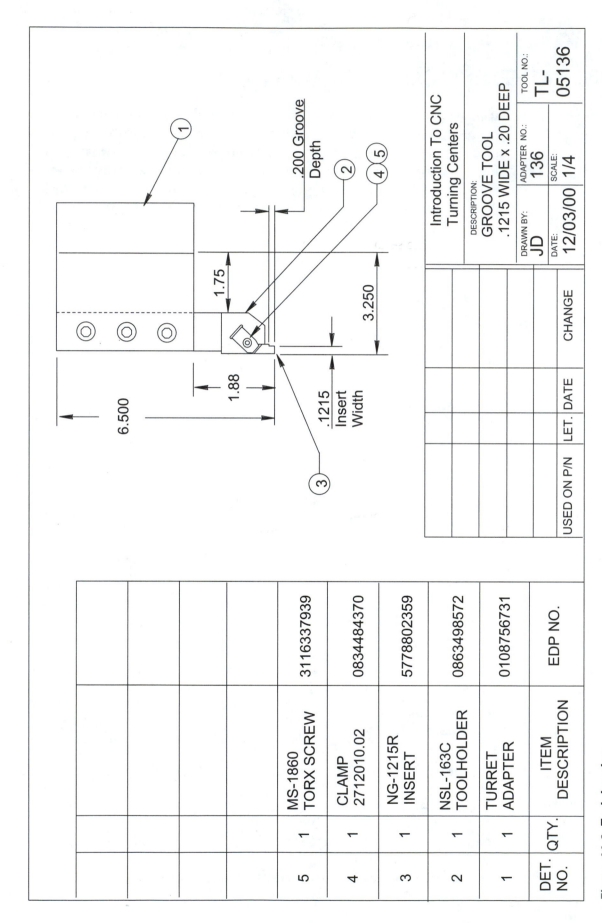

DET. NO.	QTY.	ITEM DESCRIPTION	EDP NO.				
5	1	MS-1860 TORX SCREW	3116337939				
4	1	CLAMP 2712010.02	0834484370				
3	1	NG-1215R INSERT	5778802359				
2	1	NSL-163C TOOLHOLDER	0863498572				
1	1	TURRET ADAPTER	0108756731				

Introduction To CNC Turning Centers

DESCRIPTION:
GROOVE TOOL
.1215 WIDE x .20 DEEP

DRAWN BY: JD	ADAPTER NO.: 136	TOOL NO.:
DATE: 12/03/00	SCALE: 1/4	TL-05136

USED ON P/N	LET.	DATE	CHANGE

Figure 11-6 **Tool drawing**

CNC LATHE DOCUMENTATION

The CNC process documents include the setup plan, the tool list, any special tool drawings, and the CNC program manuscript. The examples in Figures 11-7, and 11-8 illustrate one method that can be used in either paper form or on a computer CRT screen. This information must be documented and relayed to the respective CNC personnel. At this point, the CNC lathe processing of the part is complete and it can now be scheduled for production.

When the part is designated for production, the setup person starts the CNC machine setup as specified by the documentation. The setup person typically performs the following items:

- Workholding device (chuck and jaws) are fastened to the spindle.

- Tools are assembled and fastened to each specified turret station.

- The CNC program is entered and stored into the control.

- The X- and Z-zero origins are set.

- The tool length offsets are stored under the specified numbers.

- The TNR offset values are stored under the specified numbers (only if using G41/G42).

- The tool tip position numbers are stored under the specified tool offset number (only if using G41/G42).

After the machine setup is completed, the CNC program proveout is performed. This is required to ensure that the CNC program does not include any errors that will cause damage due to a collision. The proveout process is typically completed with the CNC machine control Dry Run and "Single Block" switches in the "On" position. Also, the Spindle Speed Override Control and Rapid Rate Override Control knobs are slowed down to allow the operator time to react to possible collisions. It is sometimes a common practice to initially proveout a new CNC program without a workpiece in the chuck.

Once the CNC program is debugged and edited (proved out), and the tools are adjusted to size, a workpiece can be clamped and the machine can be run in automatic mode. The next time that the same part program is run, the CNC machine is setup and it is usually proved out in Single Block only—Dry Run is not necessary.

The production run may be short or long depending on the number of parts required by the production schedule. However, the CNC setup person and CNC operator must maintain the tool cutting process to ensure that the desired quality is achieved. Process control, which includes the cutting process, typically includes the following items:

- The part is measured and produced to the drawing specifications and tolerances.

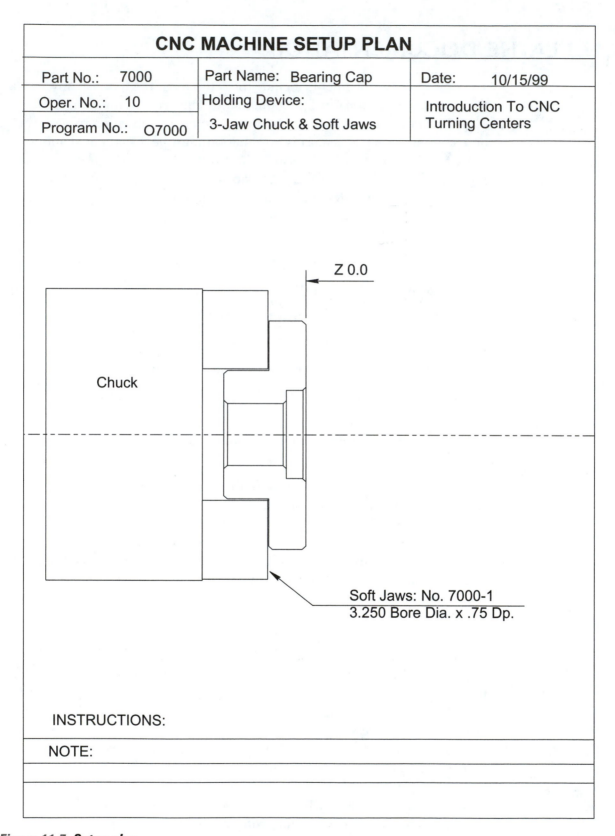

CNC MACHINE SETUP PLAN

Part No.: 7000	Part Name: Bearing Cap	Date: 10/15/99
Oper. No.: 10	Holding Device:	Introduction To CNC
Program No.: O7000	3-Jaw Chuck & Soft Jaws	Turning Centers

Z 0.0

Chuck

Soft Jaws: No. 7000-1
3.250 Bore Dia. x .75 Dp.

INSTRUCTIONS:

NOTE:

Figure 11-7 Setup plan

CNC TOOL LIST AND OPERATION DESCRIPTION

Part No.: 7000	Prog. No.: O7000	Page 1 of 1
Part Name: Bearing Cap	Oper. No.: 10	Date: 10/15/99

Station No.	Tool Offset	Tool Tip	Tool Tip Radius	Tool No.	Tool Description and Operation
T01	01	03	.031	8001	Holder No. DCLNR-164D Rough face & turn 1.25 dia. Insert: CNMG-432 C 5 .031 nose radius
T02	02	03	.031	3002	Holder No. DDJNR-164D Finish face & turn 1.25 dia. Insert: DNMG-432 C 5 .031 nose radius
T03	03			1015	.500" Dia Drill Drill (4) holes

Figure 11-8 Tool list

- The size (wear) offsets are adjusted as determined by the part inspections during the production run.

- The original CNC program file, if edited, is relayed to the CNC programmer for revision.

- The original CNC setup plan and tool list documents, if edited, are relayed to the CNC programmer for revision.

As stated previously, the proper process determination and documentation are an essential part of the CNC setup and production process. The level of CNC efficiency is dependent on both providing and using the CNC documentation required to machine a part. Therefore, it is important to identify and properly use all the CNC document information available for each CNC program.

There are various items that are required for CNC setup and machining. Table 11-1 describes some typical CNC setup items or information that can be obtained from each type of CNC document.

■___Table 11-1 *CNC documentation information*

DOCUMENT NAME	DESCRIPTION OF INFORMATION
Engineering drawing	The engineering drawing specifies what features are to be machined. It specifies the feature dimensions, sizes, tolerances, and finish requirements. Also included are the part number, revision level, and part material specifications. The drawing is typically in absolute dimensioning format so that the program coordinate values and the drawing dimensions match. Also, the drawing typically includes GD&T, which describes the part requirements by the use of geometric symbols and datum surface locations.
Setup plan	The setup plan specifies how the part is held and orientated on the machine spindle/chuck. It also specifies where the part zero origin is located. It may also specify the type of chuck, jaws, or fixturing used. If a hydraulic chuck is used, the chucking pounds per square inch (PSI) value is usually noted.
Tool list	The tool list specifies which tools have been selected to perform the machining of the part. Typical information includes the tool station number, the offset number, the TNR compensation, and tool tip direction number. Tool descriptions that are specified include turn tool type and geometry, drill diameters, tool lengths, insert identification number, and insert grade. Other items that may be specified are tool materials such as HSS, coated carbide, uncoated carbide, cobalt, diamond, and ceramic.
CNC program manuscript	The CNC program manuscript is the document where all the CNC codes needed to machine the part are specified. It includes all the specific codes to control the miscellaneous machine functions, the coordinates that control machine axis movement, and other auxiliary *command* functions. Typical information includes tool indexes, spindle speeds, and feedrate values.
Tool library	The tool library is a company's catalog of all the CNC tools that are available for machining a part. Tool types and dimensions are detailed on tool drawings for ready access and to reduce redundant tool inventory (see appendix L).

There are various forms and methods of CNC documentation used to record the CNC process. The type of form or method used is determined by the needs and resourses of each individual company. However, in order to achieve long-term success, it is essential to implement a system that adequately documents all the items required to effectively utilize CNC technology.

CHAPTER SUMMARY

The key concepts presented in this chapter include the following:

- The advantages of using CNC turning machines.
- The description of the basic CNC lathe.
- The description of the basic CNC lathe component options.
- The description of the basic CNC vertical turret lathe.
- The description of the dual spindle CNC turning center.
- The description of live tooling for CNC turning machines.
- The description of the basic CNC automatic screw machine.
- The safety rules for CNC turning machines.
- The CNC lathe processing.
- The CNC lathe documentation.

REVIEW QUESTIONS

1. Explain the difference between conventional lathes and CNC lathes.
2. List four types of CNC turning machines.
3. The CNC automatic screw machine is designed for what type of production parts?
4. List the advantages of using CNC lathes in manufacturing.
5. List the safety items that a person must wear in the machine shop.
6. List the general order of operations that can be performed on CNC lathes.
7. List the documentation that is typically required for CNC turning machines.
8. List the typical items that are documented on the CNC tool list for CNC turning machines.

9. List the typical items that are documented on the CNC setup plan for CNC turning machines.

10. List the typical items that are documented on the CNC program manuscript for CNC turning machines.

11. List the typical proveout steps for CNC turning machines.

12. List the CNC lathe setup items that are typically performed by the setup person.

13. List the process control items that are typically maintained by the setup person and CNC operator.

chapter 12

CNC LATHE FUNDAMENTALS

Chapter Objectives

After studying and completing this chapter, the student should have knowledge of the following:

- **CNC lathes**
- **CNC lathe components**
- **CNC lathe Cartesian coordinate system**
- **Calculating X-Z coordinates**
- **CNC lathe program format**
- **CNC lathe codes**

THE CNC LATHE

CNC turning machines (CNC lathes) and conventional lathes (manual lathes) each have the same basic components, such as the main motor, the spindle, the bed, the tool turret, the headstock, the tailstock, the cross-slide, the carriage, and the way system. However, CNC turning machines, in addition, are outfitted with a computerized control and servomotors to operate them (see Figure 12-1). Thus, the CNC lathes are numerically controlled lathes that position and cut material automatically. Manual lathes can be very versatile and productive machine tools, but when coupled with a CNC, they become the "high tech production turning machines" of the machine shop. Complex contours such as tapers and radii can be easily and accurately machined on CNC turning machines. Additionally, repetitive lathe operations such as turning, facing, and boring are ideal machining applications for CNC turning machines.

CNC turning machines can also be designed to include a variety of optional components and features. These options are basically designed to reduce setup time, part handling, and cycle time. For instance, one option for minimizing part handling time is to add a part loading magazine or a robot arm. The maximum spindle speed (RPM) of a CNC turning machine can also vary depending on the need or application. The factors that determine the spindle speed range selected can include the type of part materials, the part sizes, and the annual part production quantities to be machined. Another factor that influences the spindle speed range is the type of tool cutter material that will be used, such as HSS, coated carbide, diamond, and ceramic.

High-production turning machines can be designed with dual chucks that enable a part to be machined on both ends. Also, live tooling can be incorporated that allows milling and drilling at various angles. Other options may include a second or third turret, a bar feeder, a parts catcher, a steady rest follower, and a tool setter.

CNC LATHE COMPONENTS

The main components of the CNC lathe include the bed, the headstock, the cross-slide, the carriage, the turret, the tailstock, the ways, the servomotors, the ball screws, the hydraulic and lubrication systems, and the machine control unit (MCU). The two primary axes of a CNC turning center are the X-axis and the Z-axis. The X-axis, which controls the cross-slide, moves the cutting tool to control the workpiece diameter. The Z-axis, which controls the carriage, moves the cutting tool lengthwise, to control the workpiece length (see Figure 12-1).

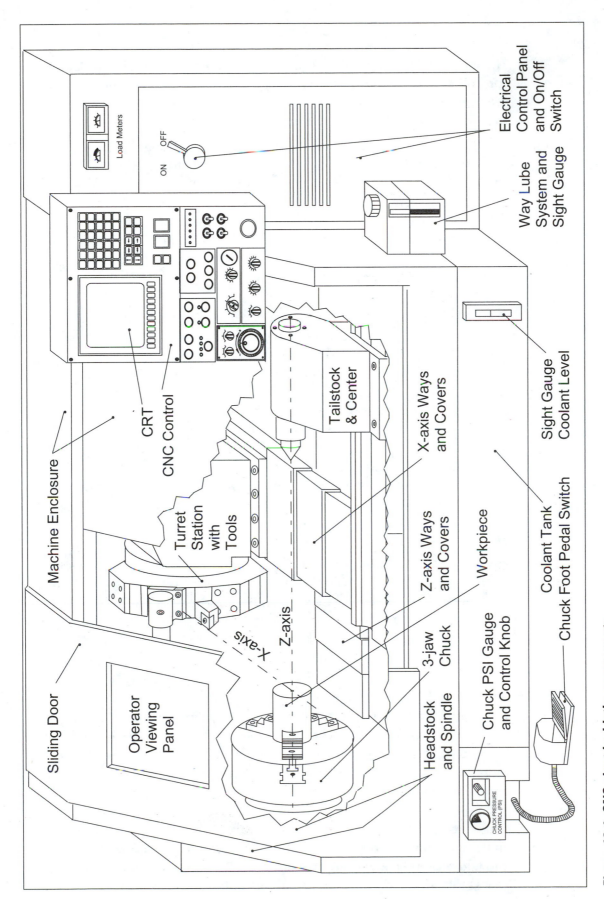

Load Meters

ON OFF

Electrical
Control Panel
and On/Off
Switch

Way Lube
System and
Sight Gauge

Machine Enclosure

CRT

CNC Control

Turret
Station
with
Tools

Tailstock
& Center

X-axis Ways
and Covers

Sight Gauge
Coolant Level

X-axis

Z-axis

Z-axis Ways
and Covers

Workpiece

Sliding Door

Operator
Viewing
Panel

3-jaw
Chuck

Headstock
and Spindle

Chuck PSI Gauge
and Control Knob

Coolant Tank

Chuck Foot Pedal Switch

CHUCK PRESSURE
CONTROL (PSI)

Figure 12-1 CNC slant bed lathe components

CNC Lathe Control

The modern CNC turning machine tool is software driven. The computer controls are programmed instead of hardwired. The control panel is where the machine operation buttons and knobs are located. The control panel typically includes the following items:

- Power On/Off button
- Cycle Start button
- Axis Select knob
- Spindle Override knob
- Load meters
- Keyboard pad
- 3.5-inch input disk drive

- Feed Hold button
- Emergency Stop button
- Axis Jog wheel
- Feed Override knob
- Turret Index button
- Mode Select knob
- Miscellaneous function buttons

CRT Display

The CRT is also located on the control panel (see Figure 12-1). The CRT display allows the operator easy visual access to CNC program and machine information. On the screen of the CNC control, the operator can view the CNC program, active codes, tool and workpiece offsets, machine positions, alarms, error messages, spindle RPM, and horsepower. Some CNC machine CRTs can also display a graphic tool path simulation of the CNC program. The tool path graphic simulation can be utilized to view and check the CNC program moves before the part is actually machined. This can minimize or eliminate programming and setup errors that could cause injury to the operator or cause damage to the cutting tools, the workpiece, and/or the machine.

Bed

The bed is designed to support and align the X-Z axes and cutting tool components of the machine. Additionally, the bed will absorb the shock and vibration associated with metal cutting conditions. Beds are designed in two ways: they either lie flat or at a slant. Most lathes have a slant-bed design, providing the operator easy access for loading and unloading parts and tools. It also allows the chips and coolant to fall away from the cutting area to the bottom of the chip conveyor.

Headstock

The headstock contains the spindle and transmission gearing or belts, which rotate the chuck and workpiece. A variable speed motor drives the headstock spindle, which is programmable in RPM. They are typically equipped with a variety of motor sizes ranging from 5 to 75 horsepower, and spindle speeds from 32 to 5500 RPM.

Chuck and Jaws

The chuck is mounted to the spindle and is equipped with a set of jaws to grip and rotate the workpiece (see Figure 12-2). There are various types of chucks depending on the manufacturing application requirements. They are

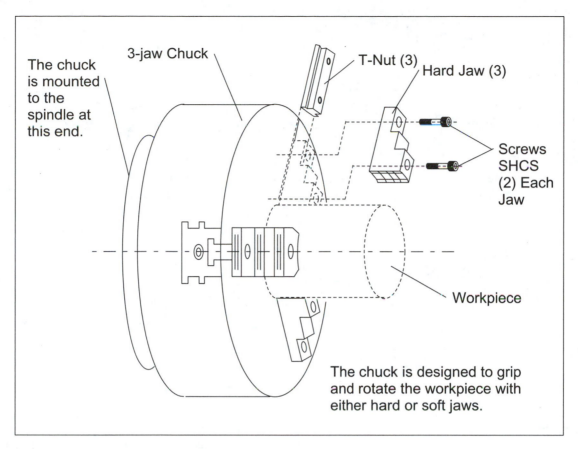

The chuck is mounted to the spindle at this end.

3-jaw Chuck

T-Nut (3)

Hard Jaw (3)

Screws SHCS (2) Each Jaw

Workpiece

The chuck is designed to grip and rotate the workpiece with either hard or soft jaws.

Figure 12-2 **The 3-Jaw chuck features**

categorized into two groups: manually operated or automatically power operated. They can also be self-centering, where all the jaws clamp and unclamp simultaneously, or independent, where each jaw clamps and unclamps separately from the other jaws. The chuck can be designed as a 2-jaw, 3-jaw, 4-jaw, 6-jaw, or a collet jaw type.

The jaws (see Figure 12-2) are either hardened (hard jaws) or mild steel (soft jaws), and are selected depending on the operation holding requirements. The hard jaws are available in various standard designs. The soft jaws are also available in various standard designs, but they require a boring operation to match the diameter that they will hold. Collet jaws are designed to grip barstock, which can be round, square, hexagon, and so on.

Tool Turret

Tool turrets come in all styles and sizes. The basic function of the turret is to hold and quickly index the cutting tools (see Figure 12-3). The CNC lathe usually is equipped with one turret that can hold from four to twelve tools. The tool turret can be square shaped or hexagon shaped. Most are hexagon, which can hold more cutting tools. Some CNC lathes are equipped with two or three turrets depending on the application. The type and number of turrets on a CNC turning center can vary with the size of the machine and the

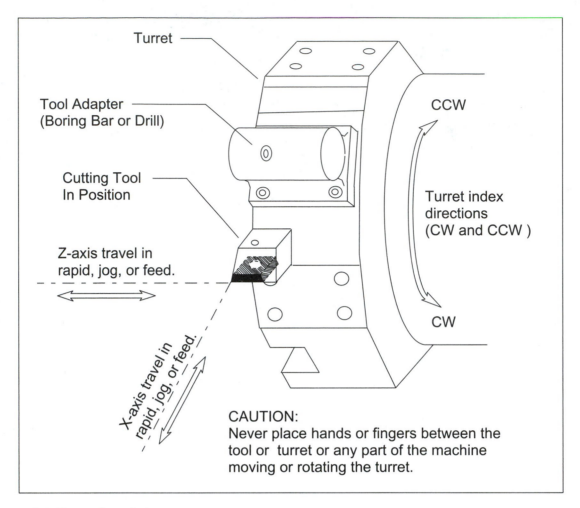

Figure 12-3 **Turret description**

manufacturer. Most turrets are capable of bidirectional indexing, and the slides on which the turrets are positioned can travel at a rapid traverse rate of approximately 400 in./min (100 m/min), which reduces noncutting time of the machine cycle.

CNC Lathe 3-Jaw Chuck Features and Descriptions

Figure 12-2 describes the 3-jaw chuck features, which include the 3-jaw chuck, the *T-Nut,* the jaws, and the screws.

The chuck is a separate component that is attached to the spindle. The operator manually changes the jaws as required by the setup plan.

Lathe chucks are designed in various styles as described as follows:

1. The 3-jaw chuck is typically self-centering to align the part radially and is designed to use either hard or soft jaws.

2. The collet chuck is designed mainly for barstock that is continuously fed to reduce handling time. Collet jaws are also available in various styles.

3. There are other chucks, which include the 4-Jaw Chuck, the 2-Jaw Chuck, and 6-Jaw Chuck.

4. The chuck can clamp the workpiece manually with a lathe chuck key, or automatically with hydraulic or pneumatic power.

CNC Lathe Turret Features and Descriptions

The illustration in Figure 12-3 describes the lathe turret features, which include the turret, the toolholders, and the turret movements.

Turret descriptions:

1. The turret is the component where the cutting tools are fastened.

2. Each tool must be indexed to the cutting location on the machine when it is required by the CNC program. Therefore, the turret is designed so that it can index radially to each tool station.

3. The turret can index in both radial directions (CW and CCW). It can be indexed both manually by the operator or automatically by the CNC program.

4. The turret has stations (usually six to twelve) which hold the cutting tool. The tools are manually loaded and fastened by the CNC setup person.

5. The turret travels in two axes, they are usually labled X and Z.

6. The CRT screen, when in the Position mode, tracks the movement and location of the turret.

Tool Indexing

Most CNC lathes have tool turrets with ten or more tool stations. This design allows a lathe tool to be fastened in each turret station so it can be indexed automatically during the CNC program cycle. Quick tool indexing is an important factor on CNC turning machines used for production purposes. Therefore, the index time is typically completed within seconds. Some CNC turning machines are equipped with quick-change toolholders, which allows the operator to manually change tools within seconds. Additionally, automatic tool change systems that can automatically clamp and unclamp tools into the turret are also used on CNC turning machines.

Tailstock

The tailstock (see Figure 12-4) is used to support workpieces that lack rigidity such as long shafts, long hollow castings, and similar parts. The tailstock can be designed to operate either manually or by CNC program commands. The tailstock basically supports one end of the workpiece with a center. The typical center (Figure 12-5) has a 60° point, which fits into the end of the workpiece for support (see Figure 12-6). Lathe centers are designed in various styles that can accommodate the various turning applications. The most

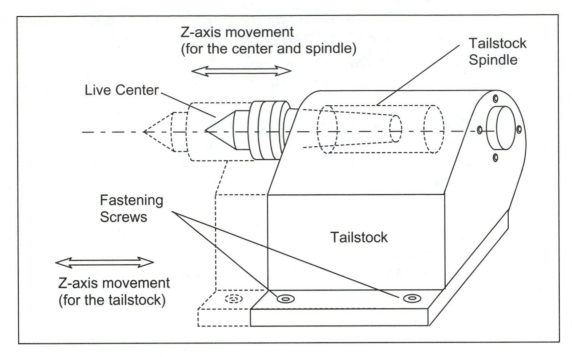

Figure 12-4 **Tailstock and center**

common is the live center, which rotates on a bearing pack to eliminate friction. The tailstock travels on its own hardened and ground-bearing ways.

Ways and Way Covers

The ways are precision-hardened rails that allow the turret to travel in rapid traverse and feed motion (X- and Z-axes). The way covers are a means of protecting the ways from damage due to scraping from metal chips, or due to dents from metal items such as tools or parts.

Way Lube System

The way lube system is designed to keep the ways lubricated, which will reduce wear from friction and eliminate early machine failure. The operator and maintenance personnel must check the way lube system reservoir level on a daily basis.

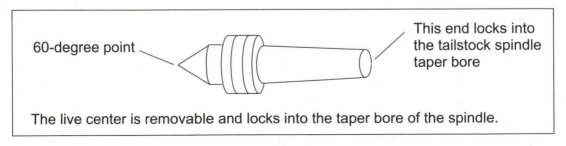

Figure 12-5 **Tailstock center**

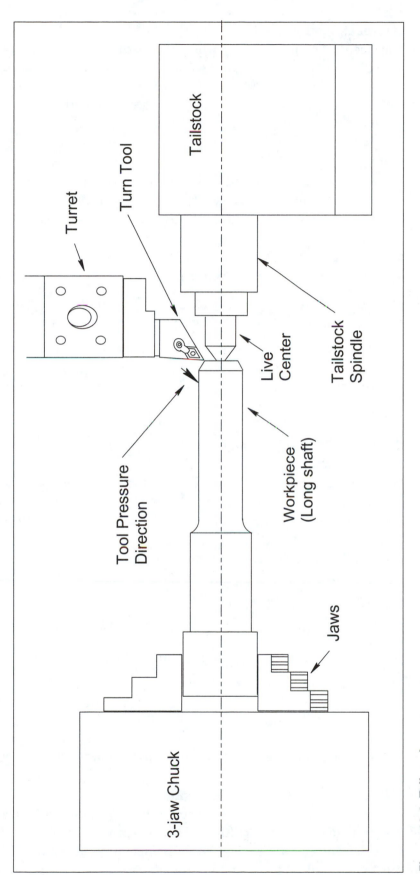

Figure 12-6 Tailstock setup

271

Electrical Control Panel

The electrical control panel is where the main Power On/Off switch is usually located. This is also where all the electric components, including fuses and reset buttons, are enclosed for safety. The electrical control panel is usually locked to prevent entry by unauthorized personnel. An authorized electrician should be contacted if entry is required.

Ball Screw

The rotary motion generated by the drive motors is converted to linear motion by recirculating ball screws. The ball lead screw uses rolling motion rather than the sliding motion of a normal lead screw. Sliding motion is used on conventional acme lead screws. Unlike the ball screw, the motion principle of an acme lead screw is based on friction and backlash. Below are some advantages of the ball screw versus the acme lead screw:

- Less wear
- Precise position and repeatability
- High speed capability
- Longer life

Additionally, CNC machine tool manufacturers have designed and incorporated an electronic pitch error, which compensates for backlash error into the ball screw system. This backlash compensator system is utilized on most modern CNC turning machines.

CNC Lathe Tailstock Features and Descriptions

The tailstock illustrated in Figure 12-4 is designed to slide in the Z-axis direction and support the workpiece. The tailstock can be manually or automatically positioned and fastened to the frame by the CNC machine operator. The center illustrated in the figure is a separate component that is locked into the spindle of the tailstock.

The typical tailstock setup usually requires a process similar to the steps that follow:

1. Unfasten the lock screws.

2. Slide the tailstock to the required position.

3. Allow for spindle retract to load and unload the workpiece.

4. Fasten the tailstock lock screws.

5. Check alignment of center to spindle.

When performing the setup or operating the tailstock, caution should be taken to avoid placing hands or fingers between any moving components.

CNC Lathe Tailstock Setup Descriptions

The illustration in Figure 12-6 illustrates the typical setup for a workpiece (shaft), which requires a tailstock. The workpiece, which must be center drilled at one end before the CNC lathe operation, is held at one end with the 3-jaw chuck and supported at the other end with the tailstock.

Due to the lack of rigidity in the workpiece structure, the tailstock and center are required to support it during machining operations. The tailstock and center basically counteract the tool pressure caused when the turn tool is cutting the workpiece. Cutting this type of workpiece without a tailstock will result in unsatisfactory results.

Servo Drive Motors

CNC machines use electric servomotors that turn ball screws, which in turn drive the different axes of the machine tool. The axes (X and Z) each have a separate servomotor and ball screw to control each axis independent of each other.

Auto Bar Feeder

This option can be added to reduce the handling time required for loading the workpiece material into the chuck. The bar feeder can be designed as a single tube type or magazine type. The purpose of the bar feeder is to quickly load more stock automatically at the end of the CNC machine cycle.

Parts Catcher

The purpose of the parts catcher is to catch the part immediately after it is cut off and avoid damage to the part, tools, and/or machine components. This option is typically included with bar-feed-type machines.

Secondary Turret

The main turret and the secondary turret are designed to work independently of each other. This design of the turrets allows simultaneous cutting with two tools to reduce the cycle time. The main turret is a two-axis design. The assigned axes for the primary turret are X and Z. The secondary turret is then assigned some other axis labels such as X1 and Z1.

Subturret

Some CNC turning machines such as the CNC automatic screw machine typically include a main turret and a subturret, which are designed to work independently of each other. This design of the turrets also allows simultaneous cutting with two tools to reduce the cycle time. The main turret, which is a two-axis design typically, uses the X- and Z-axes travel. The subturret is typically a single-axis design, which uses the B-axis travel.

Tool Setter

The tool setter is a sensing device on the machine that automatically references each tool in a setup. The operator manually moves each tool in both X- and Z-axes, as required, to the tool setter and touches-off. The control then automatically records the distance values in the offset storage memory. This optional device can minimize setup downtime and improves the quality of the parts produced.

Dual Spindle/Chuck

The dual spindle and chuck design allows the CNC machine to automatically machine the first operation, then transfer the workpiece to the second chuck and machine it simultaneously. This process is designed to complete all the turning operations and thus minimize lead time and WIP of the parts.

Live Tooling

The CNC lathe is capable of drilling, tapping, and reaming on center (X0.0) parallel to the machine spindle. This option allows the CNC lathe, which typically only performs turning-type operations, to also perform milling/drilling-type operations. The live tooling refers to the turret adapter, which holds and is powered to rotate tool cutters of the milling and drilling types. This option eliminates additional setups, part handling time, and downtime, thereby increasing productivity of the CNC lathe. The additional cost of this option is typically justified by the decrease in downtime, which yields a greater increase in productivity.

Chip Conveyor

The chip conveyor is designed to remove the metal chips produced from the workpiece machining operations from the CNC lathe work area. This machine feature, which is operated manually by the operator, reduces the amount of time required to clean and maintain the work area of the CNC lathe. When operating, the chip conveyor typically transfers and deposits the metal chips into a dumpster. This also controls the handling of metal chips for recycling. Most CNC lathes used for production are usually equipped with a chip conveyor.

Open-Loop Systems

The stepping motor is an electric motor that rotates a calculated amount every time the motor receives an electronic pulse from the MCU. The stepping motor's rotary motion is converted into the linear motion of the machine axes through the use of lead screws. There is no sensing or feedback system to verify and monitor the machines actual positioning. The open-loop system is simpler and rarely used for CNC machines that require accuracy and repeatability.

Closed-Loop Systems

Servomotors permit automatic operation of the machine and the closed-loop control system verifies that the machine accurately positions the CNC program commands (see Figure 12-6). The computer makes it possible to continuously monitor the machine's position and velocity while it is operating. The advantages of the servomotor are increased accuracy and repeatability. CNC machines can generally position to an accuracy of ±0.0002 inch and have a repeatability of ±0.00006 inch.

Point-to-Point and Continuous Path

Most CNC turning machines rapid-position (G00 rapid traverse) in point-to-point mode, which means that the tool does not follow an exact path to a coordinate point. However, when the tool feeds or cuts (G01 linear interpolation), it then operates in a continuous path mode.

Input Media

CNC lathe programs may be input in a variety of ways, such as punched tape, floppy disk, diskette, or wires connected to the control, called direct numerical control or *distributed numerical control* (DNC). They are also loaded through the CNC machine control panel or with the use of microcomputers and workstations. Another method of loading CNC programs is to use a remote device that connects to the CNC control for reading, storing, and writing information from or to a diskette.

The commonly used input methods on modern CNC turning machines are the DNC and the floppy disk methods. The punched tape method, which is less efficient than the DNC and floppy disk methods, is rarely used. Additionally, storing and retrieving CNC program files using the DNC and floppy disk methods are easier and faster than the other outdated methods.

CNC Lathe Closed-Loop System Descriptions

The illustration in Figure 12-7 describes the CNC lathe features, which are included in the closed-loop system.

CNC LATHE CARTESIAN COORDINATE SYSTEM

The Cartesian coordinate system is applied to CNC turning machines, which includes the CNC lathe, for all machine axis movements. This coordinate system is illustrated and described in Figures 12-8 and 12-9. These figures illustrate the labeling of each axis movement (X and Z), including the positive and negative signs, which specify the direction of the coordinate movement. The point where all the axis lines intersect is labeled the origin or X-zero and Z-zero point (see Figure 12-10). This coordinate system is divided into four quadrants (Figure 12-10), which designate the sign value of each axis. It is always necessary to specify the negative sign; the positive sign is usually the default sign and therefore it is not usually specified.

The Cartesian coordinate system can be applied to CNC machine tools in various ways. The most common application for CNC lathes is the X-axis and the Z-axis. On a CNC turning machine the X-axis specifies the diameter coordinate movement, and the Z-axis specifies the length coordinate movement. These axis movements are also illustrated in Figures 12-8 and 12-9. Another common application for CNC lathes is the U-axis and the W-axis, which also specify the diameter coordinate movement (U-axis) and the length coordinate movement (W-axis) (see Figure 12-11).

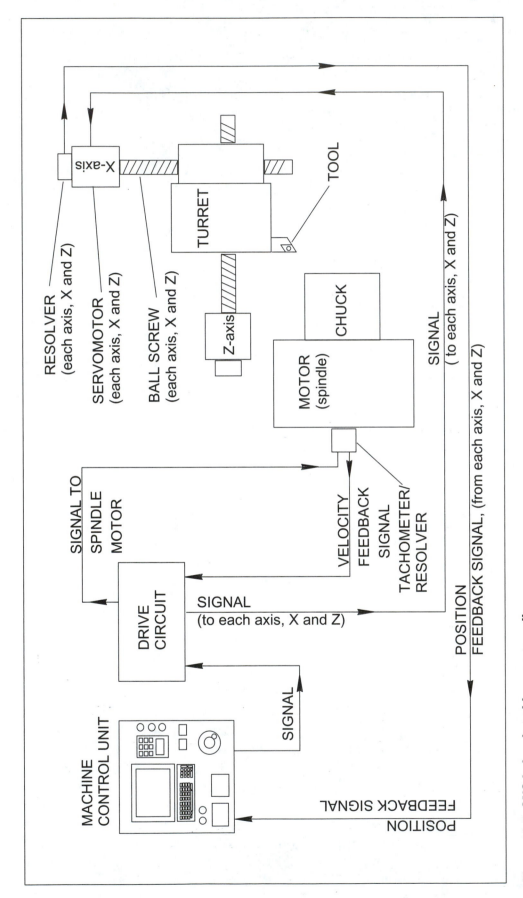

Figure 12-7 CNC lathe closed-loop system diagram

The X-zero is always the center of the part because it is also the center of the spindle axis.

The Z-zero is typically set at the front face of the part. However, Z-zero can be at any position that is appropriate for the part or the CNC program.

The X-Z zero intersect point is called the part origin. The origin must be documented so that it can be setup in the location that is related to the CNC program coordinates. The origin location is typically documented on the setup instructions.

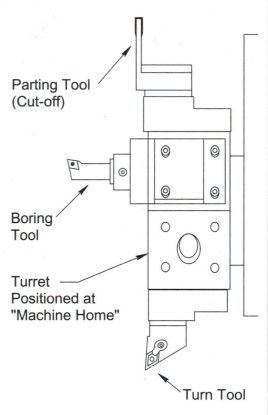

Parting Tool (Cut-off)

Boring Tool

Turret Positioned at "Machine Home"

Turn Tool

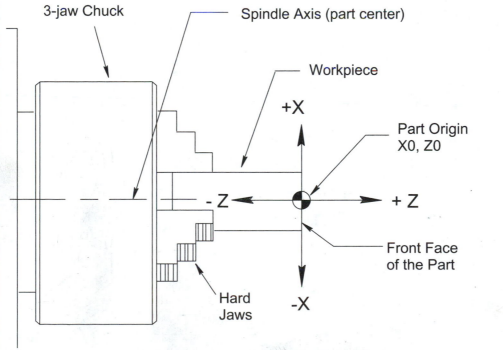

3-jaw Chuck

Spindle Axis (part center)

Workpiece

+X

Part Origin X0, Z0

-Z   + Z

-X

Front Face of the Part

Hard Jaws

Figure 12-8 Cartesian coordinate system for CNC lathes

This illustration describes the X-axis and Z-axis labels and directions. It also describes a typical X-Z zero location for the CNC vertical turret lathe (VTL).

The X-and-Z zero intersect point, which is the part origin, must be documented on the setup plan instructions.

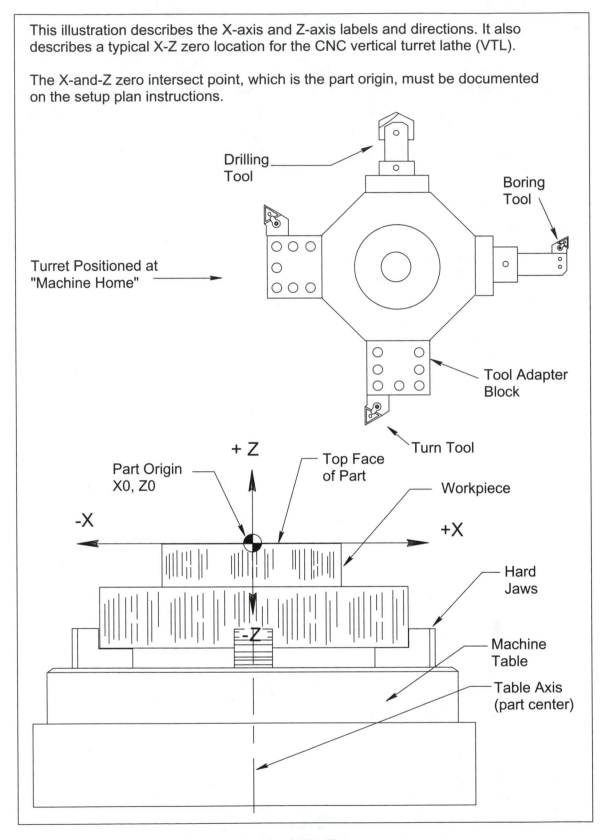

Figure 12-9 *Cartesian coordinate system for CNC VTLs*

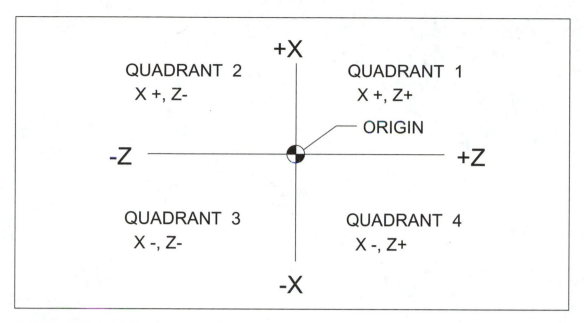

Figure 12-10 **CNC lathe Cartesian quadrants**

The other motion applied to CNC turning machine tools is rotary motion (see Figure 12-12). The rotary axis movement, which is typically applied to the spindle rotation, is the C-axis. The direction of rotation (CW or CCW) is specified by either a positive or negative sign. As stated previously, it is always necessary to specify the negative sign, and because the positive sign is usually the default sign, it is not usually specified.

It is important to note that there are some similarities between the various turning machines that are used in most manufacturing plants. For example, the length or spindle travel is always the Z-axis of the machine. The

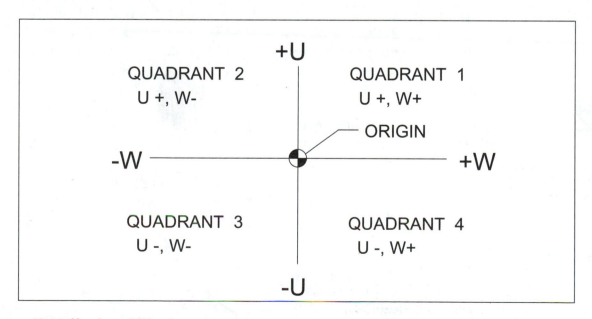

Figure 12-11 **U-axis and W-axis**

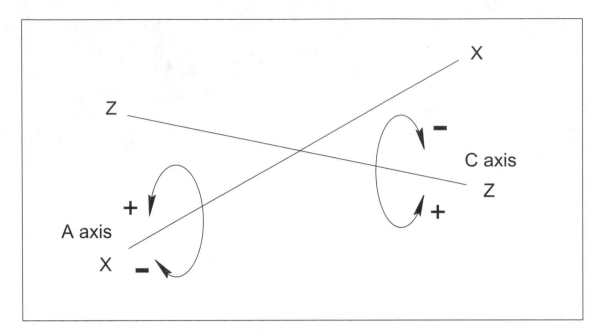

Figure 12-12 **Cartesian coordinate system for rotational axis movement**

diameter movement is always the X-axis. The X-axis value is usually expressed as a diameter; however, there are some CNC turning machines that express the X-axis value as a radius. Another item that is common to all turning machines is the cutter or toolholder and the turret. Figures 12-8 and 12-9 illustrate and describe these features associated with CNC turning machines including the turret, the cutting tools, the chuck, the jaws, the part origin, the spindle centerline axis, and the part orientation.

Most CNC turning machines have an X- and Z-axis reference point, or Machine Home position. The Machine Home position is a fixed location that calibrates each axis independently. The part origin is separate from the machine home position. The part origin position will vary depending on where the part Z-zero is located on the part. The X-zero is always the center of the part, which is also the spindle centerline axis.

Absolute and Incremental Programs

The absolute and incremental format choice, which has been previously described, is usually not applied to CNC turning machines. Most CNC lathes are only programmed in absolute mode. Absolute programs specify a coordinate position or tool path from the workpiece coordinate zero, or origin. The positive and negative signs are derived from the part origin (see Figure 12-8).

CNC Lathe Cartesian Quadrants

Figure 12-10 illustrates which sign (plus or minus) the X- and Z-coordinate points are assigned for quadrants 1–4 in absolute mode. Note that all the X and Z values in quadrant 1 are (plus). In quadrant 2 all the X values are (plus), and all the Z values are (minus). In quadrant 3 all the X and Z values are (minus). In quadrant 4 all the X values are (minus) and all the Z values are (plus).

Figure 12-11 illustrates the U-axis and W-axis that are assigned to CNC lathes. Quadrants 1–4 also indicate the (plus or minus) sign for each axis (U and W).

Rotary Cartesian Coordinate System for CNC Lathe Movement

The illustration in Figure 12-12 describes the A-rotational axis and the C-rotational axis, which are applied to CNC turning machines.

The A-rotational axis and the C-rotational axis are used to control circular motion about the X-axis and the Z-axis, respectively. One typical application is to control axis operations associated with using live tooling. Another application is for positioning a rotary table on a CNC turning center.

CNC LATHE PROGRAM FORMAT

The CNC program format for CNC lathes basically requires that operational functions be executed in an order that is consistent with how a conventional lathe operates. The following is a list of some common lathe functions:

- The appropriate RPM value must be selected before the spindle is turned ON.
- The turret must be indexed to the appropriate tool before cutting.
- The appropriate feedrate must be selected before the feed mode is executed.
- The coolant spray hose is turned ON at the appropriate time.
- The feed mode is stopped at the specified point.
- The axis slides are moved in the appropriate sequence.
- The spindle is stopped after the cutting is completed.
- The coolant spray hose is turned OFF after cutting is completed.

Additionally, there are other codes that are required strictly for the CNC program. Figure 12-13 describes the basic functions and codes that are required in a CNC lathe program.

CNC Lathe Program Example

The sample CNC program in Figure 12-13 describes the format and main components typically included in a CNC program. Note that each line of the CNC program contains various codes and coordinate numbers. These items are explained in more detail in the sections that follow. For Figure 12-13 the items on the left are the CNC program codes and the explanations are on the right side.

CNC LATHE COMMAND CODES

The CNC lathe receives a series of commands and coordinates via the CNC program as described in Figure 12-13 through the MCU. These commands are structured using letters and numbers. Each letter and number combination

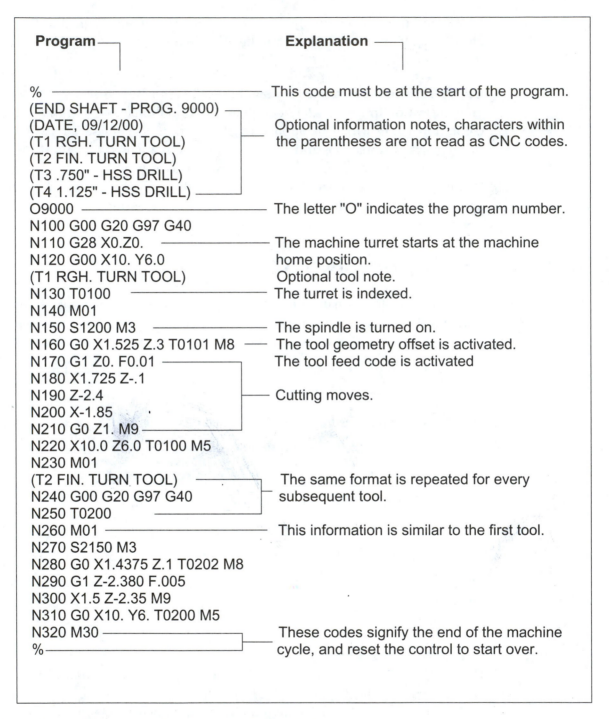

Program	Explanation
%	This code must be at the start of the program.
(END SHAFT - PROG. 9000)	
(DATE, 09/12/00)	Optional information notes, characters within
(T1 RGH. TURN TOOL)	the parentheses are not read as CNC codes.
(T2 FIN. TURN TOOL)	
(T3 .750" - HSS DRILL)	
(T4 1.125" - HSS DRILL)	
O9000	The letter "O" indicates the program number.
N100 G00 G20 G97 G40	
N110 G28 X0.Z0.	The machine turret starts at the machine
N120 G00 X10. Y6.0	home position.
(T1 RGH. TURN TOOL)	Optional tool note.
N130 T0100	The turret is indexed.
N140 M01	
N150 S1200 M3	The spindle is turned on.
N160 G0 X1.525 Z.3 T0101 M8	The tool geometry offset is activated.
N170 G1 Z0. F0.01	The tool feed code is activated
N180 X1.725 Z-.1	
N190 Z-2.4	
N200 X-1.85	Cutting moves.
N210 G0 Z1. M9	
N220 X10.0 Z6.0 T0100 M5	
N230 M01	
(T2 FIN. TURN TOOL)	The same format is repeated for every
N240 G00 G20 G97 G40	subsequent tool.
N250 T0200	
N260 M01	This information is similar to the first tool.
N270 S2150 M3	
N280 G0 X1.4375 Z.1 T0202 M8	
N290 G1 Z-2.380 F.005	
N300 X1.5 Z-2.35 M9	
N310 G0 X10. Y6. T0200 M5	
N320 M30	These codes signify the end of the machine
%	cycle, and reset the control to start over.

Figure 12-13 *CNC program format and code descriptions*

gives the control a specific command or coordinate position. For example, a common turret tool index command always starts with a letter "T" code followed by a numeric value that represents the station index position. Likewise, a coordinate position command always starts with a letter "X and/or Z" code followed by a numeric value that represents the travel-to coordinate point.

Letter Address Commands

Most letter address codes fall into two categories: modal or nonmodal. Nonmodal codes are command codes that are only active in the block in which they are specified and executed. Modal codes are command codes that once executed will remain active throughout the program until another code in the same group overrides or cancels it. Table 12-1 lists and describes some common address letter codes used in CNC programs to control the CNC turning machines.

M Code Commands

Table 12-2 lists and describes some common M codes that are used in CNC programs to control the CNC turning machine tool miscellaneous functions. The list also specifies if the code is modal or nonmodal.

■___ *Table 12-1* **Word address codes and descriptions**

LETTER CODE	DESCRIPTION
O	Identifies the CNC program number
N	Identifies a line number of a CNC program
X	Positions the turret coordinate diameter axis
Z	Positions the turret coordinate length axis
U	Secondary (optional) turret coordinate diameter axis; also specifies a dwell time with (G04)
W	Secondary (optional) turret coordinate length axis
A	Rotary position coordinate for X-axis
C	Rotary position coordinate for Z-axis
F	Specifies the feedrate of an axis (IPR)
S	Specifies the speed of the spindle (RPM)
T	Specifies the turret/tool position selected for indexing
H	Not assigned
D	Specifies a canned cycle parameter
I	Specifies X-axis circular interpolation (G02 or G03) coordinate
J	Not assigned
K	Specifies Z-axis circular interpolation (G02 or G03) coordinate
R	Specifies circular interpolation (G02 or G03) radius value; also specifies canned cycle (G81–G89) retract point
L	Specifies subprogram line numbers
P	Specifies subprogram number, or canned cycle parameters
Q	Specifies canned cycle parameters
M	Specifies a miscellaneous function command (see Table 12-2)
G	Specifies a preparatory function command (see Table 12-3)

■ _Table 12-2_ *Miscellaneous function M code descriptions*

M CODE	DESCRIPTION	MODE
M00	Program stop	Nonmodal
M01	Optional program stop	Nonmodal
M02	Rewind/end of program	Nonmodal
M03	Spindle start CW	Nonmodal
M04	Spindle start CCW	Nonmodal
M05	Spindle stop	Nonmodal
M06	Automatic tool change	Nonmodal
M07	Mist coolant on	Modal
M08	Flood coolant on	Modal
M09	Flood coolant off	Modal
M10	Splash guard open	Nonmodal
M11	Splash guard close	Nonmodal
M12	Bar feeding	Nonmodal
M15	Turret indexing, CCW	Nonmodal
M16	Turret indexing, CW	Nonmodal
M17	Chuck open	Nonmodal
M18	Chuck close	Nonmodal
M23	Pull-out threading on	Modal
M24	Pull-out threading off	Modal
M25	Signal for counter	Modal
M28	High-pressure coolant on	Modal
M29	High-pressure coolant off	Modal
M30	End of program and reset to the program start	Nonmodal
M31	Air blow	Nonmodal
M32	Bar pusher advance	Nonmodal
M33	Bar pusher retract	Nonmodal
M35	Part catcher forward	Nonmodal
M36	Part catcher retract	Nonmodal
M38	Chip conveyor on	Nonmodal
M39	Chip conveyor off	Nonmodal
M40	Spindle low range	Modal
M41	Spindle high range	Modal
M61 thru M66	Subturret positions 1, 2, 3, 4, 5, and 6, respectively	Nonmodal
M93	Simultaneous spindle start forward and coolant on	Nonmodal
M94	Simultaneous spindle start reverse and coolant on	Nonmodal
M95	Simultaneous spindle stop and coolant off	Nonmodal
M98	Subprogram call	Nonmodal
M99	End subprogram and return to main program	Nonmodal

G Code Commands

Table 12-3 lists and describes some common G codes that are used in CNC programs to control the CNC turning machine preparatory functions. The list also specifies if the code is modal or nonmodal.

The codes and coordinates that are programmed in each line (block) of a CNC lathe program are typically arranged in an orderly structure between the "N" sequence code and the semicolon (;) code. The CNC program should be organized in a consistent structure that will increase program efficiency. The CNC program order is typically arranged as follows:

- The letter **O,** program number, must be at the start of the CNC program.

- The parenthesis, **( ),** Information notes code can be placed next.

- The letter **N** and a number **005** is the first code of a block. Each line of information is called a sequence block.

- The letter **T** code specifies the tool change.

- The letter **M,** miscellaneous code, is usually last; only one M code per block is allowed.

- The letter **S** code with a number value is usually next.

- The letter **G,** preparatory code, is usually next; multiple G codes per block are allowed.

- The coordinate letters **X** and/or **Z** are next in this same order and they also require a number with a decimal point after the letter.

- The letter **F** code with a number value for feeding usually follows.

- The last code for each block is the semicolon **(;)** EOB code.

- The end of the CNC program is specified with an **M,** miscellaneous code.

Some CNC programs do not include N block sequence numbers for every line. However, they typically include a block sequence number for each line that starts a new tool. The M codes are basically designed to perform miscellaneous machine functions such as spindle start, spindle stop coolant on and off, and program stop.

A common lathe function is the turret tool index. This command code, which is initiated with the letter "T" and the numeric index position of the turret, is programmed as follows: T0200

When the turret index command is executed, the control automatically unclamps the turret, rotates it to the specified position, and then clamps it. A typical CNC block to perform a turret index is described below:

N010 T0202;

The (T0202) command above also automatically activates the tool offset compensation in X and Z. The first two numbers after the T code indicate the turret station number; the third and forth numbers after the T code indicate the offset number. When the tool is finished cutting, the tool offset compensation is canceled by changing the third and forth numbers to zeros (T0200).

Table 12-3 Preparatory function G code descriptions

G CODE	DESCRIPTION	MODE
G00	Rapid traverse positioning	Modal
G01	Linear positioning at a feed rate	Modal
G02	Circular interpolation CW at a feed rate	Modal
G03	Circular interpolation CCW at a feed rate	Modal
G04	Dwell	Nonmodal
G20	Inch programming format	Modal
G21	Metric programming format	Modal
G28	Zero or machine home return	Nonmodal
G40	TNR compensation cancel	Modal
G41	TNR compensation left	Modal
G42	TNR compensation right	Modal
G43	Not assigned	Modal
G49	Not assigned	Modal
G50	Maximum RPM and X-Z preset	Modal
G54–G59	Work coordinate presets	Nonmodal
G70	Finish cutting canned cycle	Nonmodal
G71	Rough cutting canned cycle	Nonmodal
G72	Stock removal cutting canned cycle	Nonmodal
G73	Pattern repeat cutting canned cycle	Nonmodal
G74	Longitudinal pecking canned cycle	Nonmodal
G75	Cross pecking canned cycle	Nonmodal
G76	Thread cuting canned cycle	Nonmodal
G80	Canned cycle cancel	Modal
G81	Drilling canned cycle	Modal
G82	Counterbore canned cycle	Modal
G83	Peck drilling canned cycle	Modal
G84	Tapping canned cycle	Modal
G85–G89	Boring canned cycles	Modal
G90	Turning canned cycle	Nonmodal
G91	Turning canned cycle	Nonmodal
G92	Threading canned cycle	Nonmodal
G94	Facing canned cycle	Nonmodal
G96	Constant SFM mode	Modal
G97	Constant RPM mode	Modal
G98	Feed rate per minute mode	Modal
G99	Feed rate per revolution mode	Modal

Another common lathe function is the spindle speed (RPM), which is controlled with the S code followed by up to four digits that represent the RPM. For example, a typical CNC block to rotate the spindle at 1575 RPM is described below:

N010 S1575 M03;

Since the S code above is only used to specify the RPM value, either an M03 CW or an M04 CCW is also required to turn the spindle on.

The G codes set preparatory machine functions such as rapid traverse mode and feed mode, and activates tool length offsets. A typical CNC block to perform a rapid move in the Z-axis is described below:

N010 G00 Z-1.543;

A typical CNC block to perform a feed move in the Z-axis is described below:

N010 G01 Z-1.543 F.012;

A typical CNC block to perform a feed move in the X&Z axes is described below:

N015 G01 X1.125 Z-2.0323 F.01;

The G96 and G97 codes are another example of G codes that are commonly used to set the spindle rotation mode. When the spindle is turned ON, it can rotate at a constant RPM or it can automatically slow down or speed up as the tool X-axis position changes. These two modes are described below:

The CNC program block below would command the control to rotate the spindle at 1575 revolutions per minute (RPM) at all times.

N010 G97 S1575 M03;

The CNC program block below would command the control to rotate the spindle at 700 SFM at all times.

N010 G96 S700 M03;

When this 700 SFM command is executed, the control calculates the RPM based on the current X-axis position and recalculates it each time the X-axis moves.

As stated previously, the axes of the CNC turning machine are controlled by the letters X and Z (see Figures 12-8 and 12-9). The X and Z letters require either a (minus) or (plus) sign to specify direction and a coordinate value to specify location. When a sign is not specified in a CNC program, the control defaults to a plus sign. The example below describes how a typical CNC block of information is interpreted by the machine control.

N005 G00 X1.75 Z-0.75 M08;

N005	is the sequence block number
G00	is the rapid mode preparatory code
X1.75 Z-0.75	is the rapid-to coordinate location
M08	is the coolant on miscellaneous code
;	is the EOB code

The CNC program, as illustrated in Figure 12-13, is basically a combination of all the blocks of information required to perform the machine functions and lathe cutting operations required to machine the workpiece.

CHAPTER SUMMARY

The key concepts presented in this chapter include the following:

- The names and descriptions of the main CNC lathe components.
- The CNC lathe Cartesian coordinate system.
- The CNC vertical turret lathe Cartesian coordinate system.
- The CNC lathe Cartesian quadrants.
- The names and descriptions of the CNC lathe axes.
- Lathe CNC program format.
- The description of lathe CNC word address codes.
- The description of lathe CNC miscellaneous function M codes.
- The description of lathe CNC preparatory function G codes.
- Lathe CNC program code arrangement structure.

REVIEW QUESTIONS

1. List the main components of the CNC lathe.
2. List the items that are displayed on the CRT.
3. What is the main purpose of the turret?
4. What is the function of the lube system?
5. When is the tailstock used on the CNC lathe?
6. Which type of tailstock center is more commonly used?
7. What is the difference between live tooling and regular lathe tooling?
8. What is the function of the tool setter?
9. What is the difference between SFM G96 and RPM G97?
10. What is the purpose of the G50 code?
11. List the various types of chucks.
12. How are the jaws fastened to the chuck?
13. List the types of chuck jaws that are mounted to the chuck.
14. What is the function of the ways and way covers?
15. How does the CNC program index the turret?
16. How does the CNC program activate the tool offset?

chapter 13

CNC LATHE CUTTING FUNDAMENTALS

Chapter Objectives

After studying and completing this chapter, the student should have knowledge of the following:

- **CNC lathe cutting terms and features**
- **CNC turning operations**
- **CNC lathe cutting tools**
- **Carbide insert technology**
- **CNC lathe machining applications**

CNC LATHE CUTTING TERMS AND FEATURES

There are various terms and features that are associated with turning machines such as CNC lathes (see Figure 13-1). In order to communicate effectively in the CNC machine shop, it is important to identify these terms and features commonly used. Some of these terms and features are also associated with other types of machines. Therefore, the commonly used CNC lathe terms and features are identified and described below as they directly apply to CNC lathes.

RPM

This is the term used to identify the revulotions per minute of the machine spindle, which uses a holding device (chuck) to hold and rotate the workpiece. The RPM value is determined by various factors entered into the RPM formula and then calculated. These factors include the type of workpiece material, the type of tool material, the type of cutting operation, the type of tool, and the diameter at the cut.

Feedrate

This is the term used to identify the rate of travel of the machine axis slides. The feedrate amount is usually expressed in IPR (see Figure 13-1). Feedrate amounts are determined by various factors, which include the DOC, the surface finish requirement, the type of workpiece material, the type of tool material, the type of cutting operation, and the type of cutting tool used.

Depth of Cut

This term is used to identify the amount of material that is machined per side of the workpiece diameter (see Figure 13-1). The amount per side is multiplied by two to determine the amount of material "on-the-diameter" that is removed. There are basically two categories for the DOC amount selection. The first DOC category is "rough cuts" and the second is "finish cuts." In some cases, where high precision or fine surface finish is required, a "semifinish cut" is also usually added to the machining process. The DOC amount is determined by various factors, which include the size tolerance requirement, the surface finish requirement, the type of workpiece material, the type of tool material, the type of cutting operation, and the cutting tool geometry that is used.

Chuck

The chuck, which is mounted to the machine spindle face, is the lathe feature used to rotate the workpiece during the machining operations (see Figure 13-1). There are various types of chucks, which basically include the self-centering 3-jaw chuck, the 4-jaw chuck, and the collet chuck. The considerations that determine the type of chuck used are the type of lathe operations to be performed, the size of the parts, and the type of the parts that will be machined.

Jaws

The jaws, which are mounted on the chuck, are used to grip and hold the workpiece securely as the part rotates during machining operations (see Figure 13-1). The jaws are available in various styles, which basically include hard jaws, soft jaws, and collet jaws. The considerations that determine the type of jaws required are the type of lathe operation to be performed, the type of the parts that will be machined, and the part feature requirements.

Lathe Cutting Terms and Features

The illustration in Figure 13-1 describes the lathe terms and features, which include the 3-jaw chuck, the jaws, the turret, the turn tool, and the workpiece. These items are identified and briefly described. It is essential to become familiar with these terms and features to fully understand the CNC turning machine (lathe) process.

Turret

The turret is the lathe component feature used to hold the cutting tools that cut the workpiece (see Figure 13-1). The turret is mounted to the machine axis slides that are designed to travel in two axes. There are various turret designs, which include square shaped, hexagon shaped, and round shaped. The considerations that determine the design type basically include the number of tool pocket stations required and the turret position on the machine.

Turn Tool

The turn tool is the lathe feature that cuts the workpiece material (see Figure 13-1). The turn tools are fastened to the machine turret pockets. There are various turn tool designs, which include OD tools and inside diameter (ID) tools. Most OD and ID turn tools are designed to cut with carbide-indexable inserts. The considerations that determine the type of turn tool selected include the type of cutting operation to be performed, the surface finish requirement, the type of workpiece material, and the geometry requirements of the workpiece features.

Insert

The indexable insert is the toolholder component that actually cuts the workpiece (see Figure 13-1). The indexable insert, which is mounted on the toolholder, is typically made of carbide. There are various insert geometry designs, which include round, square, triangle, 80° diamond, 55° diamond, 35° diamond, rectangle, trigon, and hexagon. The considerations that determine the type of insert geometry design selected include the type of cutting operation to be performed, the surface finish requirement, the type of workpiece material, and the geometry requirements of the workpiece features.

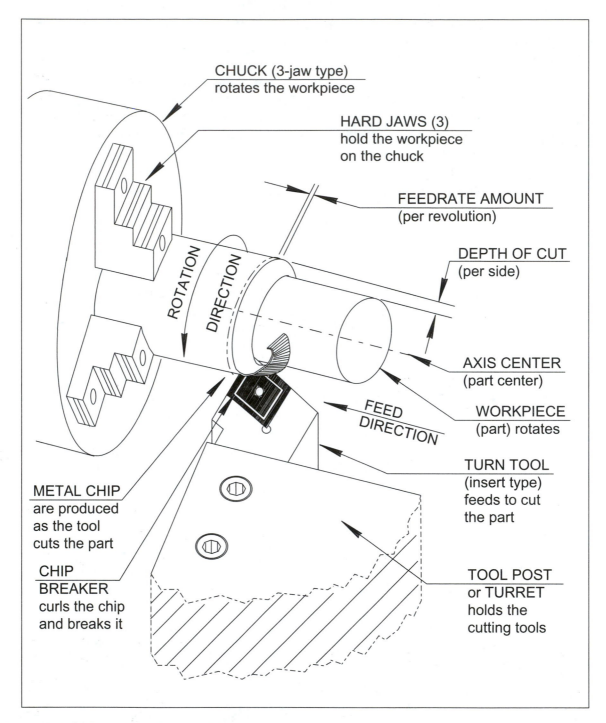

Figure 13-1 **Lathe cutting terms and features**

Metal Chip

The metal cutting process performed by the cutting tools is designed to produce small metal chips (see Figure 13-1). The preferred shape of the chip is a curl that can range from a "C-shape" to a "9-shape." These shapes can be easily handled and removed usually by a chip conveyor.

Chipbreaker

The chipbreaker is a feature of the insert that causes the metal to curl and break as it is machined by the cutting tool (see Figure 13-1). The chipbreaker is usually molded into the edge of the insert. There are various chipbreaker designs, which are designed to accommodate the various lathe cutting applications. The primary considerations that determine the design type basically include the type of cutting operation to be performed, the surface finish requirement, and the type of workpiece material.

Workpiece

The workpiece, or part, is the feature loaded into the lathe and machined to produce round-shaped items (see Figure 13-1). There are various types of parts that are selected for CNC lathe machining. They range from round-, square-, and hexagon-shaped barstock to castings of various shapes. They also typically include various types of materials such as steel, cast iron, aluminum, brass, copper, and plastics. The primary consideration that determines which parts are selected for CNC lathe machining is basically that the part has round features that require machining.

CNC TURNING OPERATIONS

There are various operations that can be performed on CNC turning machines (lathes). Therefore, it is important to identify each operation in order to understand the full capabilities of a typical CNC turning machine. It is equally important to understand the various lathe operations and terms commonly used in machine shop communications. The following pages outline and illustrate some common CNC turning operations.

Facing

The facing operation machines the end of the workpiece in order to square the face or end of the workpiece. The face cut dimensionally machines the workpiece to the specified drawing length dimensions and is typically completed in one pass. Facing is performed with the tool feeding in the X-axis direction (see Figure 13-2). Facing is also used for machining a step or "shoulder" dimension. The G96 constant SFM mode, which automatically changes the RPM as the diameter changes, is ideal for facing because the tool travels from one diameter size to another.

Turning

The turning operation machines material from the outside diameter of the workpiece. The turn cut dimensionally machines the workpiece to the specified drawing diameter dimensions and is typically completed in multiple passes. Turning is performed with the tool feeding in the Z-axis direction (see Figure 13-3). Turning is also used for machining a step diameter/s to the

This illustration describes the facing operation performed on a CNC lathe. The first view illustrates the start position of the tool and the second view illustrates the finish position of the tool.

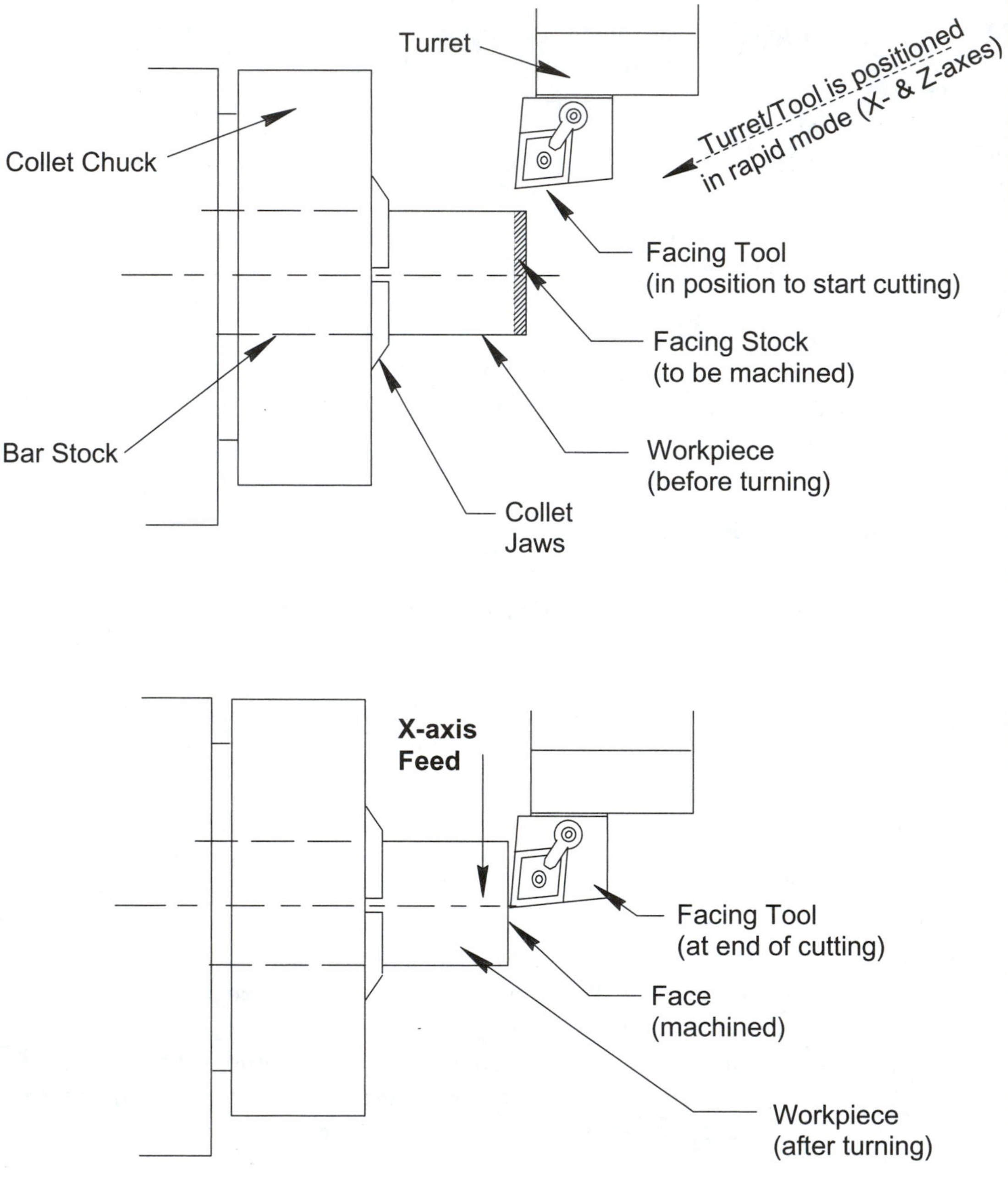

Turret

Turret/Tool is positioned in rapid mode (X- & Z-axes)

Collet Chuck

Facing Tool
(in position to start cutting)

Facing Stock
(to be machined)

Bar Stock

Workpiece
(before turning)

Collet
Jaws

**X-axis
Feed**

Facing Tool
(at end of cutting)

Face
(machined)

Workpiece
(after turning)

The facing tool feeds in the X-axis to the required X-coordinate (diameter) as illustrated above. Facing is performed in the G96 SFM mode.

Figure 13-2 Facing operation

This illustration describes the turning operation performed on a CNC lathe. The first view illustrates the start position of the tool and the second view illustrates the finish position of the tool.

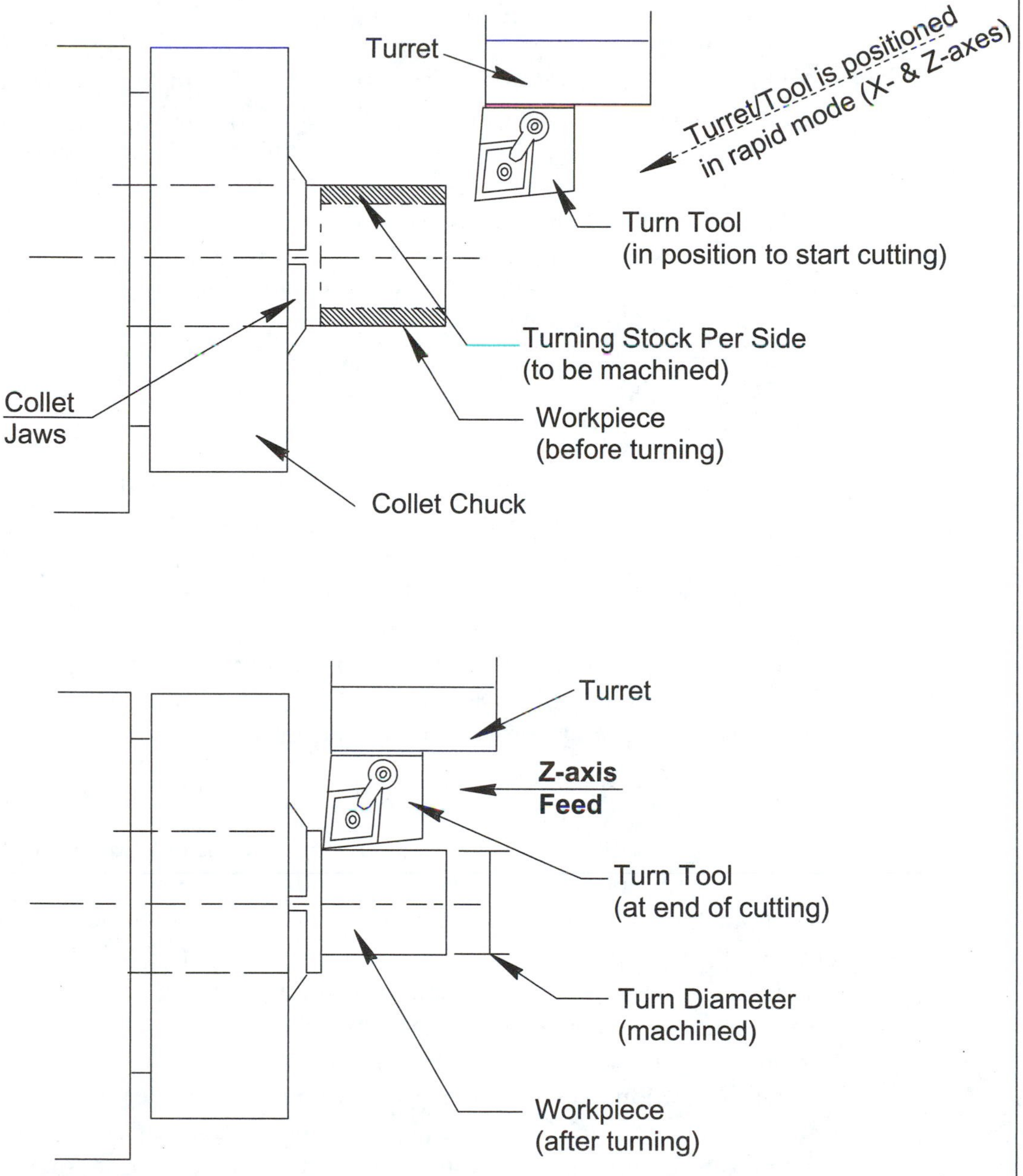

Turret

Turret/Tool is positioned in rapid mode (X- & Z-axes)

Turn Tool
(in position to start cutting)

Turning Stock Per Side
(to be machined)

Workpiece
(before turning)

Collet
Jaws

Collet Chuck

Turret

**Z-axis
Feed**

Turn Tool
(at end of cutting)

Turn Diameter
(machined)

Workpiece
(after turning)

The turning tool feeds in the Z-axis to the required Z-coordinate (length) as illustrated above. Turning is performed in the G96 SFM or G97 RPM mode.

Figure 13-3 **Turning operation**

specified drawing dimensions and is typically completed in multiple-turn passes. Typical turning operations include roughing cuts and then a finishing cut. Turning operations can be performed in either the G96 SFM mode or the G97 constant RPM mode because the tool travels in the Z-axis and the diameter position does not change.

Taper Turning

The taper turning operation also removes material from the outside diameter of the workpiece. The taper turn cut dimensionally machines the workpiece to the specified drawing diameter and taper dimensions and is typically completed in multiple passes. The taper turn cut is performed with both the X- and Z-axes feeding simultaneously (see Figure 13-4). Typical taper turning operations include roughing cuts and then a finishing cut. The G96 SFM mode is also ideal for taper turn cuts because the tool travels from one diameter size to another.

Chamfer

Chamfering is an operation that machines a 45° angle usually between 0.010 to 0.250 inch long. This operation is performed on the workpiece in order to eliminate sharp edges produced from other machining operations. The chamfer cut is also performed with both the X- and Z-axes feeding simultaneously (see Figure 13-5). Chamfers are applied to outside diameters, inside diameters, taper diameters, faces, and shoulder diameters. Typically, chamfers are performed in conjunction with the finishing face or turn cut. This blends the two surfaces together and creates a smooth edge free of sharp burrs, which are not allowed in most product design specifications.

Boring

The boring operation machines material from the inside diameter of the workpiece. The bore cut dimensionally machines the workpiece to the specified drawing bore diameter dimensions and is typically completed in multiple passes. Boring is performed with the tool feeding in the Z-axis direction (see Figure 13-6). Typical boring operations include roughing cuts and then a finishing cut. Boring operations can be performed in either the G96 SFM mode or the G97 constant RPM mode because the tool travels in the Z-axis and the diameter position does not change during the bore cut. Boring is basically performed to enlarge an existing hole to a specific diameter. Boring can typically produce holes that are accurate, straight, and have a fine surface finish.

Taper Boring

The taper boring operation also removes material from the inside diameter of the workpiece. The taper bore cut dimensionally machines the workpiece to the specified drawing inside diameter and taper dimensions. Boring is typically completed in multiple passes. The taper bore cut is also performed with

This illustration describes a taper turning operation performed on a CNC lathe. The first view illustrates the start position of the tool and the second view illustrates the finish position of the tool.

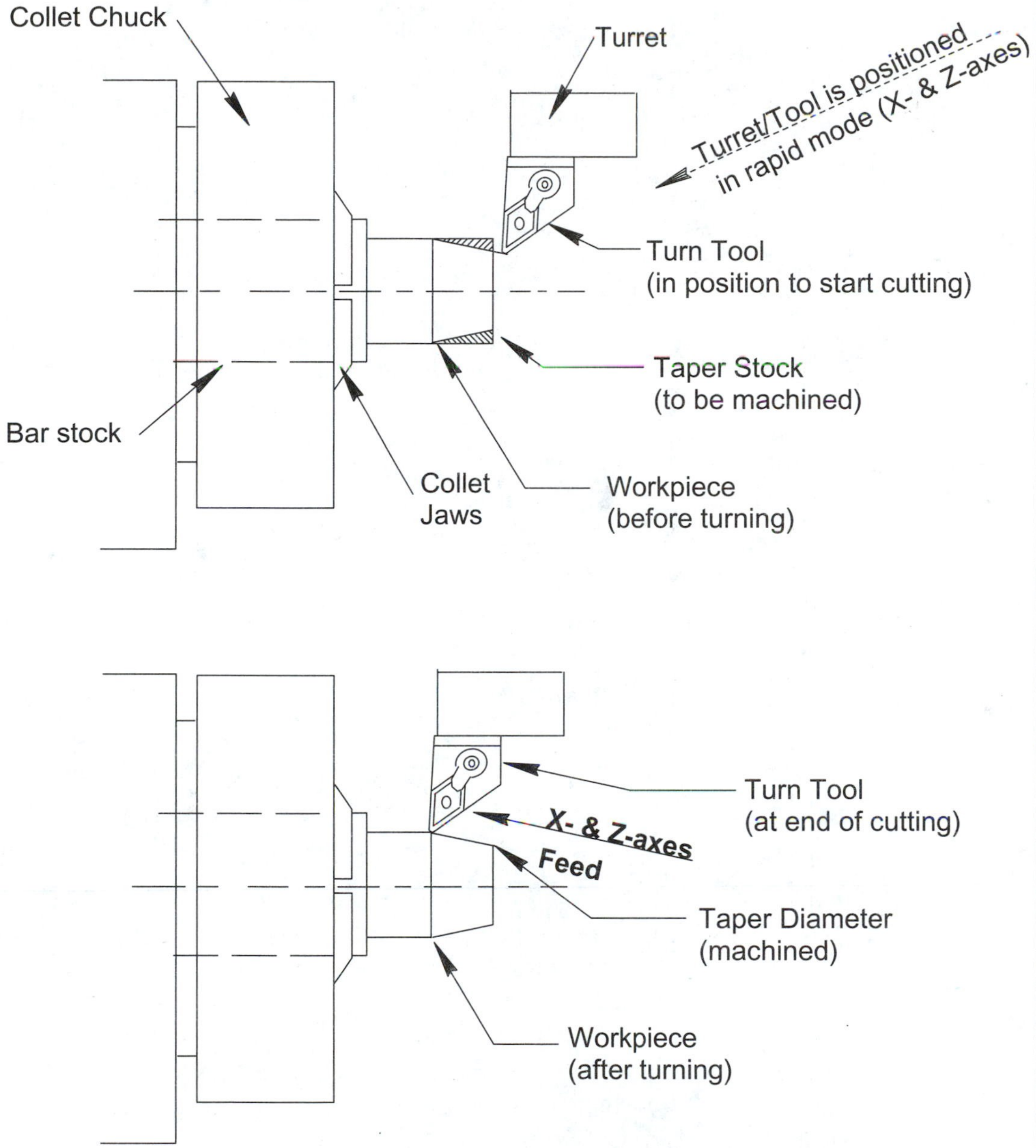

Collet Chuck

Turret

Turret/Tool is positioned in rapid mode (X- & Z-axes)

Turn Tool
(in position to start cutting)

Taper Stock
(to be machined)

Bar stock

Collet Jaws

Workpiece
(before turning)

Turn Tool
(at end of cutting)

X- & Z-axes
Feed

Taper Diameter
(machined)

Workpiece
(after turning)

The turn tool feeds in the X-axis and Z-axis simultaneously to the X-Z coordinate (diameter and length) as illustrated above. Taper turning is performed in the G96 SFM mode.

Figure 13-4 **Taper turning operation**

This illustration describes a chamfer operation performed on a CNC lathe. The first view illustrates the start position of the tool and the second view illustrates the finish position of the tool.

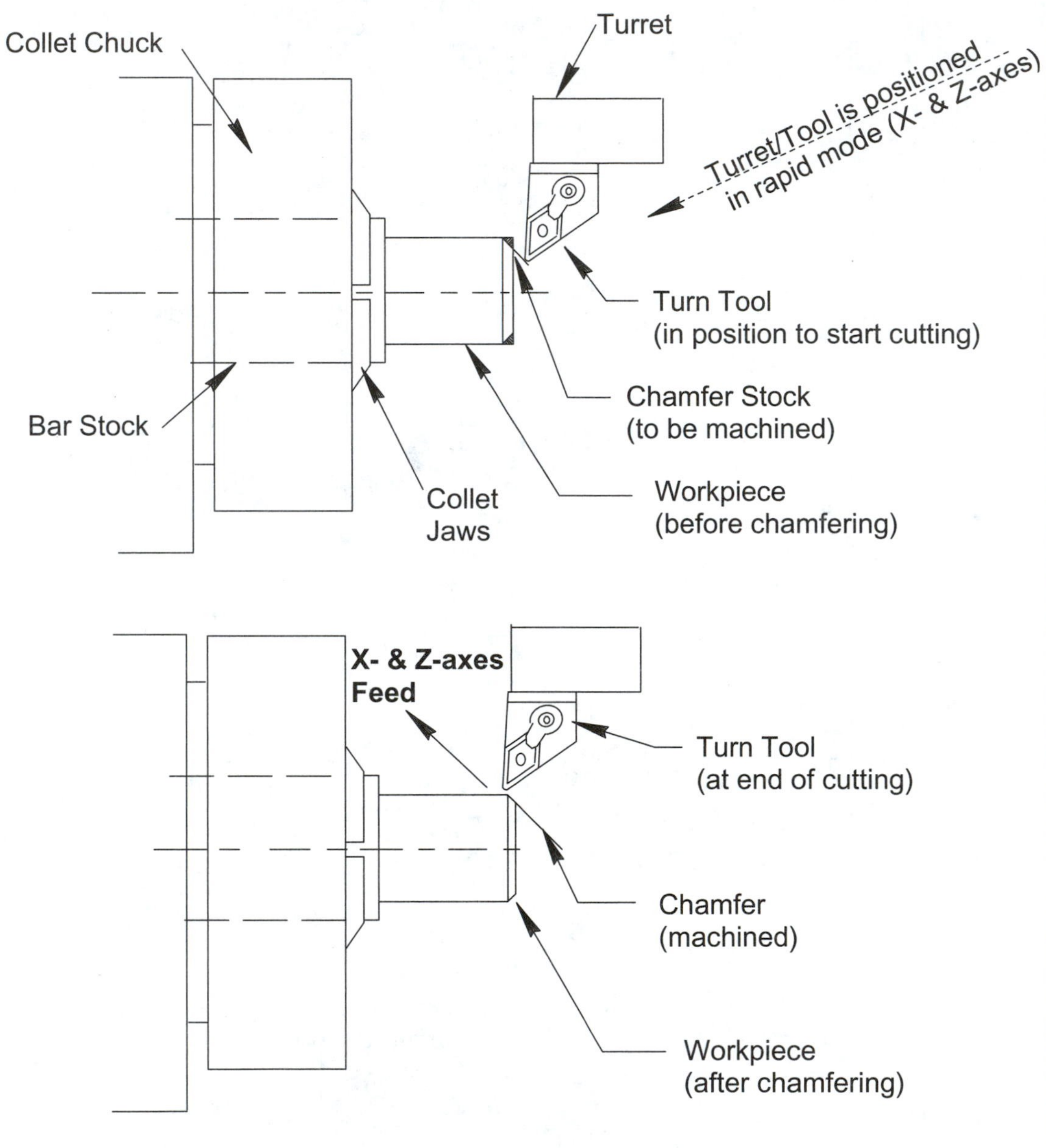

Collet Chuck

Turret

Turret/Tool is positioned in rapid mode (X- & Z-axes)

Turn Tool
(in position to start cutting)

Chamfer Stock
(to be machined)

Bar Stock

Collet
Jaws

Workpiece
(before chamfering)

X- & Z-axes
Feed

Turn Tool
(at end of cutting)

Chamfer
(machined)

Workpiece
(after chamfering)

The turn tool feeds in the X-axis and Z-axis simultaneously to the X-Z coordinate (diameter and length) as illustrated above. Chamfering is performed in the G96 SFM mode.

Figure 13-5 **Chamfering operation**

This illustration describes a boring operation performed on a CNC lathe. The first view illustrates the start position of the tool and the second view illustrates the finish position of the tool.

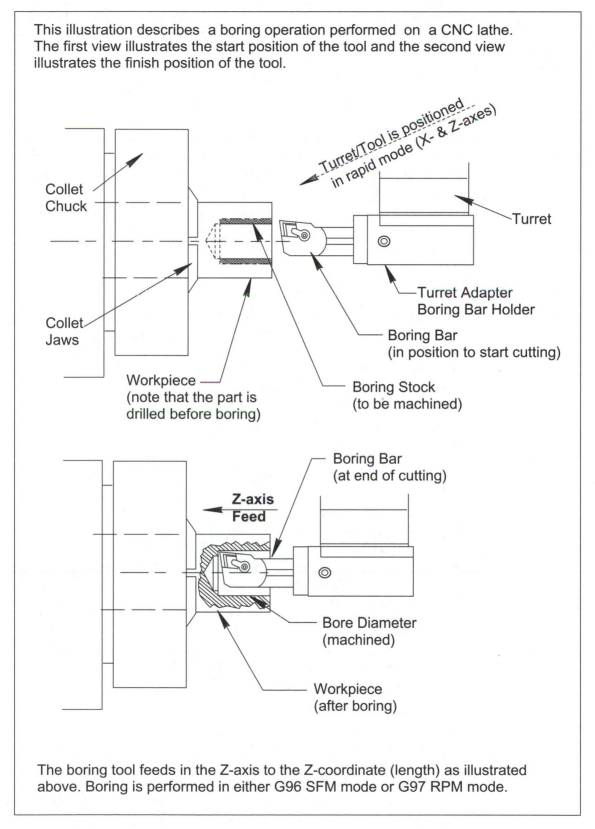

Collet Chuck

Turret/Tool is positioned in rapid mode (X- & Z-axes)

Turret

Collet Jaws

Turret Adapter Boring Bar Holder

Boring Bar (in position to start cutting)

Workpiece — (note that the part is drilled before boring)

Boring Stock (to be machined)

Boring Bar (at end of cutting)

Z-axis Feed

Bore Diameter (machined)

Workpiece (after boring)

The boring tool feeds in the Z-axis to the Z-coordinate (length) as illustrated above. Boring is performed in either G96 SFM mode or G97 RPM mode.

Figure 13-6 **Boring operation**

both the X- and Z-axes feeding simultaneously (see Figure 13-7). Typical taper boring operations include roughing cuts and then a finishing cut. The G96 SFM mode is also ideal for taper bore cuts because the tool travels from one diameter size to another.

Counterboring

The counterboring operation also machines material from the inside diameter of the workpiece. Basically, the counterbore is a step diameter of the inside diameter. The counterbore is performed with the tool feeding in the Z-axis direction (see Figure 13-8). Typical counterboring operations include roughing cuts and then a finishing cut. Counterboring machining typically requires that the step shoulder be faced at the end of the counterbore cut. Therefore, the G96 SFM mode is also ideal for counterbore cuts because the diameter position changes.

Grooving

The grooving operation can be performed on the OD, ID, and face of the workpiece. The OD and ID grooving operations are performed by the tool feeding into the workpiece at a right angle to its centerline (X-axis) (see Figure 13-9). The face grooving operation is performed by the tool feeding into the workpiece parallel to the centerline (Z-axis) of the workpiece. The G96 constant SFM mode is ideal for OD and ID grooving operations because the speed changes as the tool feeds and helps eliminate harmonics that cause chatter during the cut.

OD Threading

The OD threading operation forms a helical groove (screw thread) on the outside turn diameter of the workpiece (see Figure 13-10). Threads are usually cut with a single-point tool by feeding in the Z-axis. Single-point threading is a process performed by multiple passes that cut to the specified screw thread depth. The G97 constant RPM mode is best suited for single-point threading.

ID Threading

The ID threading operation forms a helical groove on the inside bore diameter of the workpiece (see Figure 13-11). The features described for OD threading above also apply to ID threading.

Drilling

The drilling operation is designed to machine a hole in the center of the workpiece. The twist drill, which is made of HSS, cuts the workpiece to the specific diameter of the drill and is typically completed in one pass. Drilling is performed with the tool feeding in the Z-axis direction (see Figure 13-12). Drilling operations are always performed in the G97 constant RPM mode with the drill positioned on center (X0.0). Center drills are often used to spot drill a point that helps the twist drill start straight.

This illustration describes a taper boring operation performed on a CNC lathe. The first view illustrates the start position of the tool and the second view illustrates the finish position of the tool.

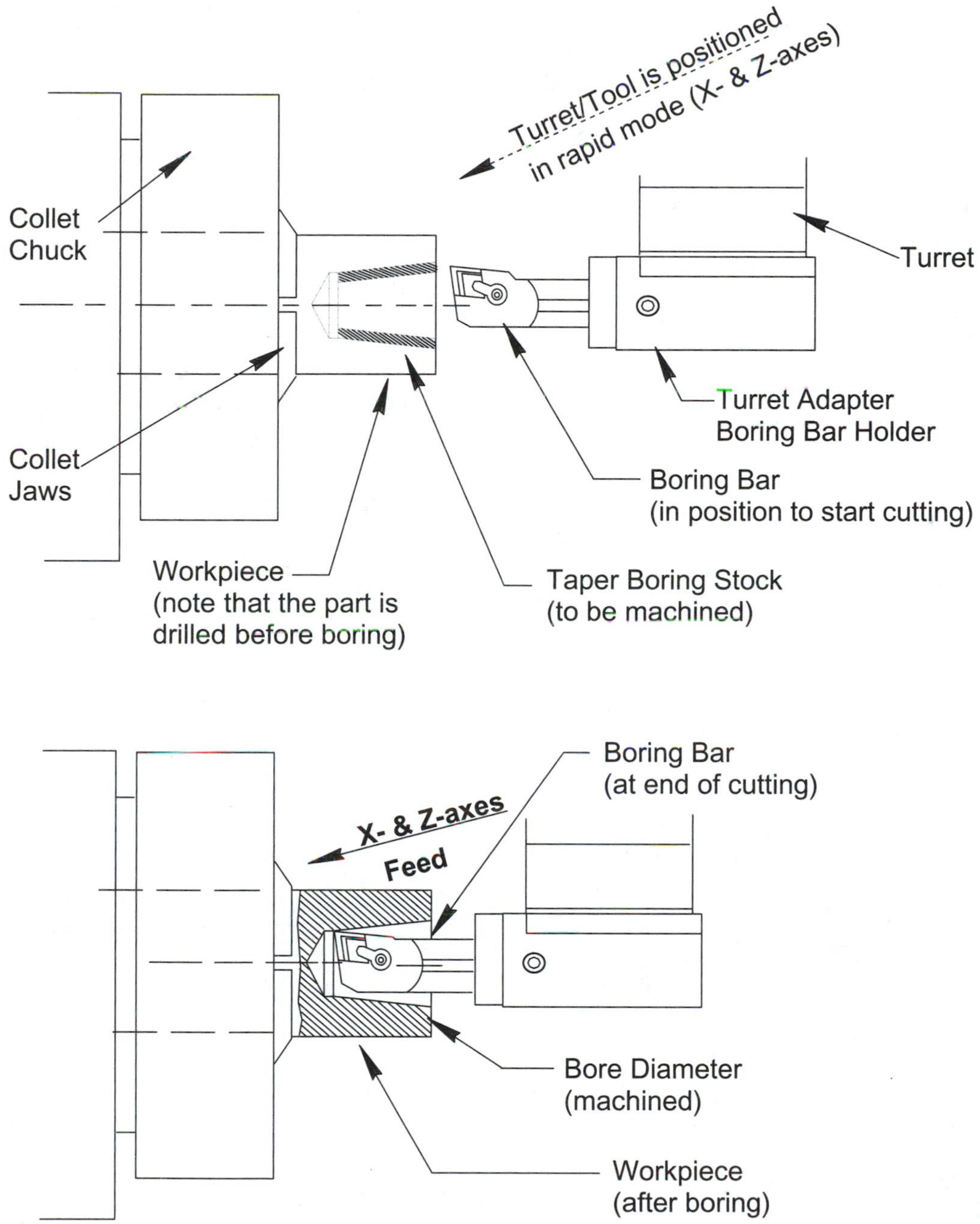

Collet Chuck

Turret/Tool is positioned in rapid mode (X- & Z-axes)

Turret

Collet Jaws

Turret Adapter Boring Bar Holder

Boring Bar (in position to start cutting)

Workpiece (note that the part is drilled before boring)

Taper Boring Stock (to be machined)

Boring Bar (at end of cutting)

X- & Z-axes Feed

Bore Diameter (machined)

Workpiece (after boring)

The boring tool feeds in the X-axis and Z-axis simultaneously to the X-Z coordinate (diameter and length) as illustrated above. Taper boring is performed in the G96 SFM mode.

Figure 13-7 Taper boring operation

This illustration describes a boring operation performed on a CNC lathe. The first view illustrates the start position of the tool and the second view illustrates the finish position of the tool.

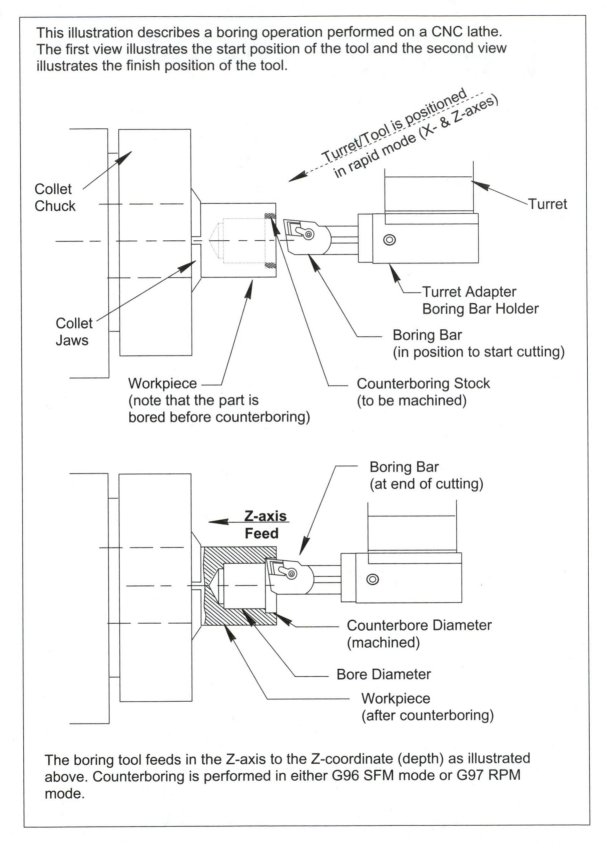

The boring tool feeds in the Z-axis to the Z-coordinate (depth) as illustrated above. Counterboring is performed in either G96 SFM mode or G97 RPM mode.

Figure 13-8 **Counterboring operation**

This illustration describes a grooving operation performed on a CNC lathe. The first view illustrates the start position of the tool and the second view illustrates the finish position of the tool.

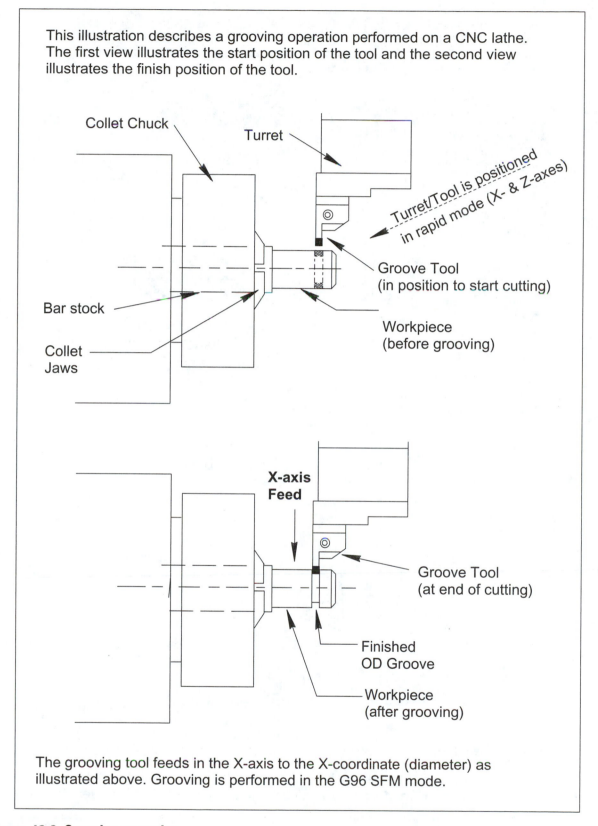

The grooving tool feeds in the X-axis to the X-coordinate (diameter) as illustrated above. Grooving is performed in the G96 SFM mode.

Figure 13-9 *Grooving operation*

This illustration describes a threading operation performed on a CNC lathe. The first view illustrates the start position of the tool and the second view illustrates the finish position of the tool.

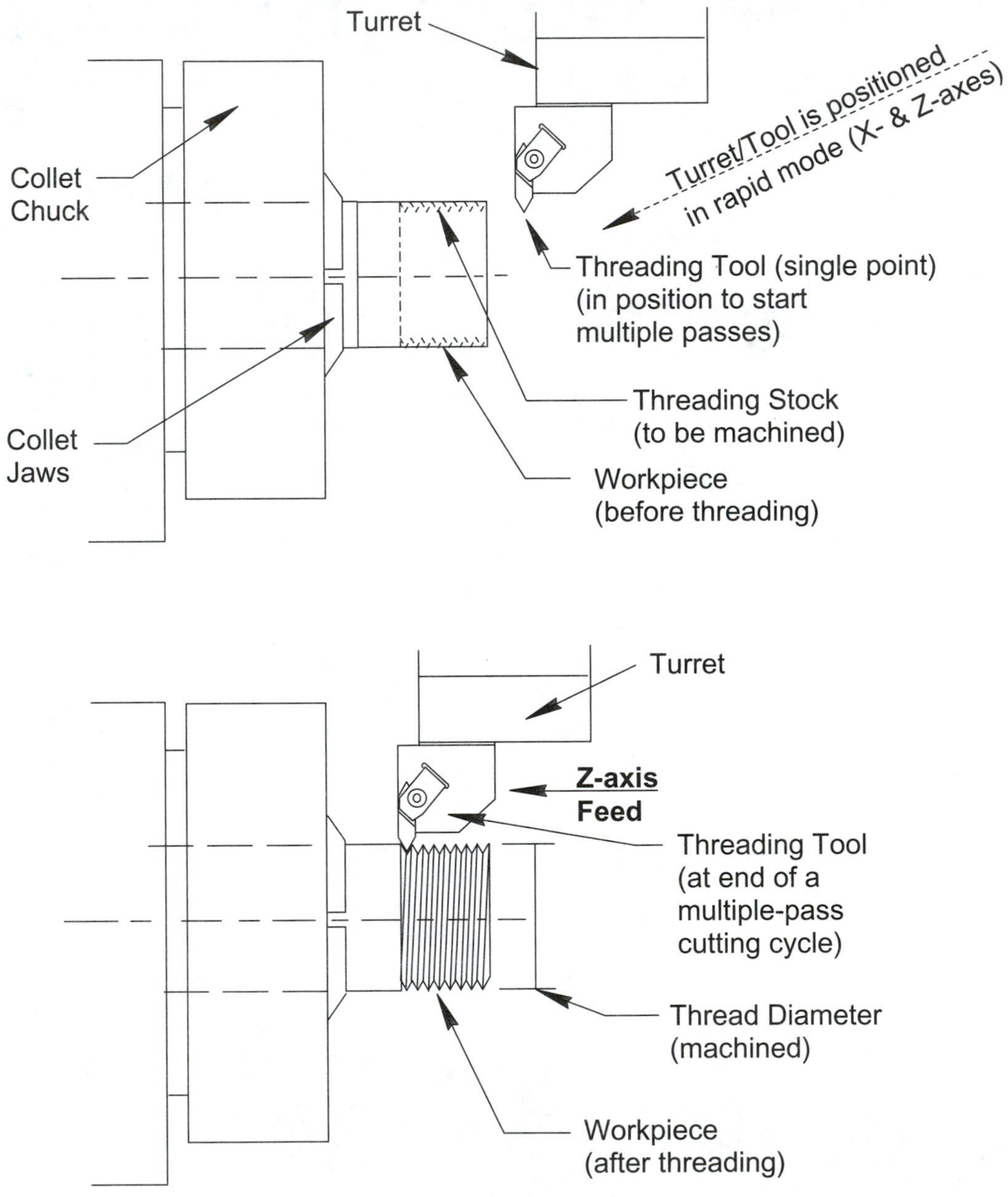

Turret

Turret/Tool is positioned in rapid mode (X- & Z-axes)

Collet Chuck

Threading Tool (single point) (in position to start multiple passes)

Threading Stock (to be machined)

Collet Jaws

Workpiece (before threading)

Turret

Z-axis Feed

Threading Tool (at end of a multiple-pass cutting cycle)

Thread Diameter (machined)

Workpiece (after threading)

The threading tool feeds in the Z-axis to the required Z-coordinate (length) as illustrated above. Threading is always performed in the G97 RPM mode.

Figure 13-10 **OD threading operation**

This illustration describes an ID threading operation performed on a CNC lathe. The first view illustrates the start position of the tool and the second view illustrates the finish position of the tool.

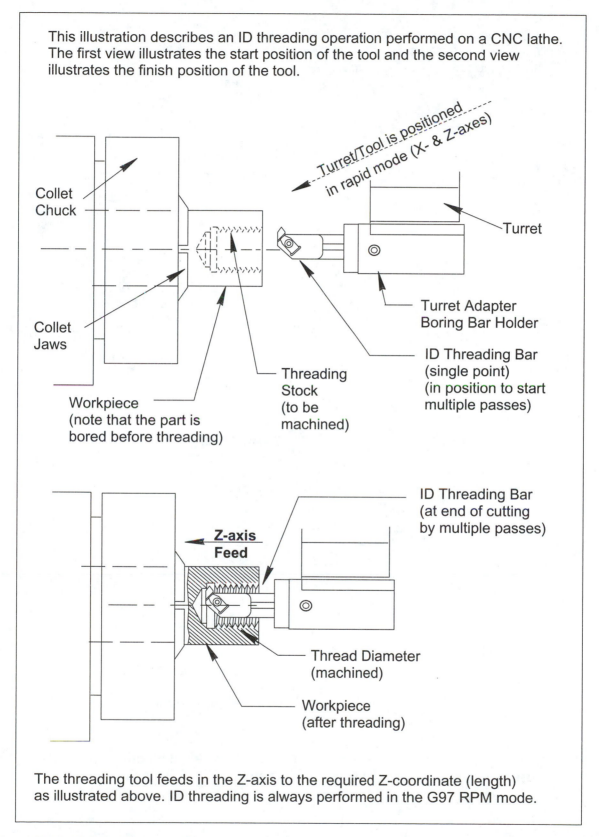

The threading tool feeds in the Z-axis to the required Z-coordinate (length) as illustrated above. ID threading is always performed in the G97 RPM mode.

Figure 13-11 **ID threading operation**

This illustration describes a drilling operation performed on a CNC lathe. The first view illustrates the start position of the tool and the second view illustrates the finish position of the tool.

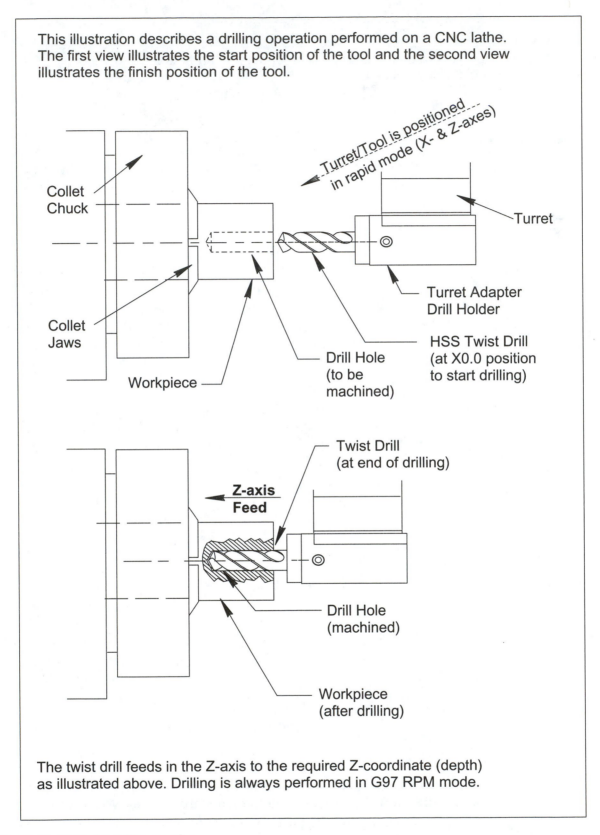

The twist drill feeds in the Z-axis to the required Z-coordinate (depth) as illustrated above. Drilling is always performed in G97 RPM mode.

Figure 13-12 **Twist drill operation**

Drilling with Carbide Indexable Insert Drills

The carbide indexable insert drill is also commonly used on CNC turning machines (see Figure 13-13). The indexable insert can drill holes at much higher rates than HSS twist drills. Insert drills are suitable for hole diameters ranging from 0.625 to 3 inches. Insert drills require higher horsepower and a high-pressure coolant system. If a hole tolerance requirement is less than ±0.003 of an inch, a secondary hole operation such as boring or reaming should be added to size the hole. Carbide indexable insert drilling operations are always performed in the G97 constant RPM mode with the drill positioned on center (X0.0). Center drilling is not required for carbide indexable insert drilling.

Spade Drilling

Spade drilling is a drilling operation that removes a larger amount of material in one pass (see Figure 13-14). Also, tool costs are usually lower with spade drilling. Therefore, spade drilling offers several advantages over drilling holes that are 1.0 inch in diameter and larger. Spade drilling operations are always performed in the G97 constant RPM mode with the spade drill positioned on center (X0.0).

Reaming

Reaming is an operation that produces holes to a high degree of accuracy and fine surface finish. The reamer is guided by the existing hole that must be machined first. Therefore, it will not correct errors in hole location or straightness. If these problems exist, it is advisable to first bore, then ream. Reaming operations are always performed in the G97 constant RPM mode with the drill positioned on center (X0.0).

Tapping

Tapping is another process that is used to produce internal threads usually for sizes smaller than 0.750 inches in diameter. This can be a delicate and sometimes troublesome process depending upon the material type and the thread size. The main problem with tapping is clearing chips from the hole and compensating for the faster feedrate required by the screw thread pitch. Tapping operations are always performed in the G97 constant RPM mode with the tap positioned on center (X0.0).

Parting

Parting, which is also referred to as "cutoff," is a machining operation that cuts the finished part off from the rough barstock. Parting is similar to grooving, except that the tool cuts to the center of the part (see Figure 13-15). Parting is typically performed with a carbide blade that is 0.125–0.250 inch wide. Parting operations are typically performed in the G96 constant SFM mode because the tool travels from one diameter size to another.

This illustration describes a carbide indexable insert drilling operation performed on a CNC lathe. The first view illustrates the start position of the tool and the second view illustrates the finish position of the tool.

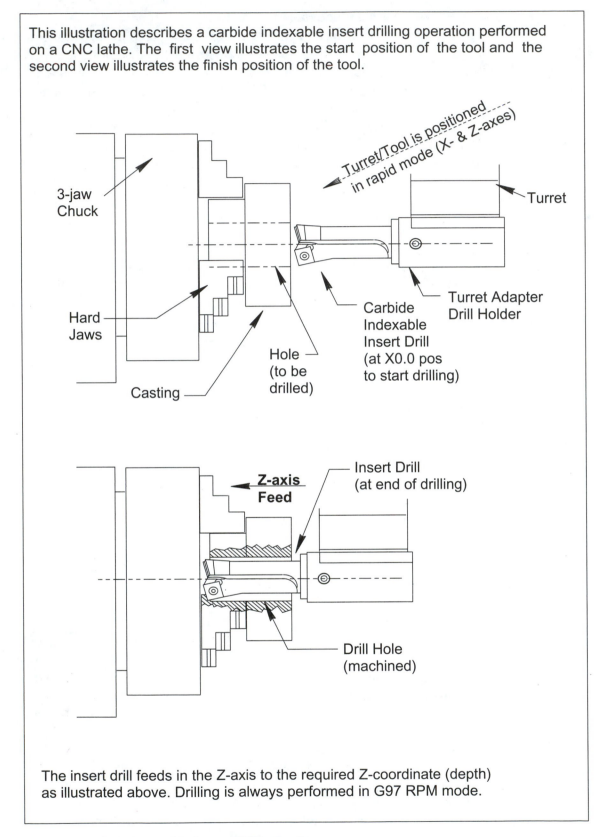

The insert drill feeds in the Z-axis to the required Z-coordinate (depth) as illustrated above. Drilling is always performed in G97 RPM mode.

Figure 13-13 *Carbide indexable insert drill operation*

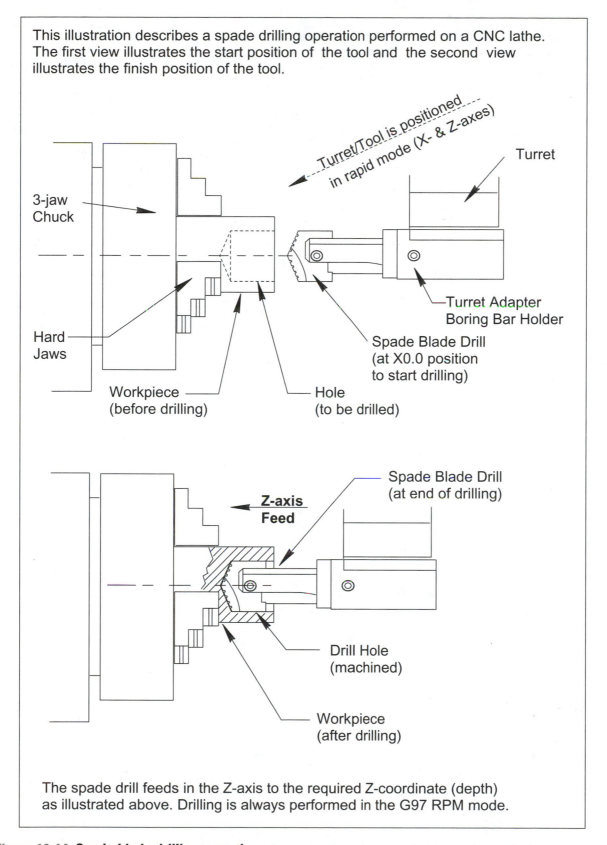

This illustration describes a spade drilling operation performed on a CNC lathe. The first view illustrates the start position of the tool and the second view illustrates the finish position of the tool.

Turret/Tool is positioned in rapid mode (X- & Z-axes)

Turret

3-jaw Chuck

Turret Adapter Boring Bar Holder

Hard Jaws

Spade Blade Drill (at X0.0 position to start drilling)

Workpiece (before drilling)

Hole (to be drilled)

Spade Blade Drill (at end of drilling)

Z-axis Feed

Drill Hole (machined)

Workpiece (after drilling)

The spade drill feeds in the Z-axis to the required Z-coordinate (depth) as illustrated above. Drilling is always performed in the G97 RPM mode.

Figure 13-14 *Spade blade drilling operation*

This illustration describes a parting operation performed on a CNC lathe. The first view illustrates the start position of the tool and the second view illustrates the finish position of the tool.

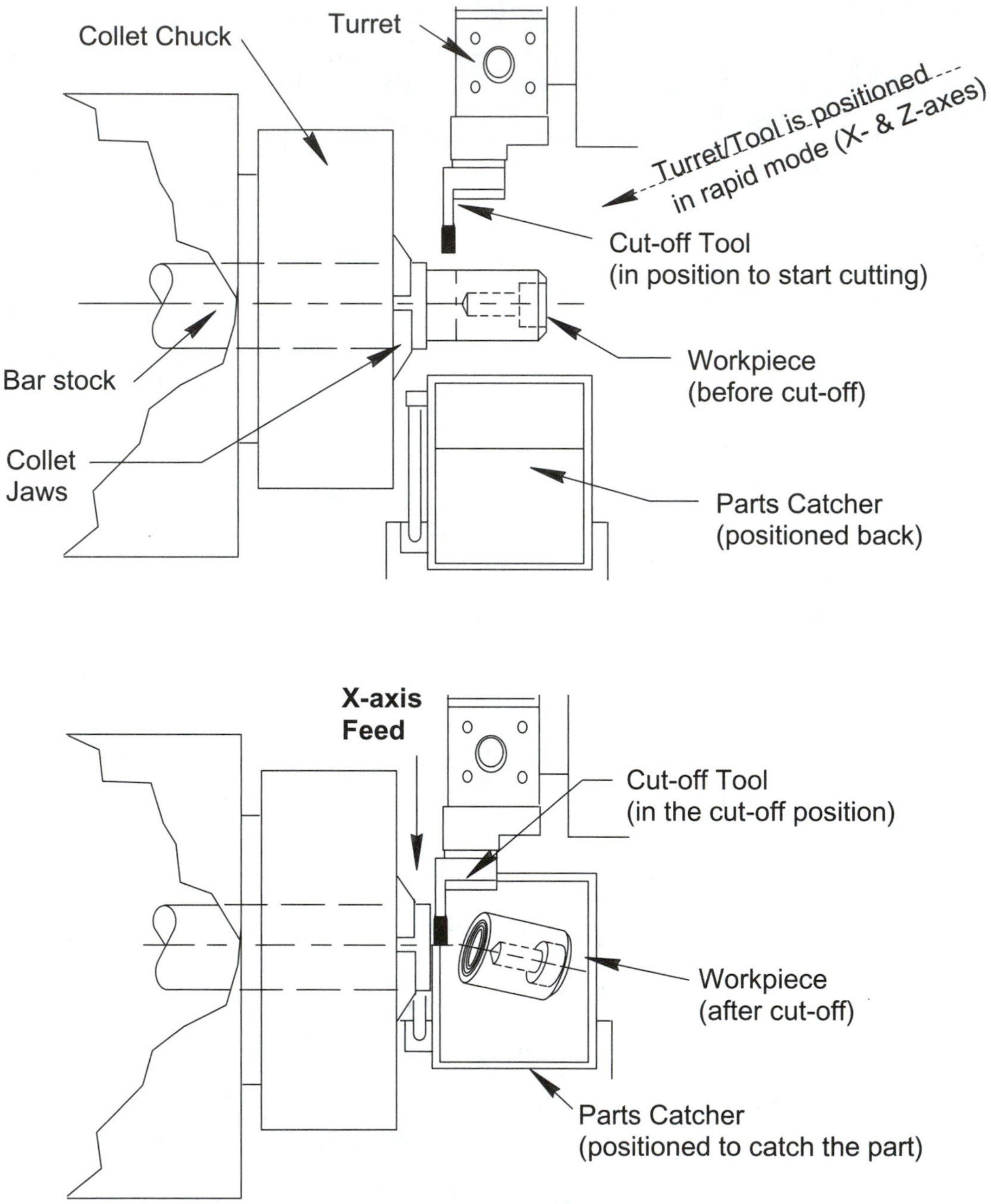

The parting tool feeds in the X-axis to the required X-coordinate (diameter) as illustrated above. Parting is performed in the G96 SFM mode.

Figure 13-15 **Parting operation**

CNC LATHE CUTTING TOOLS

The CNC lathe cutting tools are directly linked to the CNC lathe program, the CNC lathe setup, and the CNC lathe machining operations. Therefore, it is essential that the CNC setup person and CNC operator have thorough knowledge of the various types of CNC lathe cutting tools typically used on CNC lathes. This section describes and illustrates some commonly used OD and ID lathe cutting tools. It presents a view of the various types of indexable insert OD toolholders and boring bars. Also included are the grooving toolholder, the threading toolholder, the cut-off toolholder, drills, taps, and reamers. Related technical information such as "rake angles," clearance angles, tool components, and other tool characteristics are also illustrated and described. Cutting tools are usually held by some type of toolholder and then fastened to the specified CNC lathe turret station. The CNC cutting tools must be setup to match the CNC program. Tool setting is typically done with a tool tip sensor system that touches-off each tool in the X- and Z-axes to find and store the offset values.

OD Cutting Tools

The most commonly used OD cutting tools on CNC lathes are the carbide indexable insert toolholders (see Figure 13-16). These OD toolholders are designed with components that require assembly and can be replaced when worn or broken. There are two basic indexable insert OD toolholder designs—the indexable insert with a molded chipbreaker (Figure 13-17) and the indexable insert with a separate chipbreaker (Figure 13-18). The main components of the indexable insert OD toolholder are defined as follows:

- *Toolholder body* is designed to basically hold the indexable insert and all the required components.

- *Shim* is the component designed to support the indexable insert.

- *Insert lock pin* is the component designed to fasten the shim and lock the indexable insert in place.

- *Indexable insert* is the component designed to cut the material and is available in various styles and geometry shapes (see Figure 13-19).

- *Clamp screw* is the component designed to force the clamp onto the indexable insert.

- *Clamp* is the component designed to hold the indexable insert in the toolholder pocket.

- *Chipbreaker* is the component designed to curl and break the metal as it is cut by the indexable insert.

Additionally, there are basic features characteristic of all OD toolholders (see Figure 13-16). The main features of the OD toolholder are defined as follows:

- *Cutting edge* is the insert feature designed to cut the workpiece.

- *Shank* is the toolholder feature designed to be fastened to the CNC lathe turret station slot.

This illustration describes the features of the lathe insert indexable toolholder. This illustration example is of an 80° diamond toolholder commonly used on CNC lathes.

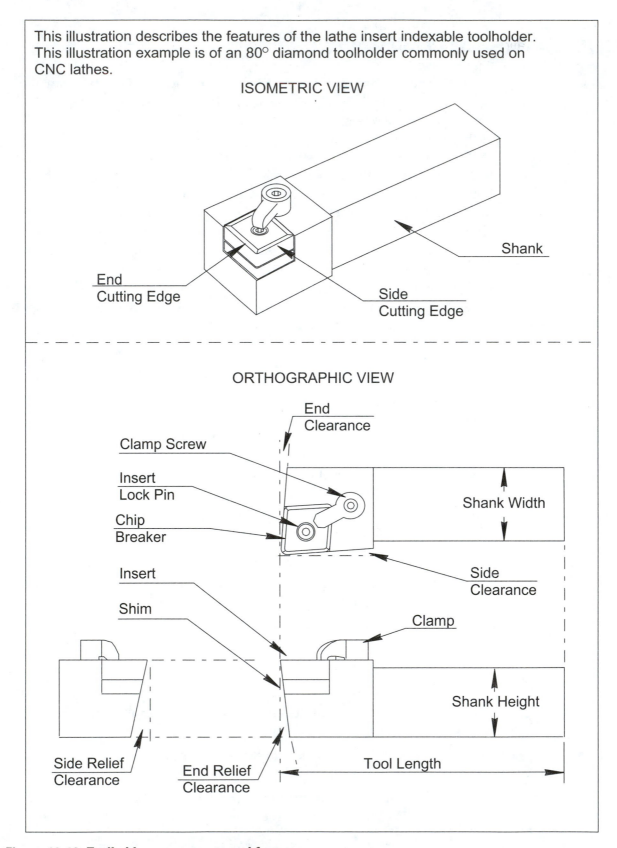

Figure 13-16 **Toolholder components and features**

This illustration is of an 80° diamond-type toolholder with the chipbreaker molded into the insert. The components are illustrated in their order of assembly.

The components of this insert indexable toolholder include the toolholder body, the shim, the insert lock pin, the insert, the clamp screw and the clamp.

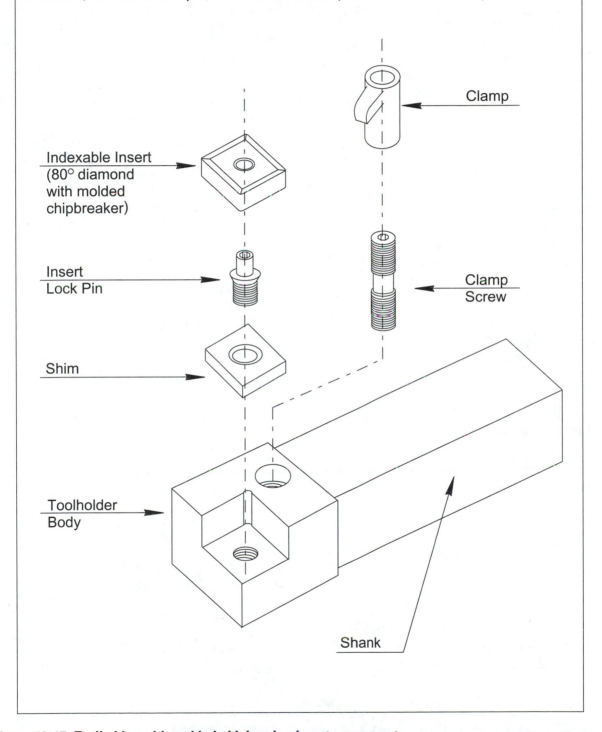

Figure 13-17 *Toolholder with molded chipbreaker insert components*

This illustration is of an 80° diamond-type toolholder with separate chipbreaker and insert. The components are illustrated in their order of assembly.

The components of this insert indexable toolholder include the toolholder body, the shim, the shim screw, the insert, the chipbreaker, the clamp screw and the clamp.

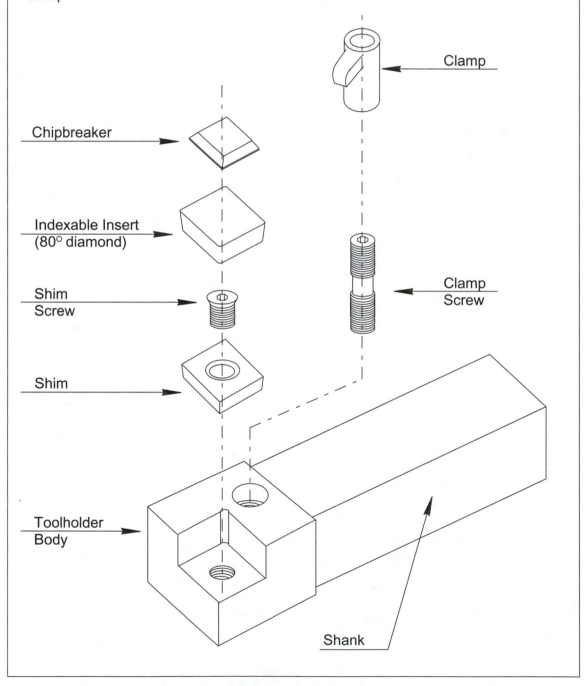

Figure 13-18 *Toolholder with separate chipbreaker and insert components*

- *Clearance angles* are toolholder features designed to prevent contact between the toolholder and the workpiece during the cut.

- *Rake angles* are toolholder and insert features designed to influence lathe machining characteristics (see Figure 13-20).

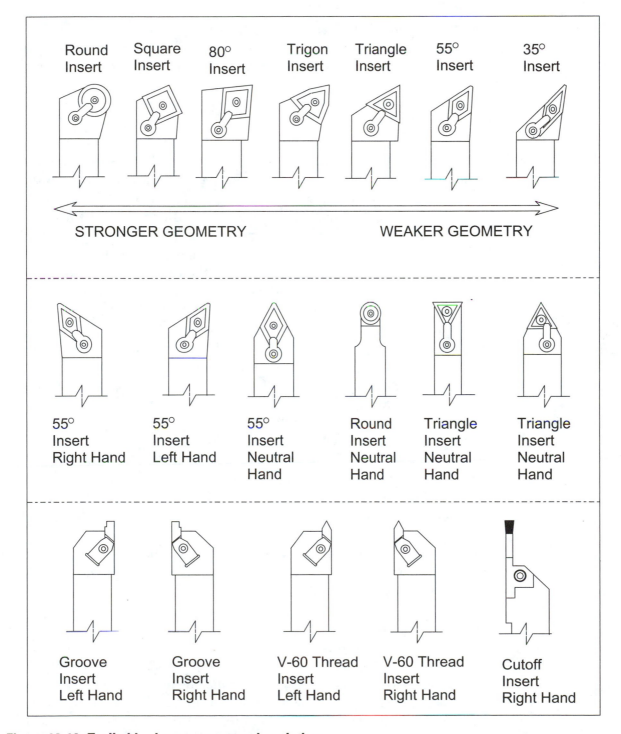

Figure 13-19 *Toolholder insert geometry descriptions*

OD Toolholder Insert Geometry Shapes and Features

The illustration in Figure 13-19 describes the various types of toolholder and insert designs typically used for CNC lathe applications. The illustration also describes which insert geometry is stronger or weaker. Additionally, the right-hand, the left-hand, and the neutral toolholder styles are described.

Insert Toolholder Rake Angle Descriptions

The illustration in Figure 13-20 describes the features and characteristics of positive, neutral, and negative top rake angles. Also included are the degrees for each rake angle.

The OD toolholders, which are available from various tool manufacturers, are identified by a standard numbering system (see Figure 13-21). The standard numbering system identifies the size of the toolholder shank, the size and shape of the insert, the toolholder cutting hand, the clamping system, and the toolholder cutting features. The shank width and height, which range in size from 1/2 to 1-1/2 inches, are usually the same size (square). The overall length (OAL) of the toolholder is usually available in one-inch increments from 2-1/2 up to 8.0 inches. Some of the insert geometry shapes typically available include the 80° diamond, the 55° diamond, the 35° diamond, the round, the triangle, the square, and the trigon. The rake angle is typically available in "positive," "neutral" and "negative." They are typically available in "right-hand cutting," "left-hand cutting," and "neutral cutting." Additionally, in order to fully use the capabilities of CNC lathes, various other lathe toolholder styles and designs are available.

Some other commonly used OD carbide-insert toolholders include the insert grooving toolholder, the insert threading toolholder, and the insert cut-off toolholder (see Figure 13-19). The OD carbide-insert toolholders range in strength and application as illustrated in Figure 13-19. Another feature of these OD insert toolholders is the orientation of the insert on the toolholder body. They are typically designed to be "right-hand" cutting, "left-hand" cutting, and "neutral" cutting, as illustrated in Figure 13-19. Generally, carbide indexable insert tools are capable of higher machining productivity than HSS tools. Therefore, modern CNC turning machines incorporate carbide indexable inserts into all cutting tool types whenever possible. However, some cutting tools such as smaller diameter drills, reamers, and taps are not suitable for carbide indexable insert designs.

ID Cutting Tools

The commonly used ID cutting tools on CNC lathes include twist drills, carbide indexable insert drills, spade drills, reamers, taps, and carbide indexable insert boring bars. The boring bars, which include boring, ID grooving, and ID threading-type inserts, are designed with components that require assembly and can be replaced when worn or broken. After facing and OD machining operations, drilling is typically the first ID machining operation performed on the CNC lathe. Drilling usually will not produce holes of high accuracy and fine surface finish. In situations where high accuracy or fine surface finish of a hole is required, other ID tools must be used after drilling. The tools typically used for those situations include either reamers or boring bars.

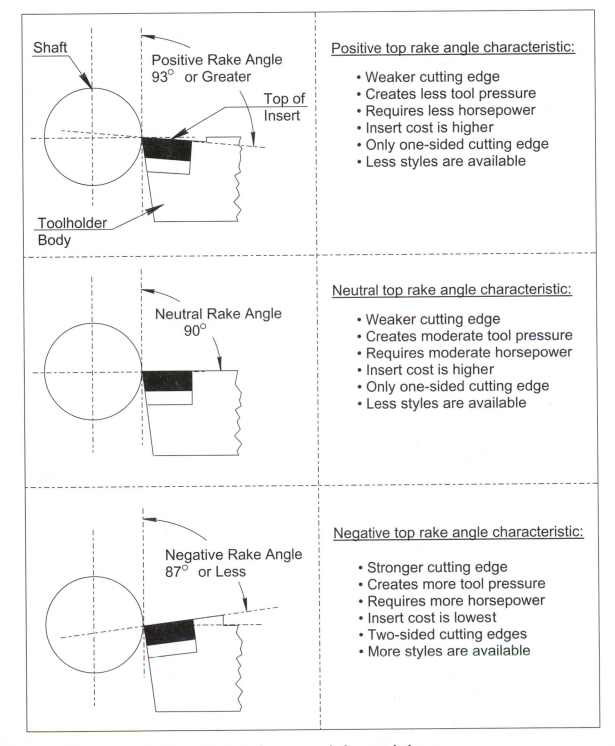

Figure 13-20 *Insert toolholder rake angle features and characteristics*

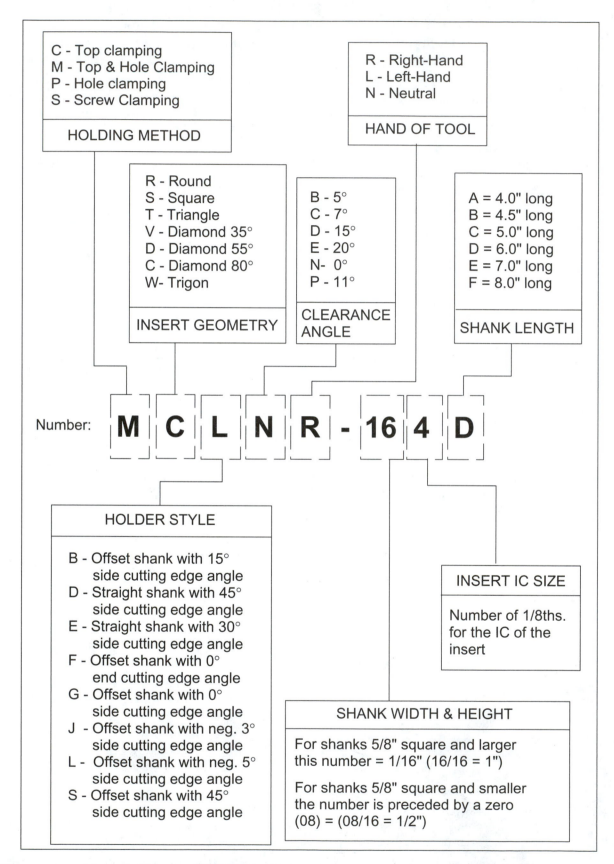

Figure 13-21 **OD toolholder identification system**

Twist Drills

For CNC lathes, the twist drill is used to produce a hole in the center of the workpiece. However, some CNC lathes are outfitted with optional live tooling, which also allows them to drill holes off center. The twist drill is made from various types of materials, which include HSS, cobalt, solid carbide, and carbide tipped. The twist drill, which is illustrated in Figure 13-22, is available in various design styles and sizes. The common design styles include jobber length, screw machine length, and taper length. Straight shanks, which are more flexible for setups, are common for drills up to 1/2 inch in diameter. Larger diameter drills are designed with either straight or tapered shanks. They are typically available in standard diameter sizes as listed below:

- Fractional diameters from 1/64 to 4.0 inches

- Number diameters from number 1 (0.228 inch) to number 97 (0.0059 inch)

- Letter diameters from letter A (0.234 inch) to letter Z (0.413 inch)

- Metric diameters from 0.15 mm (0.0059 inch) to 33.0 mm (1.2992 inch)

Twist drills are also standardized by OAL into three categories, which include the "screw machine length," the "jobber length," and the "taper length." Most drill manufacturers also produce "extra-long" sizes for deep-hole drilling applications. The OAL range for each category is listed below (English System):

- Screw machine length from 1-3/16 inches OAL to 7-3/4 inches OAL

- Jobber length from 3/4 inch OAL to 7-5/8 inches OAL

- Taper length from 2-1/4 inches OAL to 29.0 inches OAL

Metric drill lengths, which are specified in millimeters (mm), are similar to the drill lengths listed above. The tool manufacturers supply catalogs with all standard diameter and length sizes that are available.

Drilling performance and accuracy tend to decrease when either drill length or drill diameter is increased. Longer drills exhibit less stiffness and more torsional deflection. A good machining practice is to select the shortest drill length possible for any hole operation. Also, when drilling through material, a good practice is to allow one-third of the diameter ($0.3 \times$ drill diameter) of the drill plus 100 thousandths (0.100 inch) to extend sufficiently beyond the part material.

Center Drills

The center drill (Figure 13-22) is used for machining the end of a shaft (workpiece) that will be held between centers or supported at one end with the tailstock center of a lathe. Therefore, the center drill simply requires a drilling depth sufficient for the lathe center to enter and contact the end of the shaft with the 60° point. This depth (Figure 13-22) typically varies from approximately 0.125 inch deep to 0.500 inch deep depending on the diameter of the

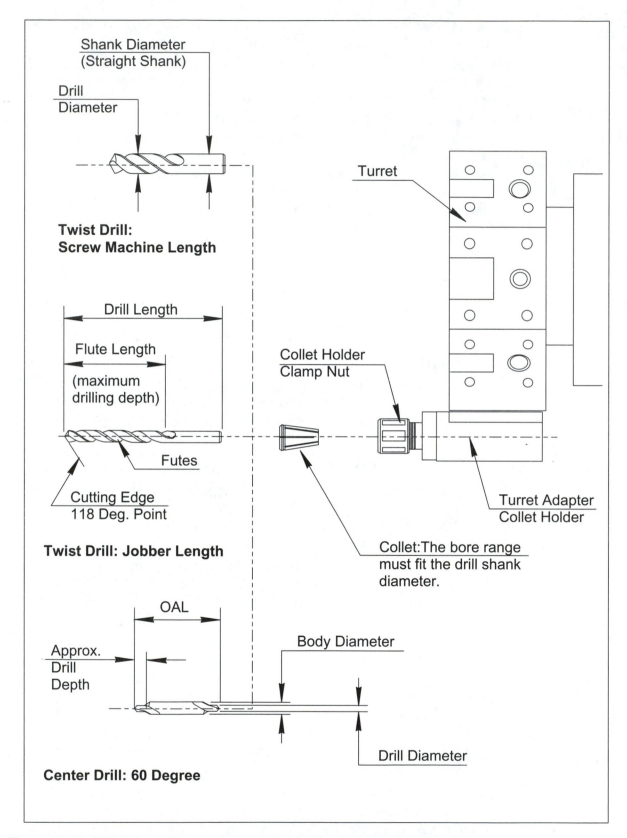

Figure 13-22 *CNC lathe drilling tools and collet holder features and descriptions*

shaft, which will be supported on the lathe center. The center drill is available in five sizes as listed below:

- Size 1: body diameter = 1/8 inch; drill diameter = 3/64 OAL = 4 inches
- Size 2: body diameter = 3/16 inch; drill diameter = 5/64 OAL = 4 inches
- Size 3: body diameter = 1/4 inch; drill diameter = 7/64 OAL = 4 inches
- Size 4: body diameter = 5/16 inch; drill diameter = 1/8   OAL = 5 inches
- Size 5: body diameter = 7/16 inch; drill diameter = 3/16 OAL = 6 inches

When center drilling it is important to align the OD of the workpiece as close to the spindle axis (X0.0) as possible. This is necessary in order to minimize runout between the center drill and the workpiece OD.

Carbide Indexable Insert Drills

Carbide indexable insert drills (Figure 13-23) represent the latest state-of-the-art advancements in CNC hole drilling. The indexable insert drill is sometimes used in place of a HSS twist drill. The indexable insert can drill holes at much higher metal removal rates than HSS twist drills. Insert drills are typically suited for hole diameters ranging from 5/8 up to 3.0 inches in diameter. Additionally, they offer the advantage of indexable or replaceable inserts that can save time during setup and tool changes. They are also capable of drilling from a solid at penetration rates of five to ten times that of HSS twist drills. The carbide inserts also allow the tool to be driven into harder materials. In most cases, insert drills will require higher horsepower from the machine and are set up with a thru-the-tool high-pressue coolant system.

Carbide-Tipped Coolant-Fed Drills

Carbide-tipped coolant-fed drills have one or two holes passing from the shank to the cutting point (see Figure 13-23). Compressed air, oil, or coolant fluid is passed through the drill as it drills the workpiece. This design enables the cutting point and work to be cooled as chips are flushed out. These drills are especially useful for drilling deep holes.

Spade Drills

The spade drill, which basically consists of a holder, a drill blade, and a clamp screw, is used for drilling larger holes in one pass (see Figure 13-23). Spade drills offer several advantages over twist drills for drilling holes 1.0 inch in diameter and up. The larger web of the spade drill ensures that during penetration less flexing occurs and thus a more accurate hole is produced. Tooling costs are lower with spade drills because standard blade holders will accommodate a variety of blade diameters, normally ranging from 5/8 up to 6.0 inches in diameter. Worn blades can be either resharpened or simply replaced with new ones. Spade drills are designed to

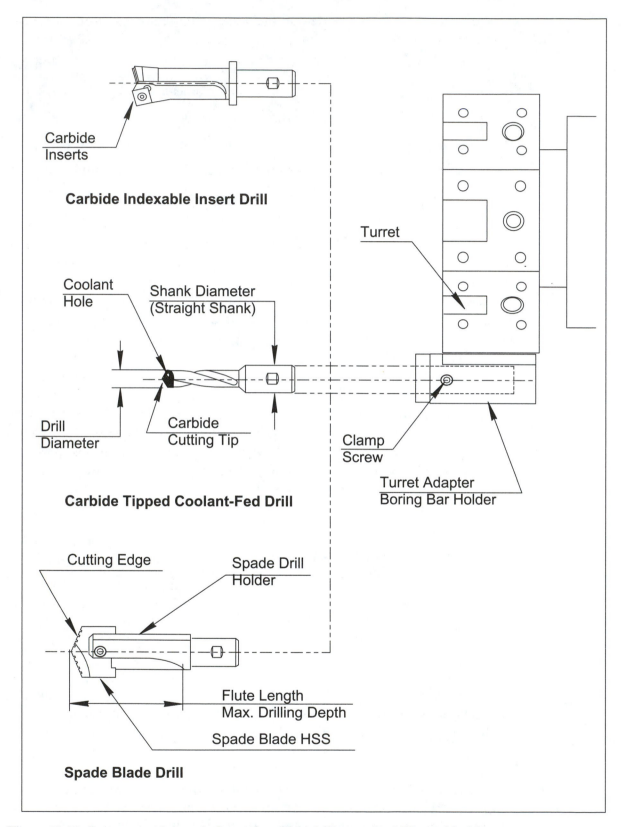

Carbide Inserts

Carbide Indexable Insert Drill

Coolant Hole

Shank Diameter (Straight Shank)

Drill Diameter

Carbide Cutting Tip

Carbide Tipped Coolant-Fed Drill

Turret

Clamp Screw

Turret Adapter Boring Bar Holder

Cutting Edge

Spade Drill Holder

Flute Length
Max. Drilling Depth

Spade Blade HSS

Spade Blade Drill

Figure 13-23 *Features and descriptions of carbide drills, spade drills, and holder*

machine a hole from the solid in one pass, eliminating the need for center drilling or multiple-pass drilling to gradually enlarge the hole size. The cutting edges of the blade incorporate chip-splitting and chip breaking action to reduce chip size and also facilitate chip removal. As with insert drills, spade drills also will require higher horsepower from the machine and are usually set up with a thru-the-tool high-pressue coolant system.

Boring Bars

As stated previously, boring bars are used in situations where high accuracy or fine surface finish of a hole is required. Additionally, boring bars typically produce better hole straightness and roundness. For CNC lathes the more commonly used boring bar tools are the carbide indexable type (see Figure 13-24). These boring bars are designed with components that require assembly and can be replaced when worn or broken. Similar to OD toolholders, there are two basic indexable insert boring bar designs: the indexable insert with a molded chipbreaker (Figure 13-17) and the indexable insert with a separate chipbreaker (Figure 13-18). The main components of the indexable insert boring bar are also similar except that boring bars do not use the shim.

As a general rule, the shortest boring bar length should be selected for any bore operation. As with all tools, the greater the length-to-diameter ratio of the tool the more flexible and error prone the boring bar will be. The finish of the surface inside the hole will also be affected because long bars tend to chatter. Boring bars, which are also available from various tool manufacturers, are also identified by a standard numbering system (see Figure 13-25).

Reamers

Reamers are also used in situations where high accuracy or fine surface finish of a hole is required. The factor that basically determines use of a reamer or a boring bar is the diameter size of the hole. A reamer is usually selected when the hole size is 0.125 inch up to 0.625 inch in diameter. The reamer, which is a cylindrical tool with straight or helical cutting edges (see Figure 13-26), is available in HSS, carbide tipped, and solid carbide. Reamers are designed to cut a small amount of material (0.005 to 0.030 inch) per side as well as the end. An important factor to consider about reaming is that reamers are guided by the existing hole, and therefore reaming will not correct errors of hole location or hole straightness. If either of these problems exist, it is necessary to first bore the hole and then ream it.

Taps

There are various thread forms, which are machined on lathes. Some common thread forms include the "Acme" thread, the "Buttress" thread, the "Square" thread, and the "Unified National" (UN) thread. The UN thread, which has a V-60 degree shape, is the most common thread form. Therefore, taps are typically designed and manufactured in the UN thread form. The UN

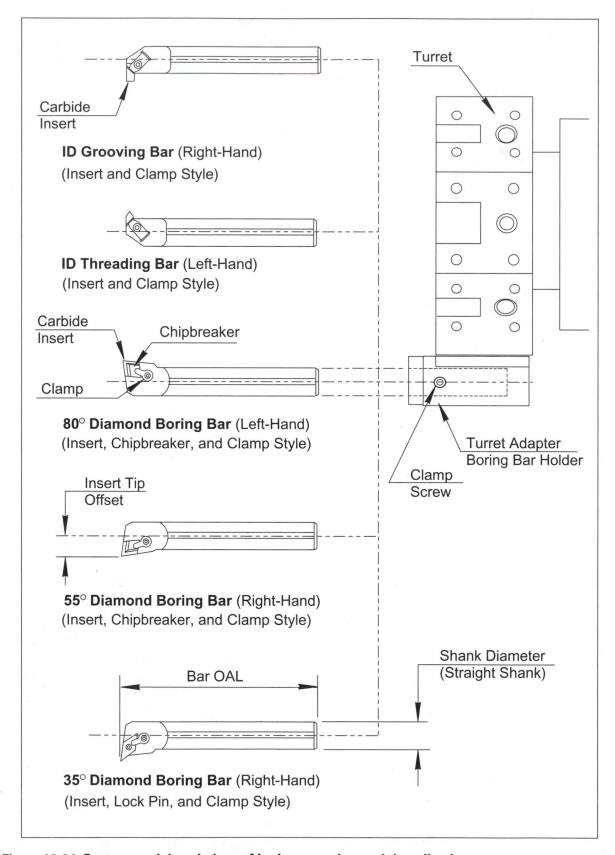

Figure 13-24 *Features and descriptions of boring, grooving, and threading bars*

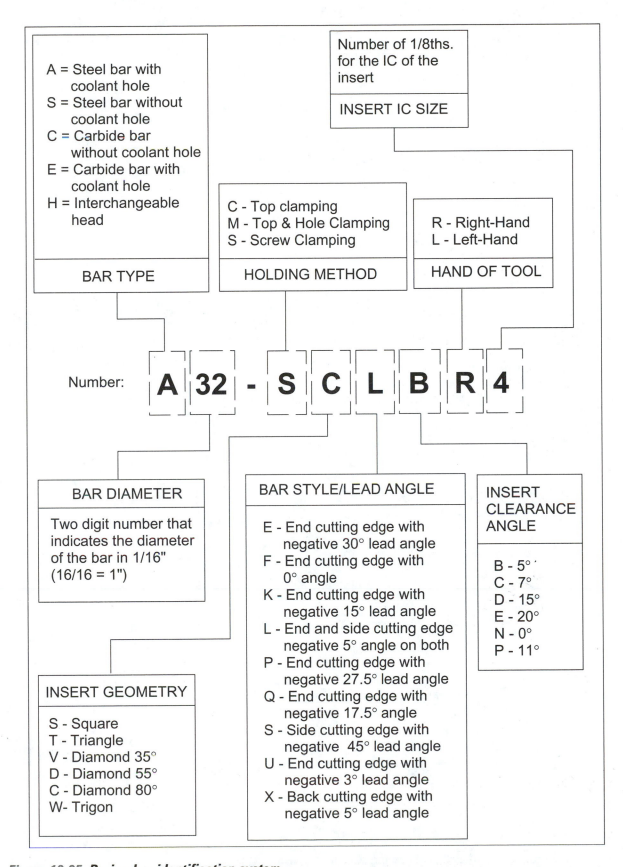

Figure 13-25 Boring bar identification system

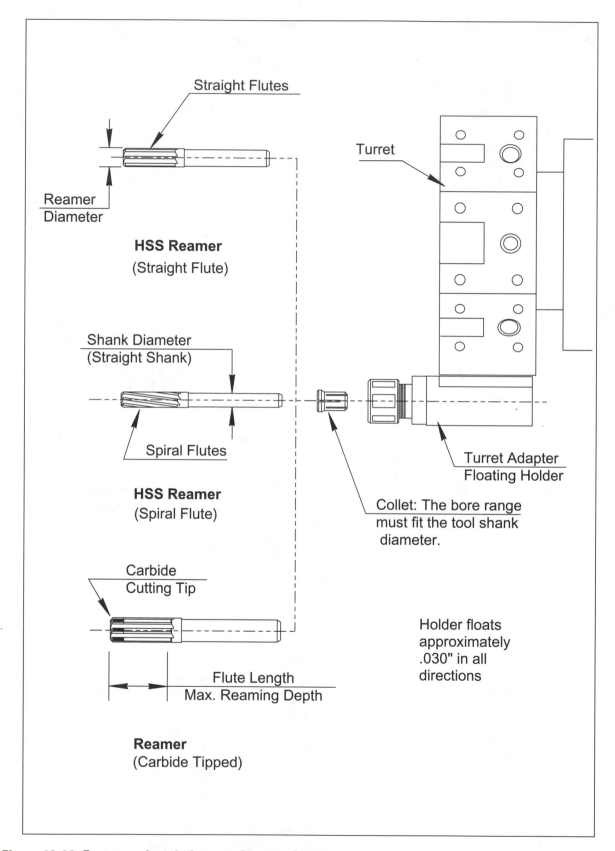

Straight Flutes

Reamer
Diameter

HSS Reamer
(Straight Flute)

Shank Diameter
(Straight Shank)

Spiral Flutes

HSS Reamer
(Spiral Flute)

Carbide
Cutting Tip

Flute Length
Max. Reaming Depth

Reamer
(Carbide Tipped)

Turret

Turret Adapter
Floating Holder

Collet: The bore range
must fit the tool shank
diameter.

Holder floats
approximately
.030" in all
directions

Figure 13-26 **Features, descriptions, and types of reamers**

thread taps are designed and manufactured in both inch and metric sizes (see appendix H). The UN thread designation, which is basically categorized according to nominal major diameter, threads per inch (TPI), class of fit, and internal or external type, is described in Figure 13-27.

Machining ID threads on CNC lathes can be performed with either a single-point ID threading bar or with a tap. The factor that basically determines use of a tap or a boring bar is the major diameter size of the threaded hole. A tap is usually selected when the tap major diameter size is 0.750 inch and smaller. Some commonly used taps, which include spiral point, plug, spiral flute, form, and bottoming, are illustrated in Figure 13-27. Due to the greater than normal cutting forces produced when threading or tapping, coolant is always applied in order to dissipate the heat and lubricate the cutting action.

It is important to note that before the thread can be machined with a tap or threading bar, the workpiece must be drilled or bored to the appropriate diameter that is designated for the thread size to be machined. For this reason, a standard "Tap Drill Size Chart" has been developed to specify the appropriate drill sizes for each tap size (see appendix H).

CARBIDE INSERT TECHNOLOGY

For CNC lathe applications, the best insert geometry to select should have strength and should be versatile. Inserts vary in strength depending on the geometric shape and size of the insert as illustrated in Figure 13-19. They are also limited to the type of operations that can be performed by the features and clearance angles of the insert. They are generally selected according to the lathe application to be performed. The 80° diamond insert best fits these criterion because it has strength and it is also versatile. Therefore, it is typically used for both rough and finish cutting, and it is also used for turning, facing, taper turning and radius contours.

Insert Identification and Grades

Indexable carbide inserts, which are manufactured to meet a variety of specific machining applications, have a standardized system of identification (see Figure 3-20). There is also a system for identifying the grades of carbide (C1–C8), which are organized according to their application. The standard ISO carbide grades are as follows:

- C-1: Roughing cuts (cast iron and nonferrous materials)

- C-2: General purpose (cast iron and nonferrous materials)

- C-3: Light finishing (cast iron and nonferrous materials)

- C-4: Precision boring (cast iron and nonferrous materials)

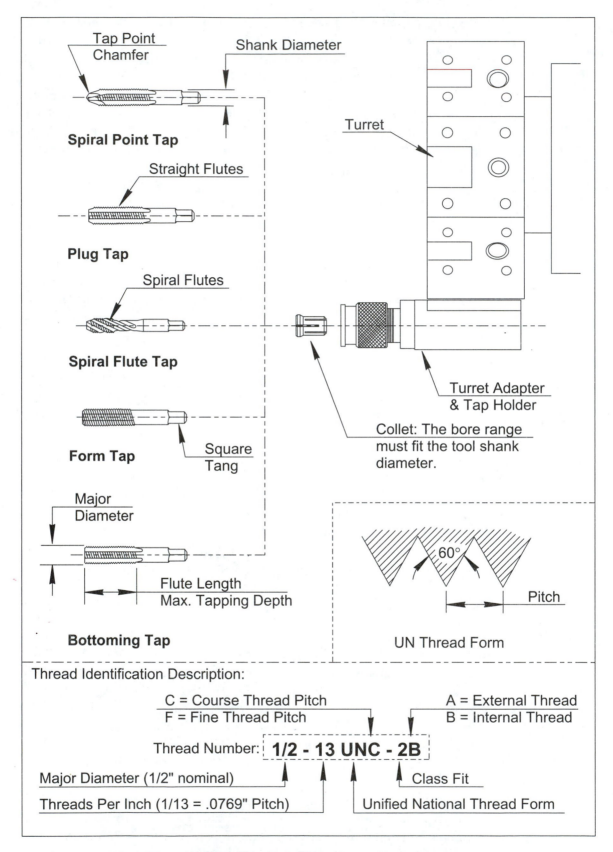

Spiral Point Tap

- Tap Point Chamfer
- Shank Diameter

Plug Tap

- Straight Flutes

Spiral Flute Tap

- Spiral Flutes

Form Tap

- Square Tang

Bottoming Tap

- Major Diameter
- Flute Length / Max. Tapping Depth

Turret

Turret Adapter & Tap Holder

Collet: The bore range must fit the tool shank diameter.

UN Thread Form — 60° — Pitch

Thread Identification Description:

C = Course Thread Pitch
F = Fine Thread Pitch

A = External Thread
B = Internal Thread

Thread Number: **1/2 - 13 UNC - 2B**

Major Diameter (1/2" nominal)

Threads Per Inch (1/13 = .0769" Pitch)

Class Fit

Unified National Thread Form

Figure 13-27 *Features, types, and identification of taps*

- C-5: Roughing cuts (steel)
- C-6: General purpose (steel)
- C-7: Finishing cuts (steel)
- C-8: Precision boring (steel)

There are various carbide insert manufacturers and they all have their own unique grade specification numbering system. They also have developed a variety of other grades that increase performance for specific machining situations. Therefore, a cross-reference chart or manufacturer's catalog should be used when seeking application information about a specific manufacturer's grade of carbide.

Insert Nose Radius Selection

Selecting the TNR of the insert is an important factor to consider because it affects the tool strength, the surface finish, and the fillet or radius that is formed on the workpiece. Generally, a TNR of 1/32 inch is selected unless it causes chatter between the tool and workpiece.

For CNC lathe programs, the TNR size selected is then used for calculating compensation amounts of any angles or radius contour cuts that are required on the workpiece. Therefore, in order to accurately machine the workpiece, the setup person must verify that the TNR size of the insert is the same as the size specified on the CNC tool list.

Insert Geometry Selection

Generally, the first consideration when selecting a geometry is the application. Then follows the type of tool, the workpiece material, the depth of cut, the finish requirement, the tolerances, the rigidity of part setup, and the machine. Therefore, the geometry that is selected must be capable of machining the workpiece and also overcome all these factors.

Other Insert Cutting Materials

Indexable inserts can be manufactured from a variety of other materials. Generally, inserts are uncoated, coated, ceramic, or diamond. The uncoated inserts are classified into the C1–C8 tungsten carbide group. The coated inserts consist of chemical vapor deposition (CVD) or physical vapor deposition (PVD). The cermet insert contains mostly TiC and TiN with a metallic binder. Ceramic inserts can be classified into two groups: alumina-base (aluminum oxide) and silicon nitride-base (sialon) compositions. Diamond inserts can also be classified into two groups: PCD and thin film diamond coatings.

Cutting Tool Pressure

Tool pressure, which is created when the cutting tool is introduced into the workpiece material, is an adverse characteristic typical of all metal cutting operations. Most machines and cutting tools are designed to be rigid enough to effectively overcome tool pressure and perform cutting operations without any adverse affects. However, certain operations, tools, and setups are more prone to create higher tool pressure, which is not exceptable for proper machining. Some situations that will probably cause higher tool pressure and adverse machining conditions are listed below:

- Machining a heavy depth of cut.

- Tools with large cutting edge surface contact.

- Negative rake cutting tools.

- Lack of rigidity in the setup.

- Lack of rigidity of the tool.

- Lack of rigidity in the structure of the workpiece.

Cutting Fluids (Coolant)

There are various types of cutting fluids (coolant), which include mineral oils, soluble oils, synthetic (chemical) fluids, and semisynthetic fluids. The most commonly used type of coolants on CNC lathes are the soluble, synthetic, and semisynthetic fluids. Coolant is usually applied to the insert or tool as it cuts metal in order to dissipate the heat and lubricate the cutting action. Additionally, when coolants are applied, pressure welding is reduced. This is especially important when machining softer metals such as aluminum, copper, bronze, and brass, which have a lower melting point and tend to build-up on the cutting edge (BUE) of the insert or tool.

The main cost objective when performing metal-removal operations is to increase productivity and decrease machining time. This is usually achieved by selecting the highest speed (RPM) that is both safe and practical while maintaining optimum "tool life" and quality of the parts. However, the result of selecting higher speeds is an increase in heat to the cutting tool and the workpiece, which can have an adverse affect on tool performance and part quality. Therefore, coolant is usually applied to most CNC lathe cutting operations, which tend to operate at higher speeds. It is important to consider that when coolant is not used, tool performance will substantially decrease. The advantages derived from using coolant are listed below:

- Cools and lubricates the tool and workpiece.

- Reduces BUE and tool wear.

- Will flush chips away from the cutting edge.

- Produces better surface finishes.

- Enables the tool to cut at optimum cutting speeds and depths of cut.

It is particularly important to apply coolant in situations that are more demanding on the tool due to higher cutting forces (tool pressure) or higher cutting friction. Some typical lathe cutting situations that are usually more demanding and should be performed with coolant include single-point threading, tapping, harder or tougher metals, softer metals, carbide drilling, spade drilling, drilling deeper than 1.0 inch in depth, and machining that removes large amounts of material.

There are also situations when it is not recommended to apply coolant for machining. Some common situations when coolant is not recommended include when machining cast iron parts and during setup or proveout when it is necessary to view the machine rapid or cutting moves to avoid collisions.

CNC LATHE APPLICATIONS

The CNC lathe has become a versatile, accurate, and productive machine tool, which is capable of producing complex parts accurately and efficiently. The CNC lathe can perform operations that previously required a number of various types of turning machines. It is important that the CNC setup person and CNC operator become familiar with the range of CNC lathe applications, which are numerous, in order to efficiently utilize CNC lathes. The typical range of applications includes turning, facing, drilling, reaming, boring, grooving, tapping, and cutoff (see Table 13-1). It is equally important to correctly match and select the cutting tools that are designed for each of these applications.

CHAPTER SUMMARY

The following key concepts that were presented in this chapter include:

- The names and descriptions of the main CNC lathe cutting terms and features.
- The names and descriptions of the basic CNC lathe turning operations.
- The CNC lathe cutting axis (X-Z) movements.
- The names and descriptions of carbide indexable insert toolholder components.
- The names and descriptions of carbide indexable insert toolholders.
- The names and descriptions of the CNC lathe cutting tool features.
- Carbide insert features and technology.
- Cutting fluid application and descriptions.
- CNC lathe applications and tooling selection.

■ *Table 13-1 CNC lathe applications*

MACHINING TYPE	OPERATION DESCRIPTION	TOOL TYPES USED
Turning		
1	Straight turning	Insert-indexable OD turn tools: 80°, 55°, square, triangle, and round
2	Step turning	Insert indexable OD turn tools: 80°, 55°, square, triangle, and round
3	Taper turning & chamfering	Insert indexable OD turn tools: 80°, 55°, square, triangle, and round
4	Contour turning (radius/angle)	Insert-indexable OD turn tools: 35°, 55°, triangle, and round
Facing		
1	End facing	Insert indexable OD turn tools: 80°, 55°, square, and triangle
2	Taper facing	Insert indexable OD turn tools: 80°, 55°, square, and triangle
3	OD shoulder facing	Insert indexable OD turn tools: 80°, 55°, square, and triangle
4	ID C'bore facing	Insert-indexable ID boring bars: 80°, 55°, 35°, and triangle
Drilling		
1	Center drilling	Center drill
2	Drilling	HSS twist drill, insert drill, spade drill, step drill, core drill, coolant-fed drill
Reaming		
1	Reaming	Reamer, taper reamer
Boring		
1	Straight boring	Insert indexable ID boring bars: 80°, 55°, square, and triangle
2	Taper boring	Insert indexable ID boring bars: 80°, 55°, square, and triangle
3	Contour boring	Insert indexable ID boring bars: 80°, 55°, square, and triangle
C'boring		
1	Counterboring	Insert indexable ID boring bars: 80°, 55°, square, and triangle
Grooving		
1	OD grooving	Insert-indexable OD grooving holder
2	ID grooving	Insert-indexable ID grooving bar
3	Face grooving	Insert-indexable face grooving holder
Threading		
1	OD threading	Insert-indexable threading holder
2	OD taper threading	Insert-indexable threading holder
3	ID threading	Insert-indexable threading bar
4	ID taper threading	Insert-indexable threading bar
Tapping		
1	Tapping	Plug tap, spiral flute tap, spiral point tap, form tap, and bottoming tap
2	Taper pipe tapping	Taper pipe tap
Cutoff		
1	Cutoff or Parting	Insert-indexable cutoff holder

REVIEW QUESTIONS

1. List the features related with turning on a CNC lathe.

2. List the considerations that determine the turn tool selection.

3. List the operations that are performed on a CNC lathe?

4. Why are clearance angles required on lathe toolholders?

5. List the top rake angles that are applied to lathe toolholders.

6. List the tool insert geometry shapes.

7. Which insert shapes are stronger and which are weaker?

8. What is the purpose of the chipbreaker?

9. List the common types of drills that are used on CNC lathes.

10. List the ID toolholders commonly used on CNC lathes.

11. List the OD toolholders commonly used on CNC lathes.

12. Which axis (X or Z) is in feed mode for facing?

13. Which axis (X or Z) is in feed mode for turning?

14. Which axis (X or Z) is in feed mode for boring?

15. Which axis (X or Z) is in feed mode for taper turning?

16. Why is G96 mode used for OD grooving operations?

17. Which spindle mode (G96 or G97) is used for threading operations?

chapter 14

CNC LATHE CONTROL AND OPERATION

Chapter Objectives

After studying and completing this chapter, the student should have knowledge of the following:

- **CNC lathe CRT and keypad panel functions**
- **CNC lathe operation panel functions**
- **CNC lathe workholding methods**
- **CNC lathe setup procedures**

This chapter identifies the CNC machine control features and operation functions. This includes descriptions and illustrations of the CRT and keypad panel, the operation control panel, CNC lathe workholding, and CNC lathe setup procedure.

CNC LATHE CRT AND KEYPAD PANEL FUNCTIONS

The CRT displays the CNC codes and operational information that are active or stored in its memory (see Figures 14-1 and 14-2). The CNC codes and information are executed by a CNC program, defaulted by the control memory, or input by the CNC operator. The latter—MDI—is performed with the keypad and Cycle Start button. Understanding the code functions, machine operations, and controls is an essential part of safely and efficiently operating a CNC machine tool.

The most common items on the CRT and keypad panel include the On and Off power for the control, the control Reset, which clears active CNC operations, and the menu keys, which are used to access various CRT screens, such as the offset screens, the CNC program directory, the axis position screens, and the alarm message screens.

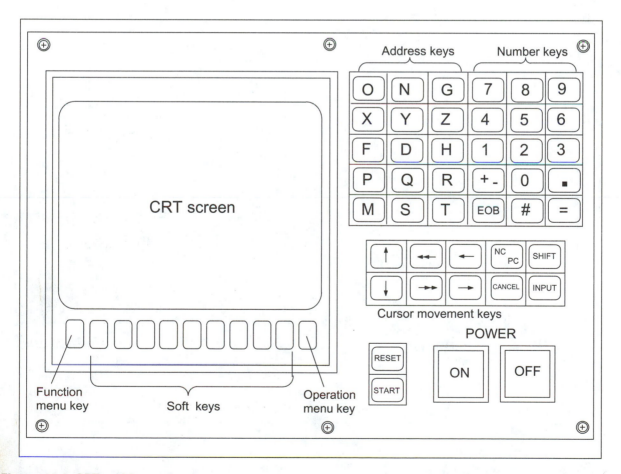

Figure 14-1 CRT and keypad

The keypad functions include entering or deleting data such as

- CNC programs
- Offset values
- Tool data
- MDI commands
- Edit commands

CRT and Keypad Descriptions

Figure 14-2 below illustrates some common CRT and keypad functions that the CNC operator must identify and understand to safely and effectively operate any CNC machine tool.

When the Function Menu key is pressed (Figure 14-2), a menu selection screen appears on the CRT. The menu items typically displayed on the CRT include.

- Position screen
- Setting screen
- Program screen
- Service screen
- Offset screen
- Message screen
- Program check

The menu list may also include other item depending on the machine and control. After selecting and pressing a menu item button, the CRT will display the selected screen. Additionally, each menu selection may have more

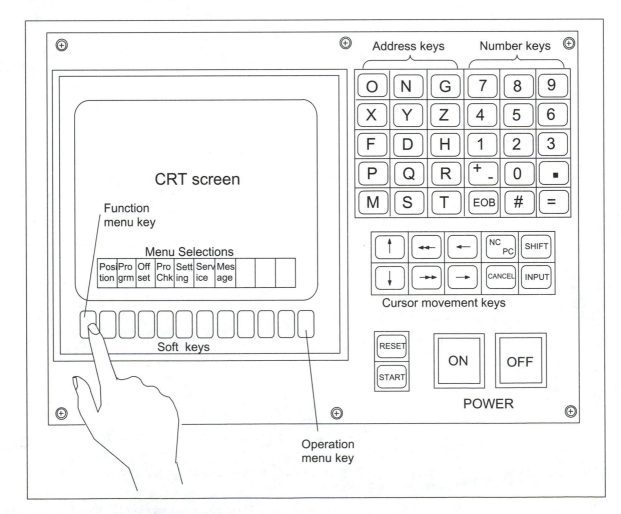

Figure 14-2 **CRT and keypad functions**

than one screen to display. The operation menu key enables the operator to select and view other screen options. The CRT and keypad are typically designed with the main feature described. However, to fully understand the CRT and keypad functions, the CNC operator should study the machine manual and receive training before operating the CNC machine.

CNC LATHE OPERATION PANEL FUNCTIONS

The CNC operator must be capable of performing a variety of functions in order to efficiently produce parts on CNC machines. Table 14-1 describes the functions that are mainly performed through the operation control panel,

Table 14-1 *Control function description*

CONTROL ITEM	FUNCTION DESCRIPTION
Alarm Lamps	When any of these lamps are turned on, the CNC control will not allow the machine to operate until the problem is corrected.
Load Meter	This meter indicates the amount of horsepower in percentage load that is being generated by the machining operation.
Spindle Speed Meter	This meter indicates the actual RPM of the spindle.
Coolant Switch	This switch operates the coolant flow (on or off).
Turret Index CW–CCW	This switch rotates the tool turret into position so that the operator can fasten and unfasten tools.
Machine Lock	This switch locks the machine axis during operation.
Tailstock Forward/Return	This pushbutton allows the operator to manually move the tailstock spindle/center into and out of the part center.
Home Return Switch	When this pushbutton is pressed, the selected axis (turret) will rapid traverse to the Home position limit switch.
Home Position Lamps X Z	When the selected axis (turret) trips the Home position limit switch and completes the move, the corresponding lamp will turn on.
Work-Grip Switch	This switch allows the operator to select either external or internal jaw gripping of the workpiece.
Spindle Jog Button	When this pushbutton is pressed, the spindle will rotate at the RPM value stored in memory until the button is released.
Spindle On Button	When this pushbutton is pressed, the spindle rotates at the RPM value stored in memory.
Spindle CW Switch	When this switch is selected, the spindle will rotate in a CW direction, also called forward.
Spindle CCW Switch	When this switch is selected, the spindle will rotate in a CCW direction; also called reverse.
Spindle Stop Button	When this pushbutton is pressed, the spindle will stop rotating.
Manual Jog + Button	When this pushbutton is pressed, the selected axis (turret) will move in a positive direction until the button is released.
Manual Jog − Button	When this pushbutton is pressed, the selected axis (turret) will move in a negative direction until the button is released.
Memory Protect	When this switch is in the ON position, the CNC program cannot be altered or deleted.

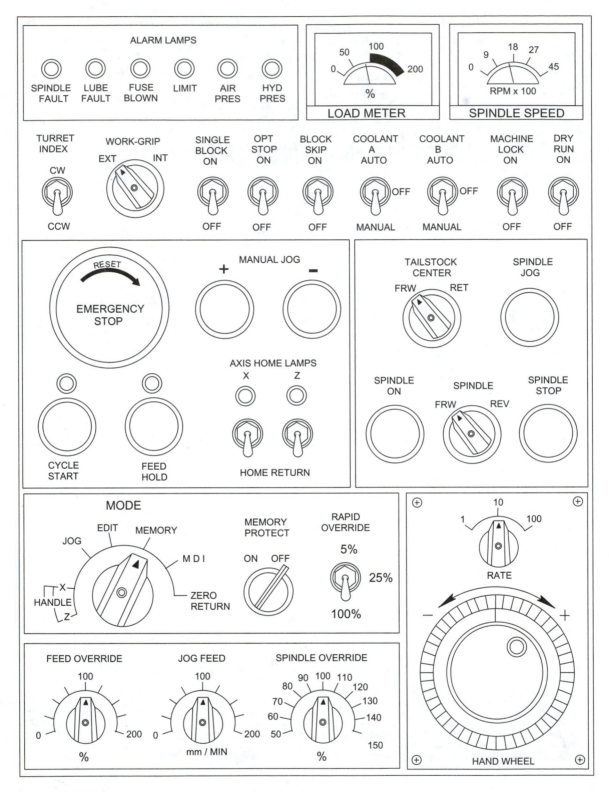

Figure 14-3 **CNC lathe operation control panel**

illustrated in Figure 14-3. Other control features are also illustrated in Figures 14-4 to 14-11.

CNC Lathe Operation Panel

The operation panel example shown in Figure 14-3 illustrates some common features that are required to operate a typical CNC lathe.

CNC Lathe Control Panel Buttons

Figures 14-4 and 14-5 are typical CNC lathe functions that are required knowledge for a CNC machine operator.

When the Emergency Stop button is pushed, all machine functions will stop operating. Usually, the button must be released by turning it clockwise. Therefore, when this button is pressed the following sequence will occur:

1. The current to the motor is interrupted.

2. The control unit assumes a reset state.

3. The fault causes must be corrected before the button is released.

4. After the button is released, each axis must be homed by the operator.

In the event of a machine emergency, all machine functions must be stopped immediately to prevent injury and damage. This is performed by immediately pressing the Emergency Stop button. The button is rotated clockwise to reset the machine.

RESET

EMERGENCY STOP

This push button is red in color.

Figure 14-4 **Emergency Stop button**

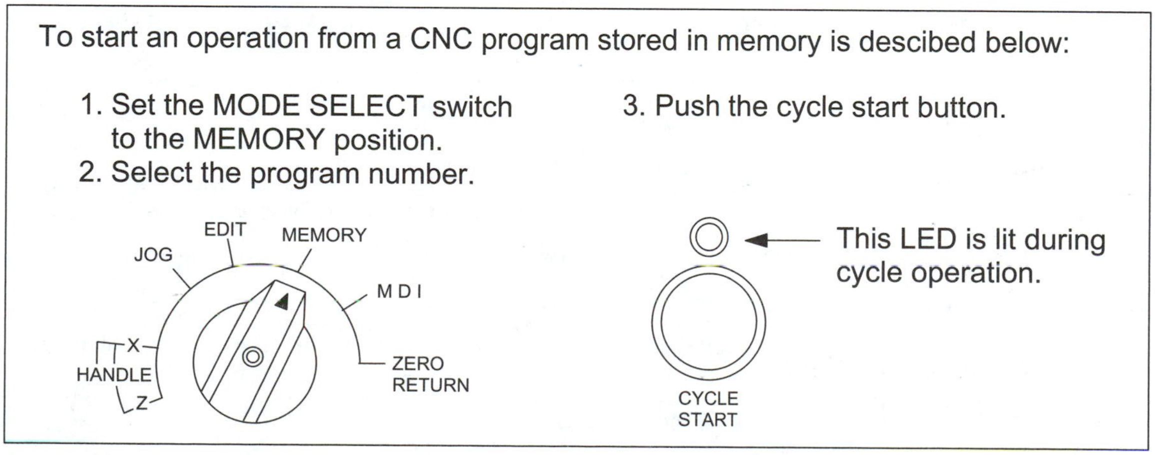

To start an operation from a CNC program stored in memory is descibed below:

1. Set the MODE SELECT switch to the MEMORY position.
2. Select the program number.

3. Push the cycle start button.

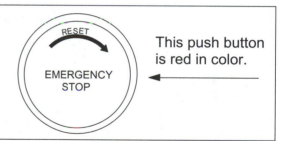

EDIT   MEMORY
JOG
                MDI
X
HANDLE
Z   ZERO
                RETURN

This LED is lit during cycle operation.

CYCLE START

Figure 14-5 **Cycle Start button**

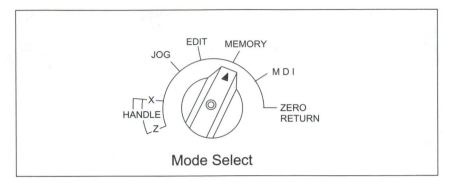

Figure 14-6 **Mode Select switch**

When the Cycle Start button is pushed, automatic operation starts and the Cycle Start LED is lit. The Cycle Start button is ignored in the following cases:

1. When the Feed Hold button is pushed.

2. When the Emergency Stop button is pushed.

3. When the Reset signal is turned ON.

4. When the Mode Select switch is set to a wrong position.

5. When a sequence number is being searched.

6. When an alarm has occurred.

Mode Selections

The CNC machine is capable of performing various functions, which the CNC operator is required to identify and understand. The Mode Select switch (Figure 14-6) is used to specify the type of machine or control function that the control and the machine will perform. Table 14-2 explains the actions typically performed with each mode selection.

■ *Table 14-2* **Mode selection descriptions**

MODE SELECTION	FUNCTIONS AND OPERATIONS THAT CAN BE PERFORMED
Edit	(a) Store CNC programs to memory.
	(b) Modify, add, and delete CNC programs.
	(c) Retrieve a CNC program from memory.
Memory	(a) A CNC program stored in memory can be executed in auto or single-block cycle.
	(b) Sequence N block number search.
	(c) CNC program number search.
MDI	Manual data input can be performed via the keypad. The operator enters CNC codes and coordinates, and then executes them using the Cycle Start button.
Handle	Handle feed for X-axis or Z-axis can be executed.
Jog	Job feed for X-axis or Z-axis can be executed.
Home	The X-Z machine axes can be reset to their home position.

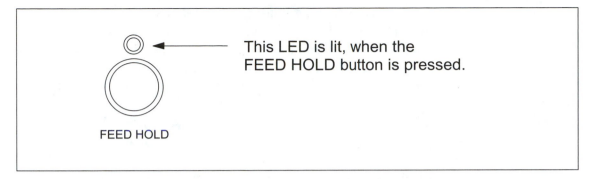

*Figure 14-7 **Feed Hold description***

Feed Hold, Dry Run, Single Block, and Optional Block Skip

The Feed Hold, Dry Run, Single Block, and Block Skip are some other typical operation functions that the CNC operator must identify and fully understand. These control operations described below are items usually performed by the CNC operator during the CNC proveout phase.

Feed Hold Button

During automatic operation or any cycle command, when this button is pressed the feed or rapid command moves are stopped. Note that spindle and other miscellaneous functions will remain on.

Dry Run

If this switch is set to ON in the cycle operation of Tape, Memory, or MDI, an "F" function specified in the CNC program is ignored, and the machine tool axes are moved at a faster rate of speed.

Single Block

When Single Block is on, the control will only execute one block of active CNC program information each time the Cycle Start button is pressed. This procedure is described below:

1. Turn on the Single Block switch.

2. Press the Cycle Start button, and a single block of information is executed.

3. When the Cycle Start button is pressed again, another block of the CNC program is read and executed. The cycle repeats until all blocks are executed.

Optional Block Skip

This is the function that allows the control to skip a block or blocks of information in which a slash is programmed before the "N," as the first character in the block. The block would appear as /N005 followed by the codes of coordinates.

Feedrate, Spindle Speed, an Rapid Traverse Overrides

The Feedrate Override, Spindle Speed Override, and Rapid Traverse Override dials enable the CNC operator to decrease or increase the programmed values. These operation control features are described in Figures 14-8 through 14-10.

Feedrate Override Dial

This dial allows the operator to reduce or increase the programmed F-code feedrate. The percent increments are usually 10% but vary depending on the type of control.

This function is active in the following cases:
1. Feed modes by G01, G02, and G03
2. Feed mode during canned cycle execution

Feed Override

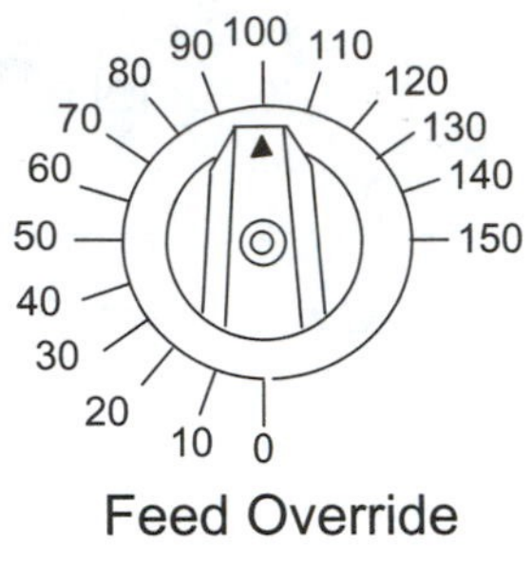

Figure 14-8 **Feedrate Override control knob**

Spindle Speed Override Dial

This dial allows the operator to reduce or increase the programmed S-code spindle speed. The percent increments are usually 10% but vary depending on the type of control.

This function is active in the following cases:
1. Spindle is ON by S-code and M03/M04
2. Spindle is ON by MDI

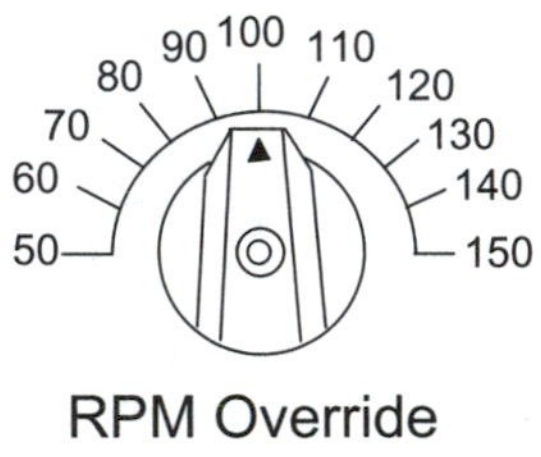

RPM Override

Figure 14-9 **Spindle speed override control knob**

Rapid Traverse Override Switch

The rapid override switch of 100%, 25% and 5% can be provided with the machine operator's panel. When the feedrate is 10 m/min and the switch is set to the position of 25%, the actual feedrate becomes 2.5 m/min.

This function is active in the following cases:
1. Rapid traverse by G00, G27, G28, G29, and G30
2. Rapid traverse during canned cycle execution
3. Manual rapid traverse operation
4. Rapid traverse Home return

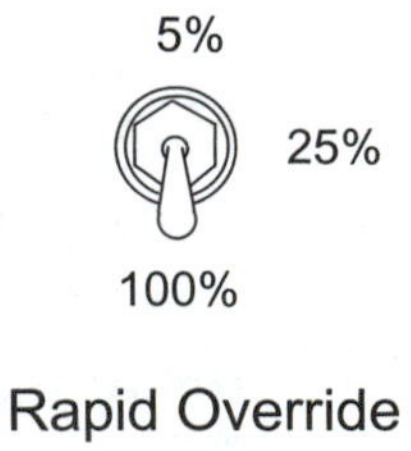

5%

25%

100%

Rapid Override

Figure 14-10 **Rapid traverse override control knob**

Manual Pulse Generator Handwheel

The CNC operator, during the process of setups and operating the CNC machine, will be required to manually move each machine axis and grip the workpiece. The "Manual Pulse Generator Hand Wheel" (see Figure 14-11) is the CNC machine control feature that enables the operator to move the machine in any axis (X or Z). While most workpieces are gripped externally, some are gripped internally. The Work-Grip Switch (see Figure 14-12) must be set accordingly.

1. Set MODE SELECT switch to the HANDLE position either the X-axis or the Z-axis.

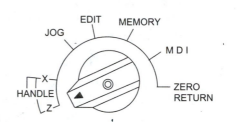

2. Set the movement amount switch. A typical example of graduations is described below:

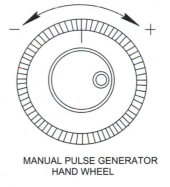

Input system	X 1	X 10	X100
Metric input	0.001 mm	0.01 mm	0.1 mm
Inch input	0.0001 inch	0.001 inch	0.01 inch

3. Rotate the manual pulse generator handle.
   Clockwise rotation.................+ direction
   Counterclockwise rotation......- direction
   (The direction varies with machine tool builder.)
   The hand wheel is either fixed or detachable for remote operation.

MANUAL PULSE GENERATOR
HAND WHEEL

Figure 14-11 **Hand wheel operation**

Work-Grip Switch

The work-grip switch enables the operator to select between external (OD) jaw gripping or internal (ID) jaw gripping.

Figure 14-12 **Work-grip switch**

CNC LATHE WORKHOLDING METHODS

Proper location and orientation of the workpiece, which ensures both accuracy and repeatability of the parts produced, are essential to the lathe machining process. In this case, accuracy is defined as machining a part's features to the specified size, form, and location. Repeatability is defined as accurately performing the machining operations for all the parts throughout the production run. Therefore, the workholding method that is selected must be capable of ensuring both the accuracy and repeatability of the CNC lathe production process.

Locating Fundamentals

To ensure accuracy and repeatability, the workpiece must be properly orientated within the workholding device (chuck). To accomplish this, the chuck must perform three primary functions, which include properly holding, supporting, and locating the workpiece throughout the CNC lathe machining cycle.

Properly holding the part, which is usually gripped with chuck jaws, requires that sufficient chucking force is applied to prevent any movement or slipping of the part during machining. Also, the chucking force should be applied with caution so that the workpiece is not distorted. Ideally, the workholding method should be designed so that the workpiece is located by stationary stops, held (gripped) in place, and oriented by the chuck jaws.

When holding and locating the workpiece on a CNC lathe, there are basically two axes (X and Z) that are considered for alignment (see Figure 14-13). The part is located radially (X-axis) so that it is concentric to the machine spindle/chuck axis. The part is also located lengthwise (Z-axis) usually to a fixed surface on the chuck (stops) or jaw steps (see Figure 14-13). Another common method of locating the part in the Z-axis is to position the turret with a stop, bump the part to it, then clamp the part (see Figure 14-14).

Methods of Holding the Part during Machining

The CNC programmer usually determines how the part will be held during machining operations. The workholding method and instructions are documented (Setup Plan) and provided to the CNC setup person. In most cases, the parts are held with a 3-jaw chuck, which is available in various designs styles. The chuck typically uses hard jaws and soft jaws to grip and hold the workpiece. In some cases where the part is complicated, special jaws or a fixture may need to be designed.

Workholding Devices

There are various types of workholding devices for CNC lathes. The most common workholding device used on turning machines is the chuck. There are various types of chucks, which include the self-centering, countercentrifugal, and collet (see Figure 14-15) that are used to fit the various workpiece configurations and lathe applications. When a CNC lathe is initially setup, it typically involves securely fastening the chuck to the spindle mounting face. The ideal chuck type and size selected should be capable of holding all if not most of the various parts to be machined. This minimizes or eliminates the setup time required to mount the chuck on the spindle face. The jaws are then mounted to the chuck. The machine operator should be sure that all workholding items are free from chips and burrs before mounting and using them to hold the part.

3-Jaw Chuck Holding Description

The CNC operator, during the process of setups and operating the CNC machine, is required to manually clamp and unclamp the workpiece. The illustration in Figure 14-13, describes the alignment required for lathe machining.

Bump Stop Procedure

The CNC operator, during the process of setups and operating the CNC machine, is required to manually clamp and unclamp the workpiece. The illustration in Figure 14-14 describes the Z-axis bump stop procedure.

Chuck Types for CNC Lathes

The CNC operator, during the process of setups and operating the CNC machine, is required to manually clamp and unclamp the workpiece. The illustration in Figure 14-15 describes the 3-jaw chuck and the collet chuck features.

Chucks

For most CNC lathe applications, the more commonly used chucks are the 3-jaw self-centering chuck and the collet chuck. These chucks are popular for CNC lathes because they are accurate, quick, and easy to operate. Most chucks on CNC lathes are also hydraulically activated by a foot pedal switch that slides the jaws in and out to automatically clamp and unclamp the part. Most of these chucks are interchangeable, which allows the use of two or more types of chucks on a single CNC lathe. Although this adds flexibility to the CNC lathe, it also requires some setup time to interchange the chucks. Therefore, the chucks are not usually changed unless it is necessary for the machining process.

Self-Centering Chuck

Self-centering chucks, which are typically available in 2-jaw, 3-jaw, 4-jaw, and 6-jaw styles, are designed to move all jaws simultaneously to center the part to the chuck/spindle axis. Self-centering chucks generally have higher gripping forces and are more accurate than other chuck types. These chucks are recommended for barstock or turned parts that are located and gripped on either an OD or an ID. Self-centering chucks are either operated manually with a chuck key or are powered by the CNC machine (see Figure 14-15).

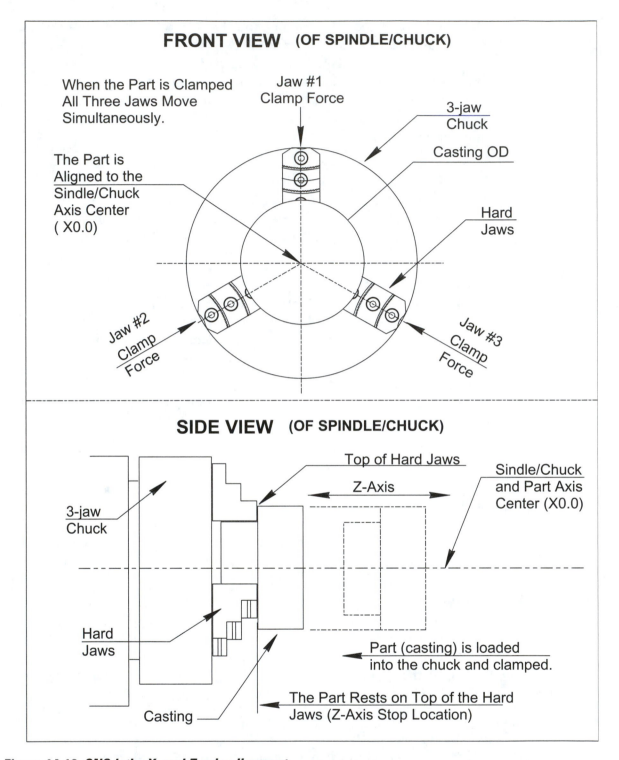

Figure 14-13 *CNC lathe X- and Z-axis alignment*

Collet Chucks

The collect is an ideal workholding device for small- to medium-sized parts where accuracy and quickness is needed. Collets are available for holding hexagonal, square, round, or custom shapes. Due to their efficiency and flexibility, they are typically used on CNC screw machines with bar feeding systems (see Figure 14-15).

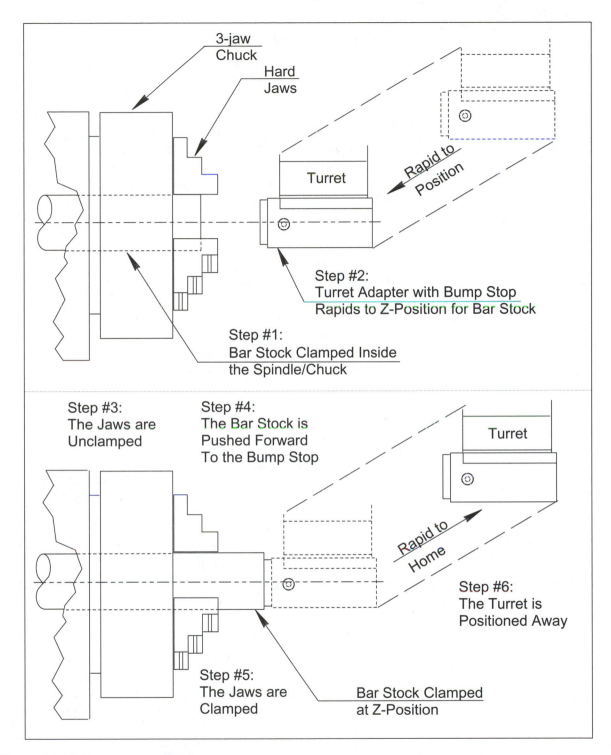

Figure 14-14 Bump stop operation

Independent 4-Jaw Chuck

The independent 4-jaw chucks, which are operated manually with a chuck key, are designed to move each jaw independent of the others to hold, locate, and center the part to the chuck/spindle axis. Independent 4-jaw chucks require more time to clamp and align the part than other chuck types. For

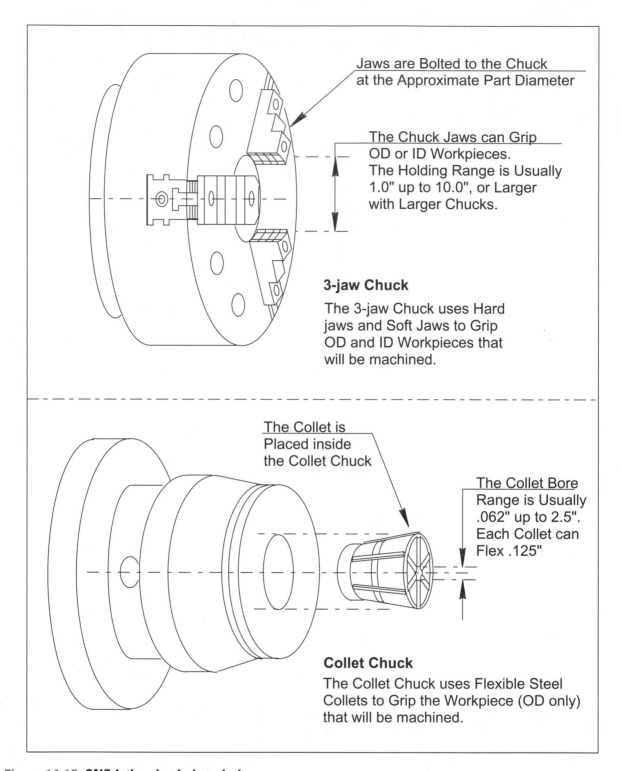

Jaws are Bolted to the Chuck at the Approximate Part Diameter

The Chuck Jaws can Grip OD or ID Workpieces. The Holding Range is Usually 1.0" up to 10.0", or Larger with Larger Chucks.

3-jaw Chuck

The 3-jaw Chuck uses Hard jaws and Soft Jaws to Grip OD and ID Workpieces that will be machined.

The Collet is Placed inside the Collet Chuck

The Collet Bore Range is Usually .062" up to 2.5". Each Collet can Flex .125"

Collet Chuck

The Collet Chuck uses Flexible Steel Collets to Grip the Workpiece (OD only) that will be machined.

Figure 14-15 **CNC lathe chuck descriptions**

this reason, they are rarely used on CNC lathes where cycle time is critical to the level of productivity achieved. These chucks are recommended for parts located and gripped on surfaces that are not concentric or evenly proportioned (see Figure 14-16).

Chuck Jaws

There are various types of chuck jaws available to accommodate the various lathe applications. The two types of standard chuck jaws that are commonly used for CNC lathes include hard jaws and soft jaws (see Figure 14-17). Hard jaws are used when maximum holding power is needed on unfinished surfaces such as castings or rough barstock. Soft jaws are used for minimizing runout between two or more part diameters and on finished surfaces where jaw marks are not acceptable. Soft jaws are typically made from mild steel and are bored to fit the exact outside diameter of the part to be held in the chuck. Soft jaws are usually machined (bored-out) with standard carbide boring bars. Hard jaws are typically used for the first operation of a part, which allows the jaw marks to be machined away on the second operation. Soft jaws are typically used for the second operation of the part. They are used to hold on the turned diameter from the first operation. They are bored-out on the chuck to fit the part diameter and to control runout between the first and second side diameters.

4-Jaw Chuck Holding Description

The CNC operator, during the process of setups and operating the CNC machine, will be required to manually clamp and unclamp the workpiece. The illustration in Figure 14-16 describes the 4-jaw alignment required for lathe machining.

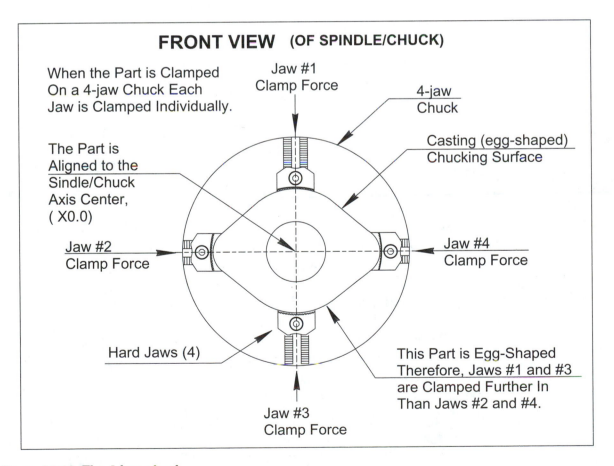

Figure 14-16 *The 4-jaw chuck*

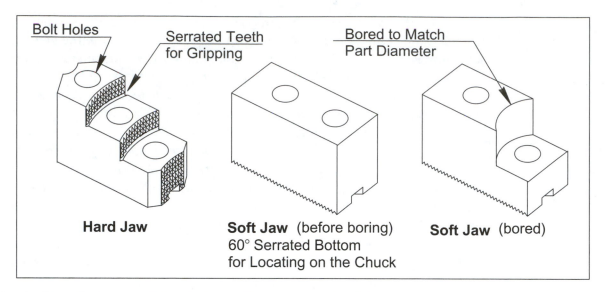

Figure 14-17 *Features and descriptions of hard jaws and soft jaws*

Chuck Jaw Pressures

The CNC operator and CNC setup person are required to adjust the chuck jaw clamping pressure depending on the machining application to be performed. The chuck jaw clamping pressure should be set with caution so as not exceed the specified maximum pressure on the warning plate. If greater pressure is used, high stress forces can be created within the chuck or the machine, which may result in personal injury. Power chucks are usually operated between 200 and 500 psi. Additionally, operating the chuck below 200 psi may not provide enough clamping force on the part to hold it securely. Therefore, the chuck clamping pressure should be set at the specified maximum PSI unless that amount of pressure can distort or damage the part.

Centrifugal Force

The amount of centrifugal force that can be sustained by the lathe chuck imposes speed limits upon all types of chucks. Centrifugal force, which increases as the speed of spindle rotation (RPM) increases, exerts an outward force on the chuck jaws. This tends to reduce the amount of jaw clamping force on the part. Every chuck is affected to some degree by the high internal stresses caused by centrifugal force. Consequently, all chuck manufacturers specify a maximum rotation speed (RPM), which should never be exceeded.

Special Workholding Methods

In situations where standard chuck jaws are not adequate for holding the part, special jaws or a fixture can be designed to fit the part requirements. Special workholding designs can range from simple to complex. Complex workholding designs, which should only be utilized as a last option, are usually costly and time-consuming to build. Additionally, the cost to design and build any workholding device must be justified in advance to ensure that it is cost effective.

Whichever workholding method is selected, the CNC setup person and CNC operator must use proper care to ensure that the workholding device is safe and adequate. The following is a list of items that should be observed and practiced:

1. The proper chuck type is selected.

2. The proper jaw type is selected.

3. The chuck must be mounted to the spindle face securely.

4. The jaws must be mounted on the chuck securely.

5. The chuck maximum speed should be identified and not exceeded.

6. The proper jaw clamping pressure should be determined and set.

7. The jaws should be evenly spaced on the chuck.

8. The jaws should clamp the part in the middle of the clamping stroke not at the end of the stroke.

CNC LATHE SETUP PROCEDURE

Machining typically requires some method of holding the workpiece and some method of cutting the workpiece. These workholding and cutting methods are dependent on various factors, which include the type of machine to use, the type of cutting operation to be performed, and the type of workpiece material. For CNC machines, these workholding and cutting methods are crucial to the accuracy and efficiency of the parts produced. Therefore, the CNC setup person should be as knowledgeable as possible about CNC setup procedures.

In order to ensure that each setup is correctly and properly completed, each part scheduled for CNC machining includes setup instructions. The setup instructions typically list the cutting tools, tool components, and all workholding information. The CNC lathe setup person reviews the setup instructions and then proceeds with the setup. Additionally, the CNC setup person must be proficient at reading and understanding the engineering drawing, the tool drawing, the tool library, and the CNC program manuscript. To fully understand the setup process, it is important to first understand how the CNC lathe accurately positions from one coordinate point to another. For this reason, it is important to differentiate between the "Machine Home" position and the "Part Origin" position.

Machine Home Position

When a CNC machine is initially turned on, it does not register an accurate position for each axis. The CNC machine needs to be homed, which moves each axis to a known position where each axis is then initialized. The

Machine Home position or machine zero is a location set initially by the machine manufacturer. The CNC machine is homed at the start of the part program setup. When homed, the machine fully retracts the turret (machine X0 and Z0) and positions to its maximum turret preset position. The machine can be homed manually by the Zero Return on the control Mode Select knob, or automatically by the G28 code in MDI mode.

Part Origin

As previously discussed, the part origin (part X0 and Z0) is the zero location of the part X- and Z-axis Cartesian coordinate system. In absolute coordinate programming, all tool movements of the CNC program are taken with respect to this part origin zero and not the machine zero. The point from which the dimensions are taken on the part print is usually considered the same as the part origin. Therefore, the setup must include finding and setting the part X- and Z-axis zero. The X-zero is always the spindle/part center axis.

Setup

Typically, the setup person starts by assembling the cutting tools in the toolholders. Then the tooling assemblies are loaded and secured into the tool turret stations according to the order outlined in the tool list (see Figure 14-18). Next, the workholding (chuck, chuck jaws, collet) items are fastened in place on the machine spindle (see Figure 14-19). The type of chuck (3-jaw, collet, or 4-jaw) usually is not changed. However, the collet size usually is changed to fit the size of barstock required. Also, the type of jaws (hard jaws or soft jaws) usually are changed to accommodate the type of operation required (see Figures 14-20 and 14-21).

Tool Assembly and Mounting Description

The setup procedure illustrated in Figure 14-18 describes the typical toolholder assembly and mounting on the CNC lathe turret. Each tool must be assembled and mounted according to the setup plan in the same manner.

Mounting the 3-Jaw Chuck and Jaws

The illustration in Figure 14-19 describes the setup procedure for mounting the 3-jaw chuck and jaws. The chuck is a separate component that is attached to the spindle face. The jaws are then mounted to the chuck as specified by the setup plan.

Hard Jaw and Soft Jaw Operations Description

The setup procedures illustrated in Figures 14-20 and 14-21 describe the typical hard jaw setup and the soft jaw setup. The setup plan specifies the type of jaw setup.

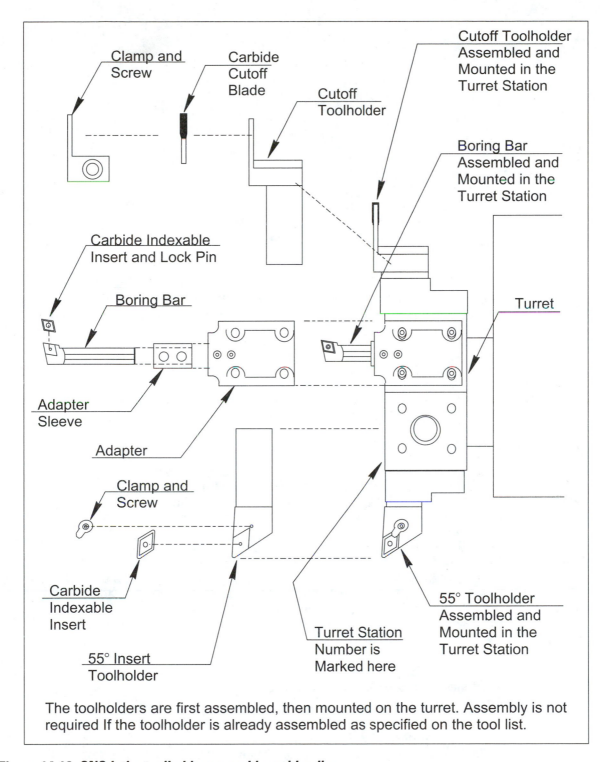

The toolholders are first assembled, then mounted on the turret. Assembly is not required If the toolholder is already assembled as specified on the tool list.

Figure 14-18 *CNC lathe toolholder assembly and loading*

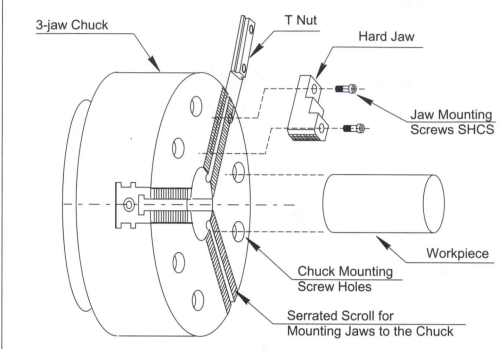

CAUTION:
Never rotate the spindle until all jaws and components are securely fastened. Always remove wrench tools after loosening and tightening.

3-jaw Chuck Setup Procedure

1. Load and clamp the chuck on the spindle—most chucks are heavy and should be lifted with a hoist. The chuck can remain fastened to the spindle if it is adequate for all parts that will be run on the lathe.
2. Select the hard jaws or soft jaws and align them radially to fit the part diameter. Each jaw must be in the same radial position to hold and rotate the part evenly (concentric) to the spindle axis.
3. Fasten each jaw securely with mounting screws.
4. Load the part with the jaws in the open position, then clamp the part.
5. Check the clamping stroke of the jaws (clamp and unclamp); the part should be clamped approximately in the middle of the stroke.
6. Visually check that the part rotates in alignment with the spindle. The part should not wobble axially or radially.

Figure 14-19 *The 3-jaw chuck setup. SHCS: socket head cap screw*

Tool Geometry Offsets

The setup person must next measure and enter the X- and Z-axis values of the tool tip length and diameter offsets for each tool. The various tool tip geometry values used for the machining operations on a CNC lathe will usually vary. Therefore, the machine control must be setup to compensate for these variations when moving the tools to the X- and Z-axis coordinates. It can only do so if it knows the initial distance between the tip of the tool and the part X0 and Z0.

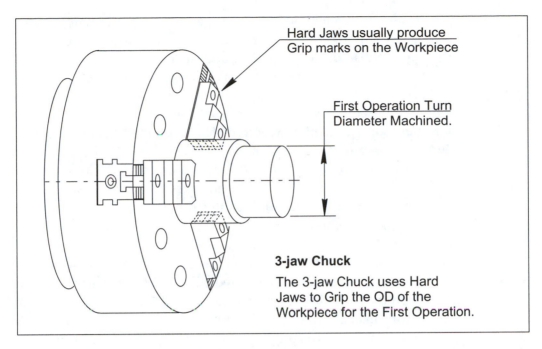

Hard Jaws usually produce Grip marks on the Workpiece

First Operation Turn Diameter Machined.

3-jaw Chuck

The 3-jaw Chuck uses Hard Jaws to Grip the OD of the Workpiece for the First Operation.

Figure 14-20 **The 3-jaw chuck setup with hard jaws**

Setting X- and Z-Axes Offsets

There are various methods of setting the X- and Z-axes offsets for the cutting tools on the CNC lathe. For instance, the tools can be set to specified lengths from the turret, they can be preset with modular tooling, they can be set automatically with a tool sensing system, or they can be manually set by the touch-off method.

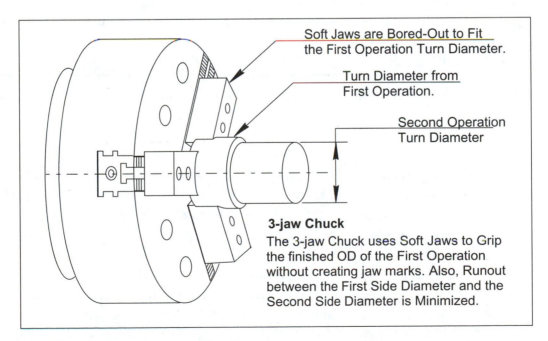

Soft Jaws are Bored-Out to Fit the First Operation Turn Diameter.

Turn Diameter from First Operation.

Second Operation Turn Diameter

3-jaw Chuck

The 3-jaw Chuck uses Soft Jaws to Grip the finished OD of the First Operation without creating jaw marks. Also, Runout between the First Side Diameter and the Second Side Diameter is Minimized.

Figure 14-21 **The 3-jaw chuck setup with soft jaws**

The manual touch-off method requires the setup person to index, position, and then physically touch-off each tool tip. The CNC setup person sets the jog switch to the lowest setting and then moves the tool in the X-axis until it touches a measured diameter of the part and the Z-axis to the Z-zero surface. The offset values are found by viewing the "Position Screen" on the CRT. The offset values are then recorded and stored in the control memory for each tool. This method requires a fair amount of machine operating skill and nonproductive time to complete.

Therefore, some modern CNC lathes are outfitted with an automatic tool setter, which allows the CNC setup person to quickly touch-off each tool tip (see Figure 14-22). Additionally, the tool setter method can be a more accurate method of finding the tool tip geometry values. This method requires that the CNC setup person index and position each tool, and then perform the touch-off with the machine control. The distance between the tool tip and the part zero is measured and stored automatically by the control. Each tool must be similarly indexed, positioned, touched-off, and then stored for each axis (X and Z). The setup person must also confirm that the offsets are correctly stored by the specified offset number in the control offset memory.

Manual Setting Z-Zero

One method of setting the part Z-zero origin is accomplished by placing a specified tool at the distance specified on the setup plan and then pressing the Origin button on the control (see Figure 14-23). When manually setting the part origin, the setup person positions the turret with the tool in the Z-axis directly at the specified part zero coordinate and zeros out the machine coordinate system on the MCU console. The setup person finds and enters the location of the part Z-zero with respect to Machine home as indicated on the setup sheet. The position values are found by viewing the Position Screen on the CRT. There are other techniques that can also be used to set the Z-zero. Whichever Z-zero setting method is selected, the CNC setup person and CNC operator must confirm that the Z-zero is properly and accurately setup.

Part X-Zero

Due to the nature of the lathe cutting process, which positions the cutting tool perpendicular to the spindle axis center, the spindle axis center is always the X-zero of the part. The cutting tool is positioned out from the center to the radius of the desired diameter to be turned. Therefore, the X-axis does not require any setup by the CNC setup person.

CNC Lathe Tool Setter Description

The CNC lathe tool setter (Figure 14-22) is ususally stored away from the machining area to avoid interferences or collisions between the workpiece and the cutting tools. When it is needed, the operator simply swings it into position as illustrated. Each tool tip must be set in the same manner.

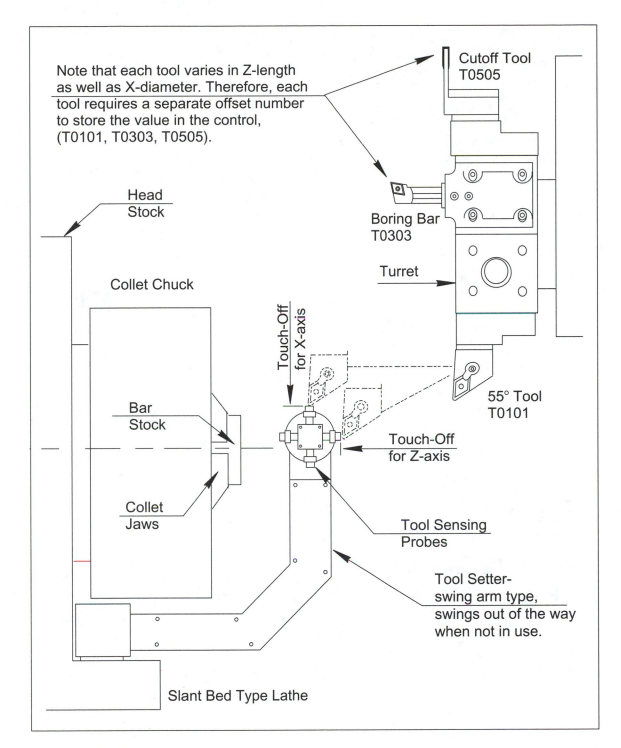

Note that each tool varies in Z-length as well as X-diameter. Therefore, each tool requires a separate offset number to store the value in the control, (T0101, T0303, T0505).

Cutoff Tool
T0505

Boring Bar
T0303

Head
Stock

Collet Chuck

Turret

Touch-Off
for X-axis

Bar
Stock

55° Tool
T0101

Touch-Off
for Z-axis

Collet
Jaws

Tool Sensing
Probes

Tool Setter-
swing arm type,
swings out of the way
when not in use.

Slant Bed Type Lathe

Figure 14-22 **CNC lathe tool setter**

Setting Z-Zero Description

The illustration (Figure 14-23) describes a setup technique for setting the Z-zero on a CNC lathe. The operator basically measures the distance from the top of the jaw, or whichever surface is specified, to the tip of the tool as specified by the setup instructions. The CNC setup person should

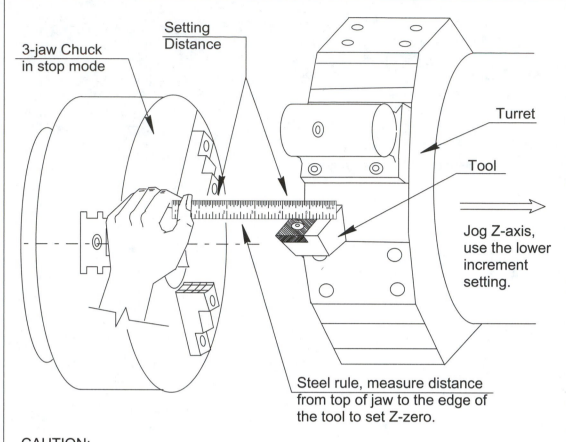

3-jaw Chuck in stop mode

Setting Distance

Turret

Tool

Jog Z-axis, use the lower increment setting.

Steel rule, measure distance from top of jaw to the edge of the tool to set Z-zero.

CAUTION:
Never place hands or fingers between the tool or turret and any part of the machine while moving any axis or rotating the spindle.

Z-zero Touch-off Procedure:

1. Spindle rotation must be stopped.
2. Index turret to specified touch-off tool.
3. Move the tool close to the top of the jaw, for example, if the Z-zero distance is 4.000" move the tool edge to approximately 3.0" to 3.5" from the top of jaw.
4. Set Z-axis increment jog to low setting.
5. Place the steel rule between the top of the jaw and the tool edge as illustrated.
6. Move the Z-axis away from the jaw until the specified distance is reached.
7. Preset the Z-axis to 00.0000 on the CRT screen.

Figure 14-23 **Setup technique for Z-zero**

always exercise caution to avoid injury or collisions when moving the turret.

Downloading the CNC Program into Memory

The next setup step is to download the CNC program into the machine's control memory and begin the CNC proveout. As previously explained, the CNC program can be entered into the machine control memory by a number of methods. The common modern methods include the "PC downloading" method and the "floppy disk" downloading method. The PC downloading method requires that a PC station be hardwired to the CNC machine control so that the CNC program can be transferred into the MCU memory. The floppy disk method requires that the machine control be outfitted with a floppy disk drive to accept the CNC program file from a floppy disk. The CNC setup person also confirms that the CNC program number in the control memory is the same as the program number specified on the CNC setup instructions.

After the CNC program is downloaded, the setup is complete and the CNC program proveout phase can begin. Most CNC machines are equipped with suffient memory to store a number of CNC programs. Therefore, each CNC program is usually kept in the machine control memory and recalled whenever the part is run again.

Setup Notes

The following is a general list of CNC lathe setup items that should be observed and put into practice:

- Always verify that the tools, jaws, and workpiece clear the swing of the machine.

- Check the workpiece, the tools, and the jaws for interferences.

- Always clamp the workpiece securely to avoid slipping or movement.

- Verify that the chuck is mounted to the spindle face securely.

- Verify that the jaws are mounted on the chuck securely.

- Check that the workpiece is not distorted from clamping.

- For holding rough surfaces, use hard jaws to prevent the workpiece from slipping during the cut.

- For holding finished surfaces, use soft jaws to avoid clamping marks.

- Whenever possible, test for suffient jaw clamping force before machining.

- When fastening the jaws, select a bolt length that will engage the entire thread length of the nut.

- Never torque a nut to the end of the bolt threads.

- Clean all chuck surfaces before mounting the jaws.

- Clean all chuck and jaw surfaces before mounting the workpiece.

- Verify that the Z-zero is correct.

- Verify that the tool geometry offsets are correct.

- Verify that the jaws are evenly spaced on the chuck.

- Verify that the jaw clamps the workpiece in the middle of the clamping stroke.

The list of setup notes covers the basic items that are typically related to setup of a CNC lathe. There are various other types of setups that may have other requirements not listed. Likewise, this chapter only covered the basics that are typically related to CNC lathe setup. The setup person should exercise similar practices with any other setup situations.

CHAPTER SUMMARY

The key concepts presented in this chapter include the following:

- The names and descriptions of the CNC lathe CRT and keypad functions.
- The names and descriptions of the CNC lathe operation panel functions.
- The names and descriptions of the CNC lathe workholding methods.
- The names and descriptions of lathe chucks used on CNC lathes.
- The names and descriptions of chuck jaws used on CNC lathes.
- The names and descriptions of CNC lathe setup requirements.
- The descriptions of CNC lathe tool geometry offset process.
- The descriptions of CNC lathe Z-axis locating methods.
- The descriptions of CNC lathe setup practices.

REVIEW QUESTIONS

1. List the items typically displayed on the CNC lathe CRT.
2. What happens when an alarm is activated?
3. How is the work gripping changed from external to internal on a CNC lathe?
4. Which control panel feature indicates that the machine is homed?
5. Which control panel feature moves the CNC lathe in the X- or Z-axes?
6. When is it necessary to press the Emergency Stop button?
7. How can the CNC operator slow or stop the programmed feed rate?
8. How can the CNC operator slow the programmed spindle speed?
9. How can the CNC operator slow the programmed rapid moves?
10. List the common workholding devices used on a CNC lathe.
11. When are hard jaws and soft jaws used?

12. Explain why chucking pressure is adjusted?

13. How does centrifugal force affect the chuck and jaw clamping?

14. How are tool geometry offsets setup on a CNC lathe?

15. Explain the advantages of using a tool setter on a CNC lathe.

16. How is the Z-zero setup on a CNC lathe?

17. List the basic setup items that should be regularly practiced.

chapter **15**

CNC LATHE TECHNICAL DATA

Chapter Objectives

After studying and completing this chapter, the student should have knowledge of the following:

- **CNC lathe machining calculations**

- **CNC turning feeds and speeds**

- **Geometric dimensioning and tolerancing applications**

- **Lathe surface finish**

- **Screw threads**

- **CNC lathe alarm codes**

CNC LATHE MACHINING CALCULATIONS

This chapter covers CNC technical calculations and other technical data related to CNC lathes. This includes explanations and illustrations that deal with CNC lathe machining calculations, GD&T, surface finish, material types, hardness rating, and basic CNC lathe alarm messages.

Calculation Example 1

The CNC lathe program includes the coordinates that position the turret and cutting tools. The X- and Z-coordinates are mainly derived from the dimensions on the part drawing. However, in some cases where the coordinates are not dimensioned on the part drawing, it becomes necessary to perform calculations. Figures 15-1 through 15-8 illustrate some typical calculations for machining CNC lathe parts.

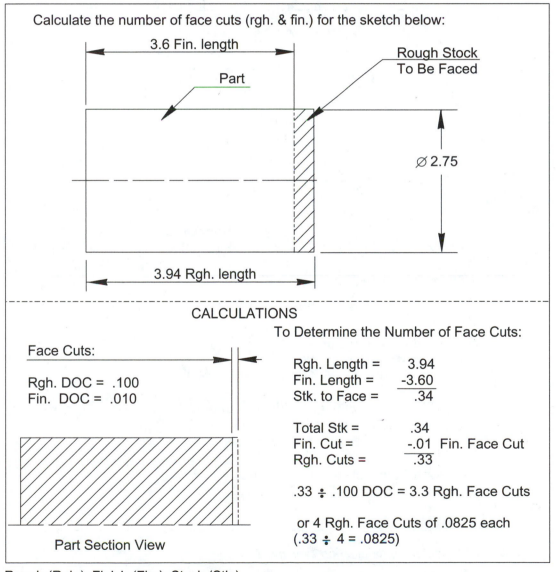

Calculate the number of face cuts (rgh. & fin.) for the sketch below:

3.6 Fin. length

Rough Stock
To Be Faced

Part

⌀ 2.75

3.94 Rgh. length

CALCULATIONS

To Determine the Number of Face Cuts:

Face Cuts:

Rgh. DOC = .100
Fin. DOC = .010

Rgh. Length =	3.94
Fin. Length =	-3.60
Stk. to Face =	.34

Total Stk =	.34	
Fin. Cut =	-.01	Fin. Face Cut
Rgh. Cuts =	.33	

.33 ÷ .100 DOC = 3.3 Rgh. Face Cuts

or 4 Rgh. Face Cuts of .0825 each
(.33 ÷ 4 = .0825)

Part Section View

Rough (Rgh.); Finish (Fin.); Stock (Stk.)

Figure 15-1 **CNC lathe machining calculation example 1**

Calculation Example 2

In Figure 15-2, some calculations are required to determine the number of turn cuts to machine the part. The finish DOC, the rough DOC, and part dimensions on the drawing are factors considered when calculating turn cuts.

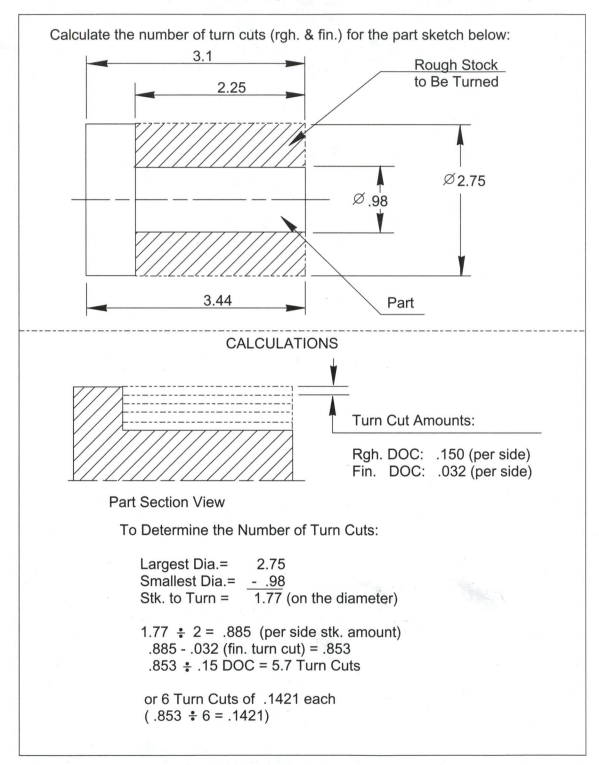

Calculate the number of turn cuts (rgh. & fin.) for the part sketch below:

Rough Stock to Be Turned

Part

CALCULATIONS

Part Section View

Turn Cut Amounts:

Rgh. DOC: .150 (per side)
Fin. DOC: .032 (per side)

To Determine the Number of Turn Cuts:

Largest Dia.=   2.75
Smallest Dia.=   - .98
Stk. to Turn =   1.77 (on the diameter)

1.77 ÷ 2 = .885 (per side stk. amount)
 .885 - .032 (fin. turn cut) = .853
 .853 ÷ .15 DOC = 5.7 Turn Cuts

or 6 Turn Cuts of .1421 each
(.853 ÷ 6 = .1421)

Figure 15-2 CNC lathe machining calculation example 2

Calculation Example 3

A typical CNC lathe operation may include turning a taper diameter as illustrated in Figure 15-3. In this situation, some calculations are required to determine the tangent point in X-Z for the start and the end of the cut. This calculation, which is detailed in Figure 15-4, is the TNR compensation. This calculation is required whenever the insert tool is programmed to cut in the X-Z axes simultaneously.

Before calculating the TNR compensation, it is important to understand how the tool tip is positioned at the start of the cut and also at the end of

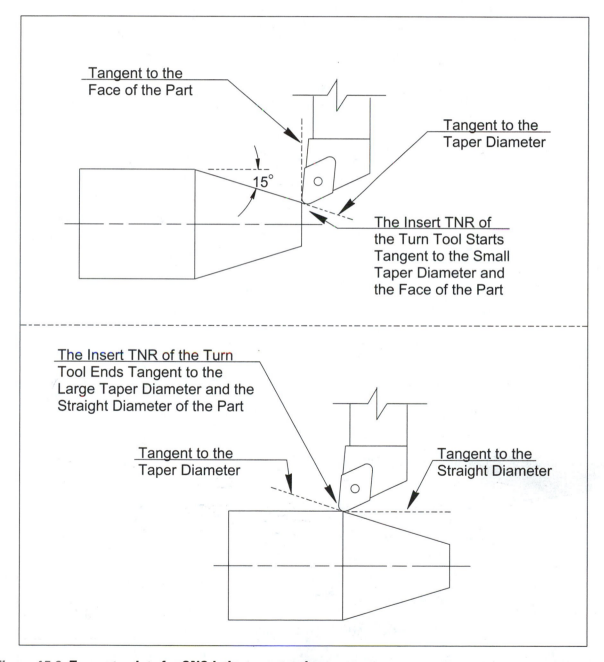

Figure 15-3 Tangent points for CNC lathe taper turning

the cut. Therefore, Figure 15-3 illustrates the start tangent point and then the end tangent point of the taper cut.

The illustrations, formulas and calculations in Figure 15-4 are required to determine the TNR compensation for a taper turn. The X-axis compensation is the Xc value and the Z-axis compensation is the Zc value. The calculations below are typically determined by CAD/CAM or by the G41/G42 TNR compensation codes.

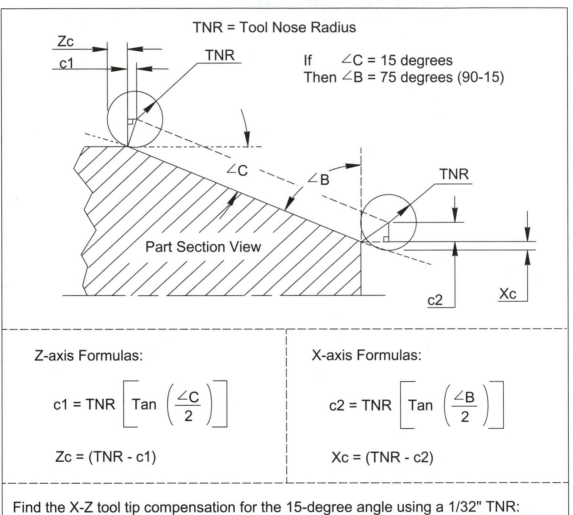

TNR = Tool Nose Radius

If $\angle C$ = 15 degrees
Then $\angle B$ = 75 degrees (90-15)

Part Section View

Z-axis Formulas:

$$c_1 = TNR \left[Tan \left(\frac{\angle C}{2} \right) \right]$$

$$Zc = (TNR - c_1)$$

X-axis Formulas:

$$c_2 = TNR \left[Tan \left(\frac{\angle B}{2} \right) \right]$$

$$Xc = (TNR - c_2)$$

Find the X-Z tool tip compensation for the 15-degree angle using a 1/32" TNR:

① c_1 = (.03125 x .131652498) = .00411

② Zc = (.03125 - c_1) = .0271 (Z-axis compensation)

① c_2 = (.03125 x .767326988) = .02398

② Xc = (.03125 - c_2) = .0073 (X-axis compensation)

*Figure 15-4 **CNC lathe machining calculation example 3***

Calculation Example 4

The illustrations, formulas, and calculations in Figure 15-5 are required to determine the TNR compensation for a radius turn of an arc that is less than 90° radially. In this case, the TNR starts tangent in the X-axis as illustrated below and therefore only the Z-axis requires compensation.

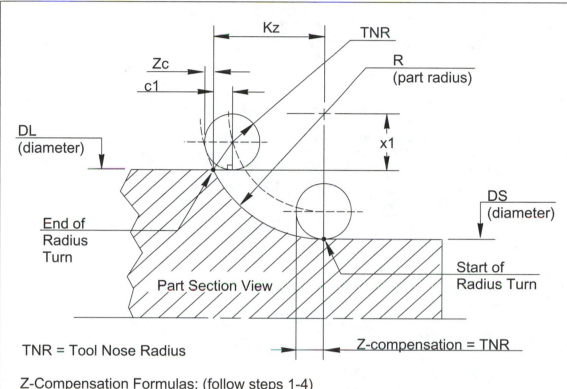

TNR = Tool Nose Radius

Z-Compensation Formulas: (follow steps 1-4)

① $x1 = R - \left[\dfrac{(DL - DS)}{2}\right]$ ③ $c1 = Kz - \sqrt{(R - TNR)^2 - (x1 - TNR)^2}$

② $Kz = \sqrt{R^2 - x1^2}$ ④ $Zc = (TNR - c1)$

Find the Z-compensation for: R = 1.500, DS = 1.125, DL = 2.500, TNR = .03125

① $x1 = (1.500 - .6875)$, which = .8125

② $Kz = \sqrt{1.5^2 - .8125^2}$, which = 1.26089

③ $c1 = 1.26089 - \sqrt{1.46875^2 - .78125^2}$, which = .01716

④ $Zc = (.03125 - .01716)$, which = .01409

*Figure 15-5 **CNC lathe machining calculation example 4***

Calculation Example 5

The illustrations, formulas, and calculations in Figure 15-6 are required to determine the TNR compensation for a radius turn of an arc that is less than 90° radially. In this case, the TNR starts tangent in the Z-axis as illustrated below and therefore only the X-axis requires compensation.

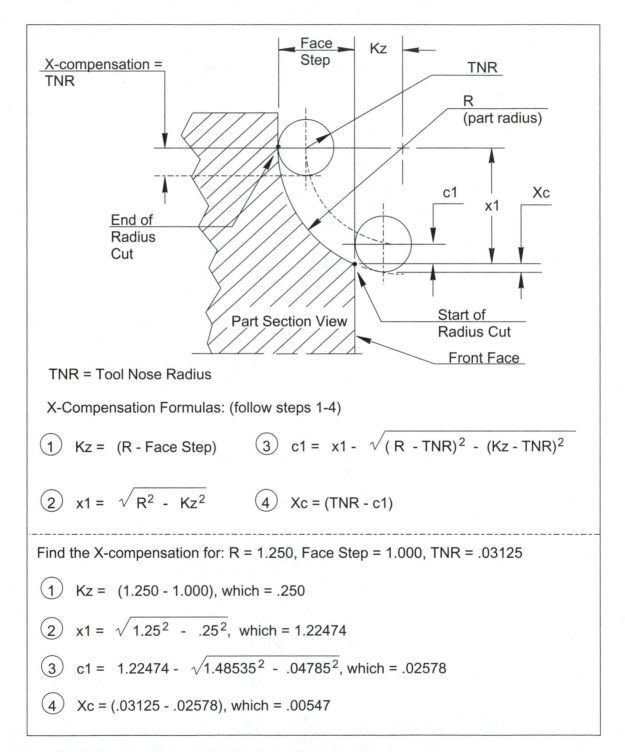

TNR = Tool Nose Radius

X-Compensation Formulas: (follow steps 1-4)

(1) $Kz = (R - \text{Face Step})$

(3) $c1 = x1 - \sqrt{(R - TNR)^2 - (Kz - TNR)^2}$

(2) $x1 = \sqrt{R^2 - Kz^2}$

(4) $Xc = (TNR - c1)$

Find the X-compensation for: R = 1.250, Face Step = 1.000, TNR = .03125

(1) $Kz = (1.250 - 1.000)$, which = .250

(2) $x1 = \sqrt{1.25^2 - .25^2}$, which = 1.22474

(3) $c1 = 1.22474 - \sqrt{1.48535^2 - .04785^2}$, which = .02578

(4) $Xc = (.03125 - .02578)$, which = .00547

Figure 15-6 CNC lathe machining calculation example 5

Calculation Example 6

The illustrations, formulas, and calculations in Figure 15-7 are required to determine the TNR compensation for a radius turn of an arc that is less than 90° radially. In this case, the TNR is tangent in the Z-axis and on the part OD as illustrated below. However, the Z-axis does require compensation.

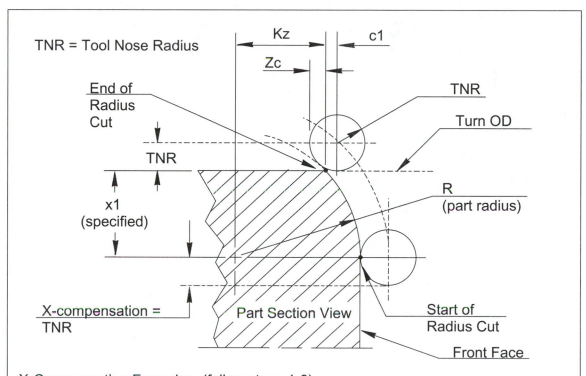

X-Compensation Formulas: (follow steps 1-3)

$x1 = $ (specified by part drawing dimension)

(1) $Kz = \sqrt{R^2 - x1^2}$

(2) $c1 = Kz - \sqrt{(R - TNR)^2 - (x1 - TNR)^2}$

(3) $Zc = (TNR - c1)$

Find the Z-compensation for: $R = 1.06$, $x1 = .750$, $TNR = .03125$

(1) $Kz = \sqrt{1.06^2 - .75^2}$, which $= .749066$

(2) $c1 = .74907 - \sqrt{1.02875^2 - .71875^2}$, which $= .01305$

(3) $Zc = (.03125 - .01305)$, which $= .0182$

Figure 15-7 CNC lathe machining calculation example 6

Calculation Example 7

The illustrations, formulas, and calculations in Figure 15-8 are required to determine the TNR compensation for a radius turn of an arc that is less than 90° radially. In this case, the TNR is tangent in the X-axis on the OD

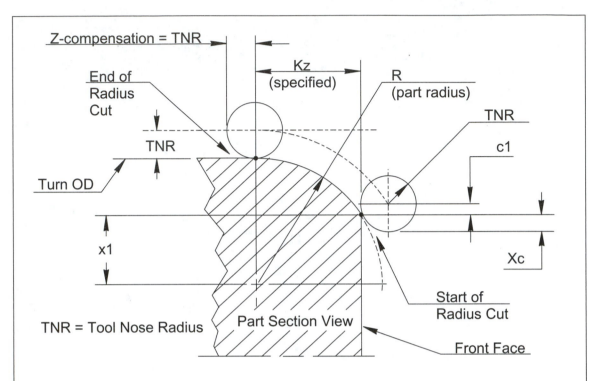

X-Compensation Formulas: (follow steps 1-3)

Kz = (specified by part drawing dimension)

① $x1 = \sqrt{R^2 - Kz^2}$

② $c1 = x1 - \sqrt{(R - TNR)^2 - (Kz - TNR)^2}$

③ $Xc = (TNR - c1)$

Find the X-compensation for: R = 1.75, Kz = 1.50, TNR = .03125

① $x1 = \sqrt{1.75^2 - 1.5^2}$, which = .901388

② $c1 = .90138 - \sqrt{1.71875^2 - 1.46875^2}$, which = .00871

③ $Xc = (.03125 - .00871)$, which = .0225

Figure 15-8 CNC lathe machining calculation example 7

as illustrated. However, the X-axis does require compensation on the front face position.

The carbide indexable insert toolholder has been instrumental in making CNC lathes very productive and capable machine tools. This is due to various factors, which include quick-index capability of the inserts, carbide cutting durability, and increased metal removal rates. The increase in metal removal rates, which reduces the cutting cycle time, is dependent on the cutting speeds (SFM) and feedrates (IPR) that are selected. Therefore, it is important to select optimum cutting speeds and feedrates that will increase metal removal rates for all CNC lathe applications. The following section describes the process and formulas required for calculating the RPM and for selecting feedrates.

CNC TURNING SPEEDS AND FEEDRATES

Cutting speeds and feedrates influence both cutting performance and tool life. There needs to be a balance between maximum cutting speed and maximum tool life. Therefore, the success of an operation in terms of accuracy of the cut, surface finish, and tool wear depends upon the proper specification of both the tool speed (RPM) and the feedrate (IPR).

Tool Speed (SFM and RPM)

The tool speed is defined as the speed of any point on the circumference of the part diameter that is being machined. It is usually expressed in SFM. The tool speed selected is then used to calculate the spindle RPM. Most tool manufacturers provide catalogs that include a technical section listing recommended SFM cutting speeds. A list of typical SFM cutting speeds for lathe machining is also detailed and described in appendix N.

The cutting speed (SFM) selected for a particular lathe operation will depend on several factors, which are listed below:

- Type of lathe operation
- Part material type to be machined
- Part material hardness to be machined
- Tool material to be used
- DOC
- Rigidity of the part to be machined
- Rigidity of the tool
- Rigidity of the lathe setup (support and gripping)

It is important to consider that selecting a cutting speed that is too high can lead to excessive dulling and burning of the tool's cutting edges. Conversely, speeds that are too low may cause the tool to wear excessively or break during operation.

Turning, facing, drilling, and boring operations, which are typically performed with carbide-insert or carbide-tipped type tools, usually require cutting speeds approximately five times faster than what is used for HSS tools. Lathe tools, which are considered single-point tools, generally have minimal cutting edge contact area and consequently cause less tool pressure. However, the probability of more tool pressure will increase as the DOC increases. Other machining factors can also have a negative affect on tool pressure. Therefore, it should be understood that the process of calculating the RPM is also determined by the actual machining experiences of the CNC setup person, CNC operator, and CNC programmer.

The following is an example that describes the process for calculating an RPM for a turning operation of mild steel barstock:

- Part material: mild steel

- Turn diameter: 2.750 inch

- Tool material: coated carbide insert

- DOC: 0.200 inch

- From appendix N, cutting speed = (300–700 SFM) = 500 middle range

- Formula from appendix M

$$RPM = \frac{4 \times CS}{DIA.}$$

The SFM cutting speed can be increased because the insert specified is coated carbide. Therefore,

$$RPM = \frac{4 \times 700}{2.750} = \frac{2800}{2.750} = 1018 \, RPM$$

The RPM calculation matrix (see Figure 15-9) has been designed to facilitate the RPM process in a step-by-step manner. This matrix can be used as a guide to ensure that all the data for calculating the RPM is applied correctly. It is important to note that this process will produce a recommended starting RPM value. Due to the various items that affect machining, the RPM values can be adjusted to the optimum level during the actual metal cutting proveout. Additionally, it should be noted that most modern CNC lathes have the capability to be programmed in either SFM mode or RPM mode.

Tool Feedrate (IPR)

The tool feedrate for lathe operations is defined as the amount (inches) at which the tool advances into the workpiece per one revolution. It is expressed as IPR. The tool IPR feedrate is the value input into the CNC

Lathe Speed (RPM) Calculation Matrix		
Step	**Description**	**Input Selections/Calculations**
1	List the **operation type**: turning, drilling, tapping, reaming, boring	
2	List the **tool material**: uncoated carbide, coated carbide, HSS, ceramic, diamond	
3	List the **part material**: aluminum, steel, alloy steel, cast iron, stainless steel, brass, copper	
4	Select the **cutting speed** (SFM), Reference Chart:	
5	List the **part diameter** (diameter being machined)	
6	RPM Formula: RPM = SFM × 4/Diameter	

Figure 15-9 **RPM calculation matrix**

program. Most tool manufacturers provide in their catalogs a technical section listing recommended IPR feedrates. A list of typical IPR recommended feedrate values for lathe machining is also detailed and described in appendix N.

Turning and boring feedrates can range from 0.0005 up to 0.030 inch IPR depending on various machining factors. Some factors include surface finish required, tolerance allowed, material hardness, and type of cut (rough or finish) to be performed. Appendix N lists some recommended IPR values for turning and boring. Drilling, reaming, and tapping feedrates are determined the same as previously described in the machining center section of this text (pp. 110 and 111), except that IPR is not converted to IPM.

Threading feedrates are derived from the pitch of the thread size to be threaded. Since the thread specification basically represents the screw thread to be machined, the feedrate of the thread must be the same as the screw thread pitch. Therefore, the feedrate is calculated by dividing one by the number of threads per inch of the screw thread. The thread designation, which is described below, is used to calculate the IPR feedrate (pitch).

The thread designation, **3/4-10 UNC-2A,** is translated as follows:

3/4	Represents the major diameter (0.750 inch nominal)
10	Represents the number of threads per inch (10)
UN	Specifies the thread series type (Unified National)

C	Specifies course; F is used for fine
2	Specifies the class fit: 1 = loose fit, 2 = moderate fit, 3 = tight fit
A	Specifies external threads; a B is used for internal threads

Consequently, for a 3/4-10 UNC-2A thread, the feedrate is calculated by dividing 1 inch by 10 threads per inch (1/10 = 0.100). This threading feed rate of 0.100 IPR is considered a heavy machining feedrate. To compensate for this heavy feedrate, the RPM for the threading tool is usually slower.

Other than threading and tapping operations, the cutting feedrate (IPR) selected for a particular lathe operation will depend on several factors. These factors, which are listed below, may or may not require an adjustment to the selected feedrate. They include the following:

- Type of lathe operation

- Part material type to be machined

- Part material hardness to be machined

- DOC

- Rigidity of the part to be machined

- Rigidity of the tool

- Rigidity of the lathe setup (support and gripping)

These are typical factors; other factors not listed may also influence the IPR selected.

The following is an example that describes the IPR selection process for an OD turning operation.

- Operation type: turning

- Part material: aluminum

- DOC: 0.062 inch

- From appendix N, IPR: (0.003 up to 0.016) = 0.0095 mean range, or 0.010

The IPR selection matrix (see Figure 15-10) can be used as a guide to ensure that all the required data is included before selecting the IPR feedrate. Due to the various items that affect machining, the IPR values can be adjusted to the optimum level during the actual metal cutting proveout.

The appropriate feedrate should be used whenever possible because improper feedrates can reduce the life of the tool. For instance, too slow a feedrate can cause problems with chip control and reduce tool life. Too fast a feedrate can cause insert chipping and breakage, which also reduces tool life. Some other factors that affect tool life include

- Small insert nose radius

- Wrong insert grade

Lathe Feedrate Selction Matrix			
Step	**Description**		**Input Selection**
1	List the **operation type**: turning, drilling, tapping, reaming, boring		
2	List the **part material**: aluminum, steel, alloy steel, cast iron, stainless steel, brass, copper		
3	Specify the **DOC**:		
4	Select the **feedrate** (IPR)		
	Reference Chart:		

Figure 15-10 ***Feedrate selection matrix***

- DOC too large for the insert

- Weak insert geometry

Speed and feed charts recommend general machining values that apply to most typical machining conditions. However, each situation has to be evaluated on an individual basis to determine the optimum feedrates and spindle speeds, which are based on the conditions that exist. Some adverse conditions may include interrupted cuts, uneven DOC, long continuous cuts, and abrasive surface scale. However, the ideal spindle speeds and tool feedrates that are selected should produce optimum metal removal rates (SFM) and still maintain optimum tool life.

LATHE GEOMETRIC DIMENSIONING AND TOLERANCING

This method of applying tolerances and dimensions to part features is referred to as GD&T and applies a datum system to establish how the part must be located for manufacturing and inspection purposes. This GD&T system is applied to each of the project drawings in chapters 16–20.

The GD&T system uses various symbols for geometric feature control instead of written notes. These symbols are easily learned because they generally reflect the geometric characteristic under consideration. The typical geometric symbols, which are used in modern engineering drawings for manufacturing, are described and illustrated in Figures 15-11 through 15-13. These illustrations describe the geometric characteristic symbol, the method

GD&T Symbol	Drawing Application	Feature Control Interpretation
STRAIGHTNESS ——— Not Related to a Datum	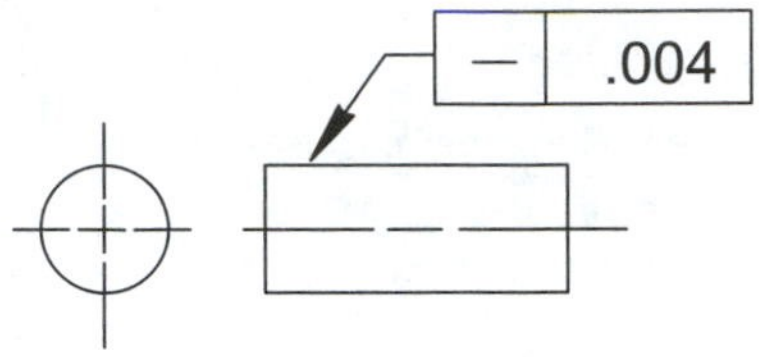 — .004	The part surface must be inside the .004" tolerance zone. 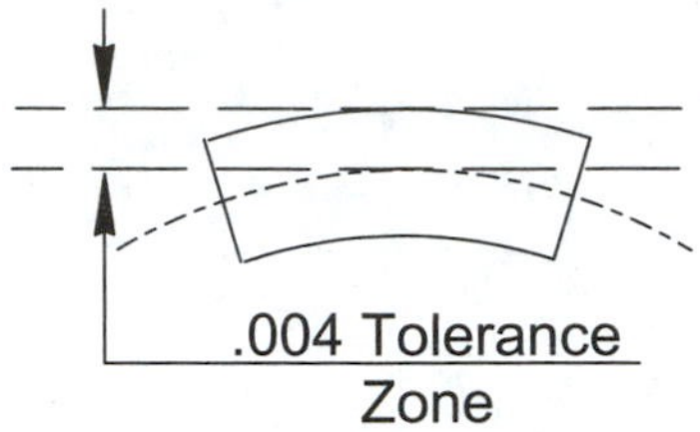 .004 Tolerance Zone
CIRCULARITY ○ Not Related to a Datum	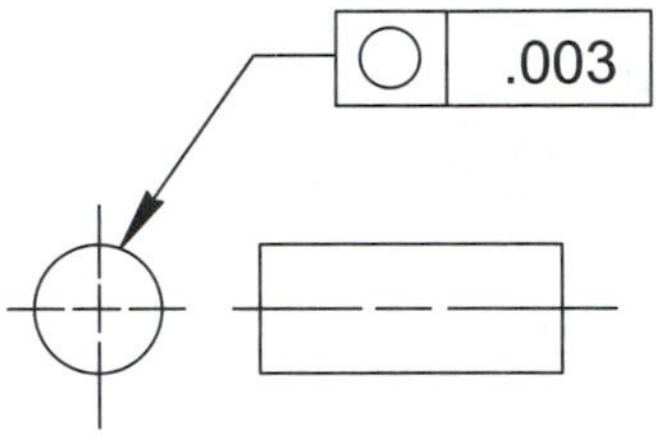 ○ .003	The part surface must be inside the .003" circular tolerance zone. 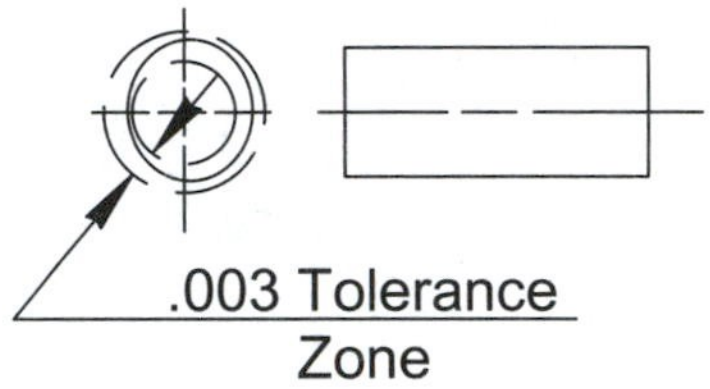 .003 Tolerance Zone
CYLINDRICITY ⌭ Not Related to a Datum	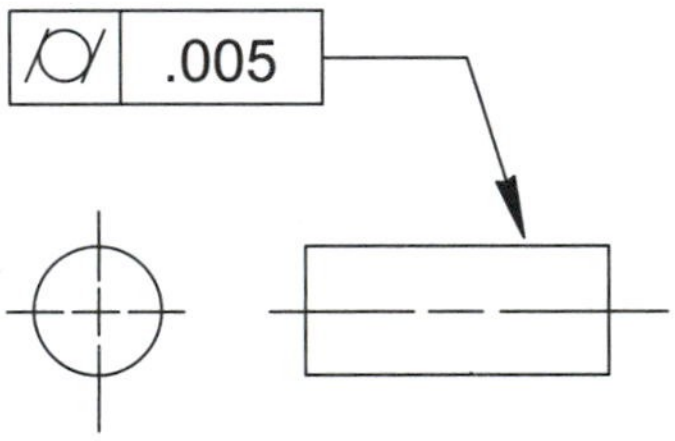 ⌭ .005	All cylindrical elements of the part must be inside both .005" tolerance zones. 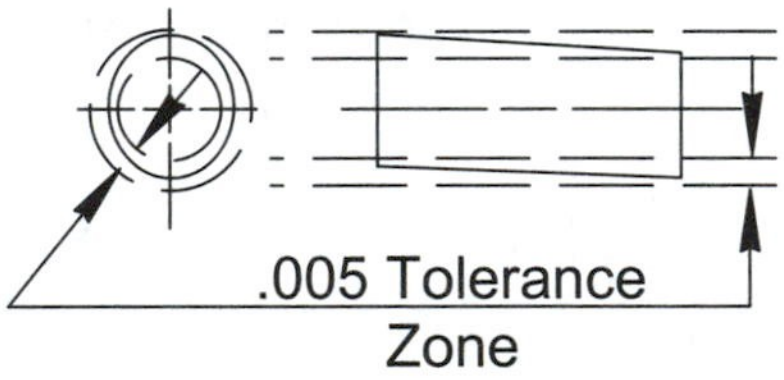 .005 Tolerance Zone
CONCENTRICITY ◎ Related to a Datum	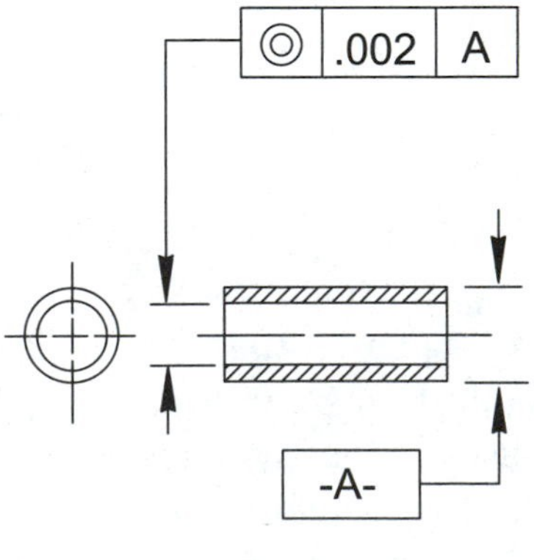 ◎ .002 A -A-	The hole centerline must be inside the .002" tolerance zone and relative to Datum -A- diameter. 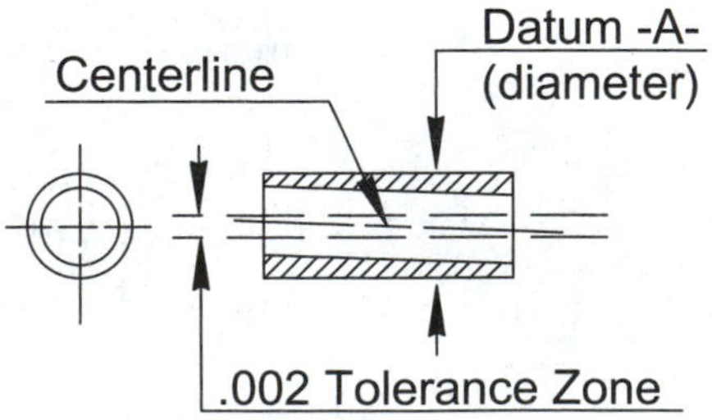 Datum -A- (diameter) Centerline .002 Tolerance Zone

Figure 15-11 *GD&T application and interpretation descriptions*

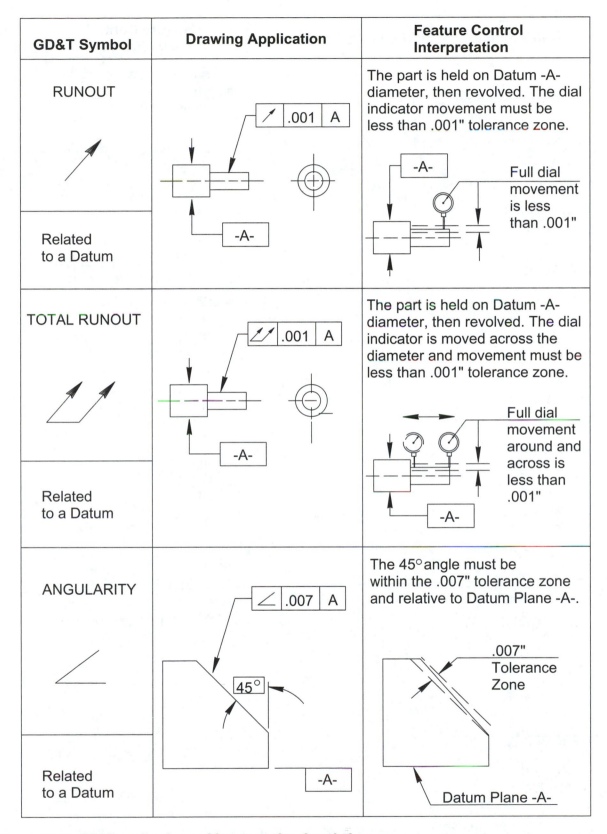

GD&T Symbol	Drawing Application	Feature Control Interpretation
RUNOUT Related to a Datum	.001 A -A-	The part is held on Datum -A- diameter, then revolved. The dial indicator movement must be less than .001" tolerance zone. Full dial movement is less than .001" -A-
TOTAL RUNOUT Related to a Datum	.001 A -A-	The part is held on Datum -A- diameter, then revolved. The dial indicator is moved across the diameter and movement must be less than .001" tolerance zone. Full dial movement around and across is less than .001" -A-
ANGULARITY Related to a Datum	.007 A 45° -A-	The 45° angle must be within the .007" tolerance zone and relative to Datum Plane -A-. .007" Tolerance Zone Datum Plane -A-

Figure 15-12 **GD&T application and interpretation descriptions**

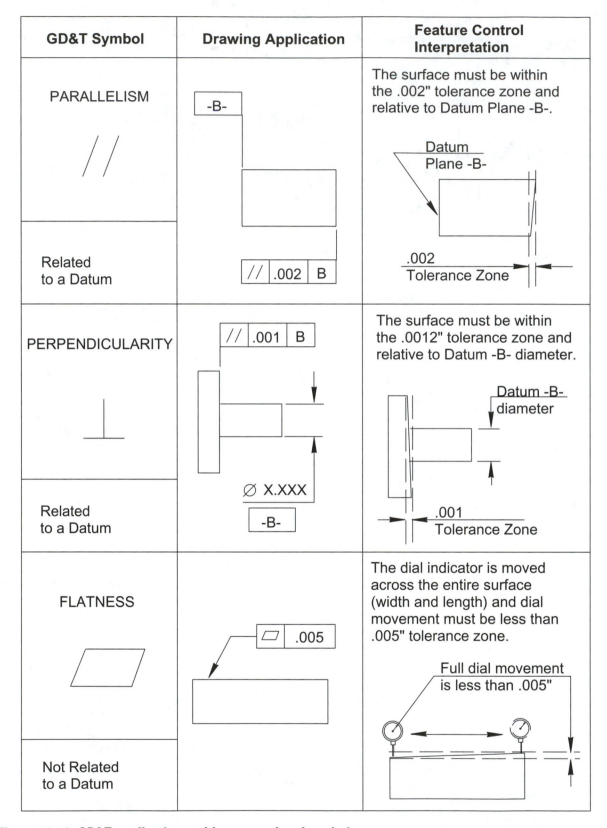

GD&T Symbol	Drawing Application	Feature Control Interpretation
PARALLELISM Related to a Datum	-B- // .002 B	The surface must be within the .002" tolerance zone and relative to Datum Plane -B-.
PERPENDICULARITY Related to a Datum	// .001 B ⌀ X.XXX -B-	The surface must be within the .0012" tolerance zone and relative to Datum -B- diameter.
FLATNESS Not Related to a Datum	▱ .005	The dial indicator is moved across the entire surface (width and length) and dial movement must be less than .005" tolerance zone.

Figure 15-13 ***GD&T application and interpretation descriptions***

of application on the engineering drawing, and the interpretation of how it controls the part features. The GD&T symbols commonly used for lathe machining applications include

- *Straightness*

- *Circularity*

- *Cylindricity*

- *Concentricity*

- *Runout*

- *Total runout*

- *Parallelism*

- *Perpendicularity*

Datum Dimensioning

The datum dimensioning method, previously described, requires that all dimensions on a drawing be placed in reference to one fixed zero point of origin. Datum dimensioning is ideally suited for absolute positioning applications and CNC machines. The GD&T method that matches the drawing dimensions with the CNC program coordinates allows effective communications between the CNC operator, the *CMM* inspectors, and any other CNC personnel.

Datum Reference

A datum is a specific surface, line, plane, or feature that is assumed to be perfect and is used as a reference point for dimensions or features. Datums are identified by letters and are shown on the part with datum feature symbols that refer to the datum reference in the feature control symbol (see Figure 15-14).

The letters B and C in the feature control symbol refers to datum B and C on the part. The datum feature symbol is shown as a letter with a dash on either side contained in a box (see Figure 15-14). All letters in the alphabet, not including the letters I, O, and Q, may be used to identify datums. These letters are not used since they closely resemble numbers. When more than 23 datums are required, double letters (AA, AB, AC, and so on) are used.

Datum references can be shown as single datums or as multiple datums. When only one letter is shown in the feature control symbol, it means the feature is related only to that single datum. Datums are arranged in their order of precedence. So, a primary datum is shown first, a secondary datum is shown next, and a tertiary datum is shown last. They are usually A, B, and C, respectively.

The lathe machining process determines and controls the surface finish characteristics produced during the machining cycle. The machining process

GD&T Symbol	Drawing Application	Feature Control Interpretation
PROFILE OF A LINE  ⌒ Can be Related to a Datum, or Not Related	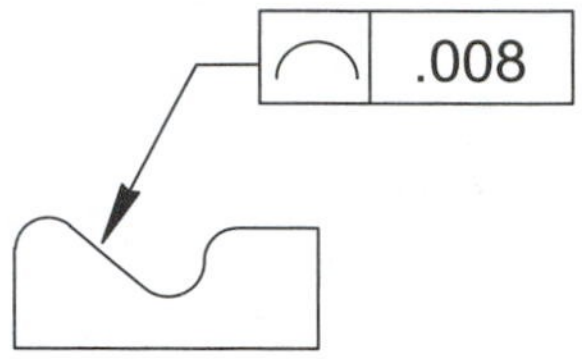 ⌒ .008	The part surface must be inside the .008" tolerance zone that is around the true profile. .008" tolerance zone around the true profile. 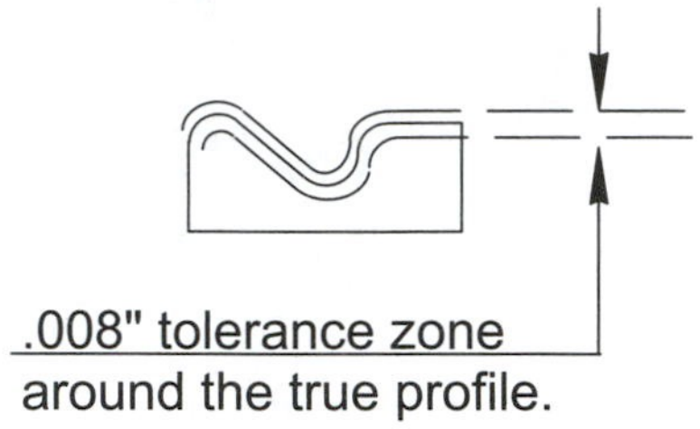
PROFILE OF A SURFACE  ⌓ Can be Related to a Datum, or Not Related	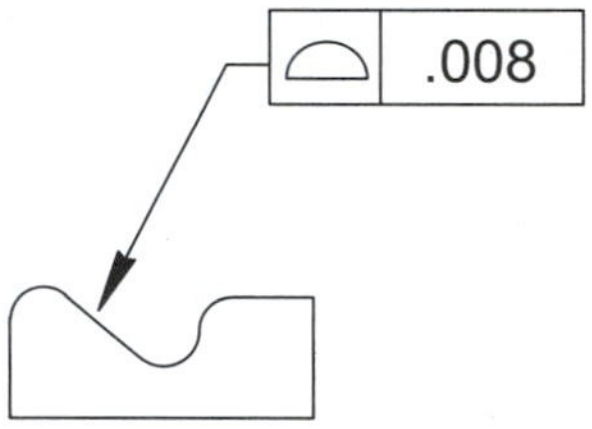 ⌓ .008	The entire part surface must be inside the .008" tolerance zone that is around the true profile. 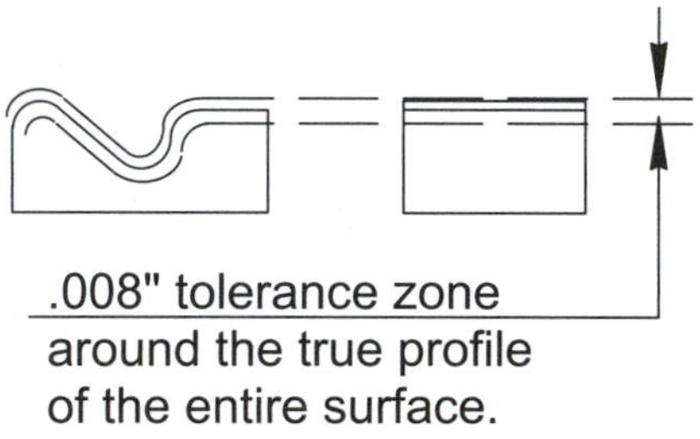 .008" tolerance zone around the true profile of the entire surface.
POSITION 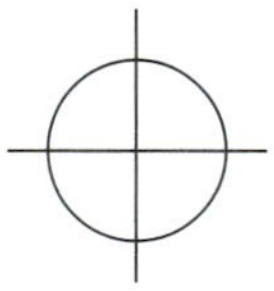 ⊕ Related to a single Datum, or multiple Datums	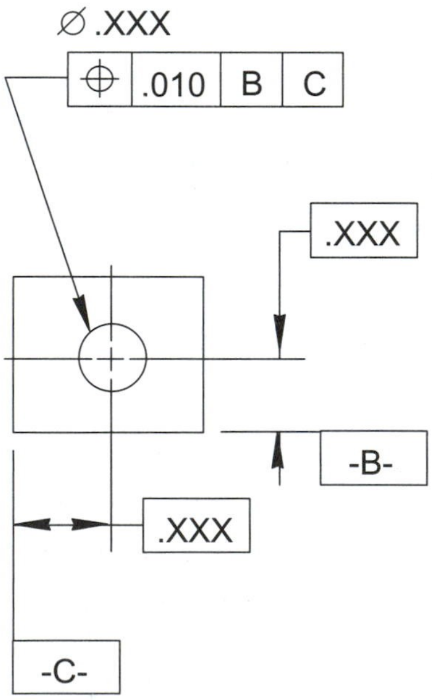 ⌀ .XXX ⊕ .010 B C .XXX -B- .XXX -C-	The hole centerline must be inside the .010" diameter tolerance zone and relative to Datum Planes -B- and -C-. The location dimensions are Basic and do not have any tolerance. 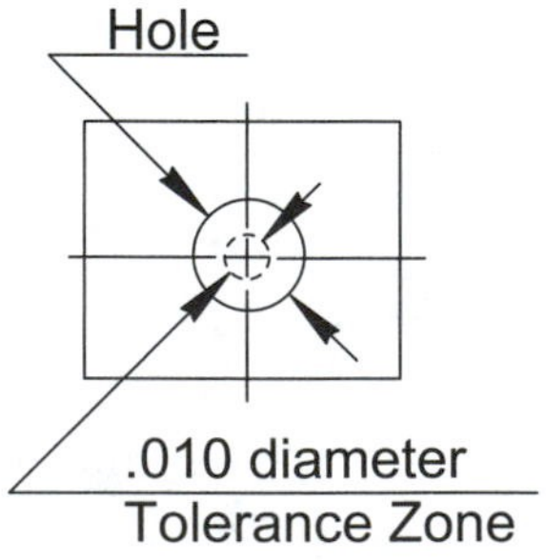 Hole .010 diameter Tolerance Zone

Figure 15-14 **GD&T application and interpretation descriptions**

includes the type of operation or operations to be performed such as facing, turning, drilling, boring, and reaming. Also included in the machining process is the number of cuts required for roughing as well as for finishing. The feedrate (IPR), which can be adjusted, is one main factor that affects the surface finish produced on the workpiece. The TNR, which can vary in radius size, is another main factor that affects the surface finish produced on the workpiece. The rigidity of both the cutting tool and the workholding device are also contributing factors in the surface finish that is produced.

SURFACE FINISH

It is important to note that overfinishing the workpiece surface adds unnecessarily to the production cost and underfinishing the workpiece surface can result in the rejection of a part or an entire production run. Surface finish or texture measurements are typically measured with either digital-type surface gages or with visual comparison-type gages within a degree of reasonable accuracy.

All machined surfaces consist of a series of peaks and valleys that when magnified resemble the surface of a phonograph record (see Figure 15-15). The characteristics of surface texture that are illustrated and described include roughness, waviness, and flaw. Surface roughness usually caused by the cutting action of the tool edges, the cutting pattern of a grinding wheel, or by the feed of the machine tool is measured over a specific distance. The measurement is typically in microinches (0.0000001 inches). Waviness, measured in inches, is similar to roughness but it occurs over a wider spacing than that used for determining roughness. Roughness may be considered as superimposed on a "wavy" surface. Flaws are irregularities, or defects, that occur at one place, or relatively infrequently on a surface. Surface finish requirements for lathe parts are typically specified on engineering drawings with a symbol and a value (see Figure 15-16).

The formula below can be applied to determine the microinch finish produced by lathe turning and boring operations. The two factors required are the insert TNR and the feedrate IPR value.

$$\text{Microinch Finish} = \frac{(\text{IPR})^2}{16 \times \text{TNR}} \times 1{,}000{,}000$$

For example, using an insert TNR of 0.03125 inch and selecting a feed rate IPR of 0.005 inch the expected surface finish to be produced is 50 microinches.

$$\frac{(0.005)^2}{16 \times 0.03125} \times 1{,}000{,}000 = 50 \text{ microinch finish}$$

There are various process methods that produce surface finishes. They range from very fine to very rough, or 2 microinches to 1000 microinches.

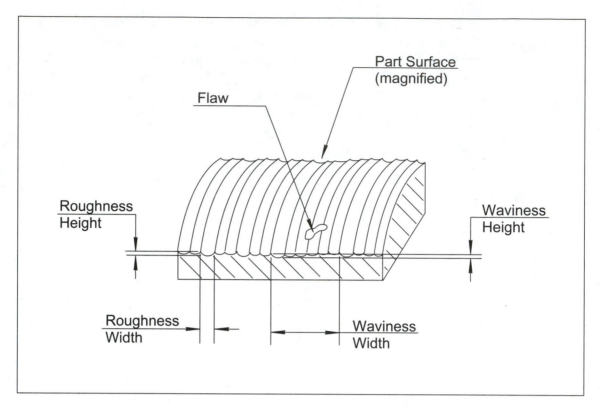

Figure 15-15 **Surface finish characteristics**

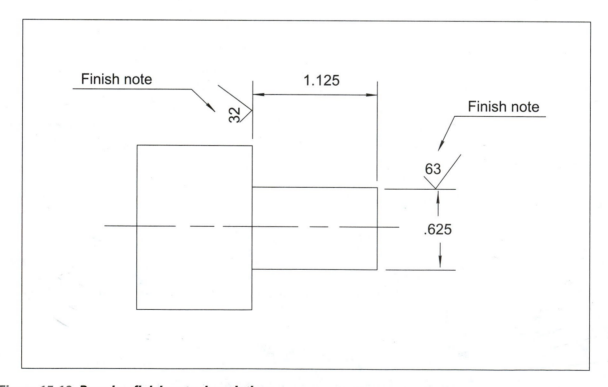

Figure 15-16 **Drawing finish note descriptions**

Lathe turning and boring operations typically produce surface finishes that range from 16 to 250 microinches (see Chart 5-1, p. 122).

SCREW THREADS

The screw thread is used universally throughout industry for the production of bolts, screws, nuts, and similar fasteners. The screw thread is also used to transmit motion and to transmit power. These bolts, screws, nuts, and fasteners are used on all types of machines and equipment for various mechanical functions. It is important to identify these common types of threads, their features, and their characteristics. This section describes the common threads machined on CNC lathes. It also describes the technical features required for machining and identifying threads.

Screw Thread Forms

The most common screw thread forms, which are machined on CNC lathes, include the American National thread, the Acme thread, the Square thread, and the Buttress thread (see Figure 15-17). The American National (UN) thread, which is the most common, is typically manufactured in English (inch) thread form, Metric thread form, and Pipe thread form (see Figures 15-18 and 15-19).

UN Threads (English)

The American National thread, which is commonly recognized as UN thread, is designed in course thread series (UNC), fine thread series (UNF), and extrafine thread series (UNEF). This refers to the number of threads per inch (TPI) that are on the screw. They are classified by how the internal and external screw threads fit together when assembled. Thread classes for external threads are 1A, 2A, 3A, and for internal threads they are 1B, 2B, 3B. The most commonly used classes are the 2A and 2B. They are manufactured in various standard sizes and thread series (see Table 15-1).

UN Threads (Metric)

The Metric UN thread is identical in form to the English UN thread. Metric threads are also designed in course and fine thread series. They are classified by how the internal and external screw threads fit together when assembled. They are also manufactured in various standard sizes and thread series (see Table 15-2).

Pipe Threads

The American National standard taper pipe thread, which is commonly recognized as NPT thread, is designed to create a tight "leak proof" fit when the

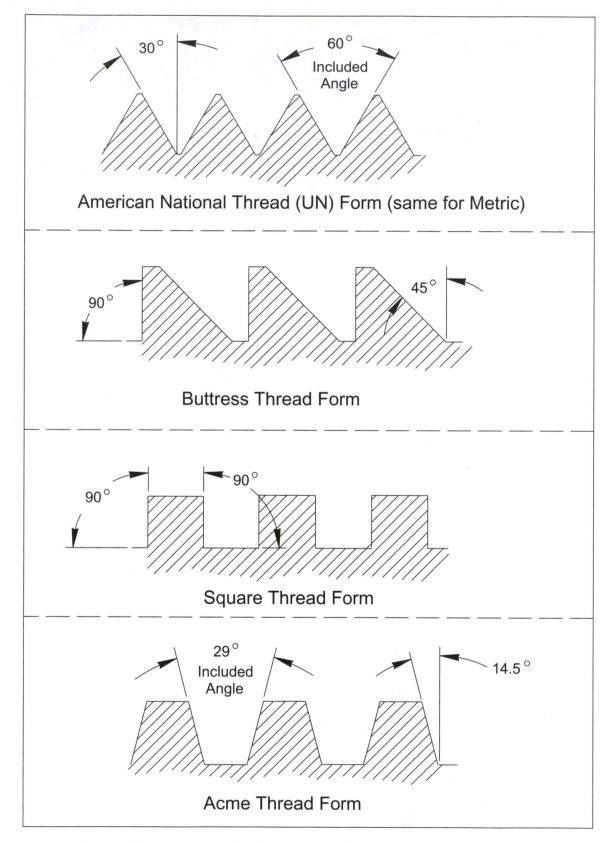

Figure 15-17 Screw thread forms

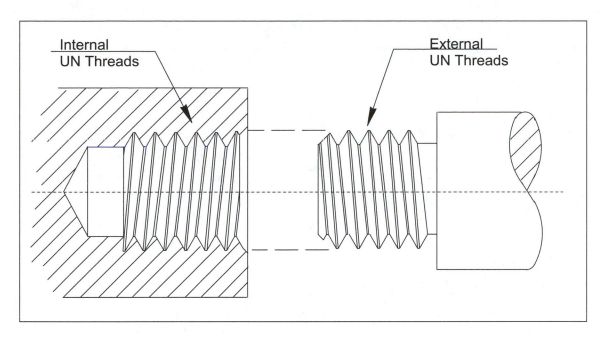

Figure 15-18 **UN external and internal screw thread**

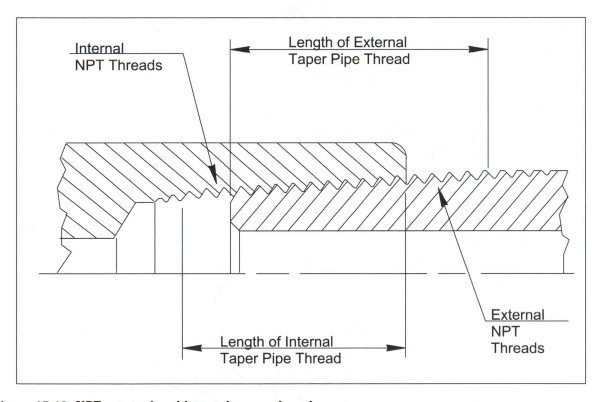

Figure 15-19 **NPT external and internal screw thread**

Table 15-1 **American National thread sizes**

0–80 UNF	1/4–20 UNC	5/8–11 UNC	1-1/4–7 UNC
1–64 UNC	1/4–28 UNF	5/8–18 UNF	1-1/4–12 UNF
1–72 UNF	1/4–32 UNEF	5/8–24 UNEF	1-1/4–18 UNEF
2–56 UNC	5/16–18 UNC	11/16–24 UNEF	1-5/16–18 UNEF
2–64 UNF	5/16–24 UNF	3/4–10 UNC	1-3/8–6 UNC
3–48 UNC	5/16–32 UNEF	3/4–16 UNF	1-3/8–12 UNF
3–56 UNF	3/8–16 UNC	3/4–20 UNEF	1-3/8–18 UNEF
4–40 UNC	3/8–24 UNF	13/16–20 UNEF	1-7/16–12 UNEF
4–48 UNF	3/8–32 UNEF	7/8–9 UNC	1-1/2–6 UNC
5–40 UNC	7/16–14 UNC	7/8–14 UNF	1-1/2–12 UNF
5–44 UNF	7/16–20 UNF	7/8–20 UNEF	1-1/2–18 UNEF
6–32 UNC	7/16–28 UNEF	15/16–20 UNEF	
6–40 UNF	1/2–13 UNC	1 inch–8 UNC	
8–32 UNC	1/2–20 UNF	1 inch–12 UNF	
8–36 UNF	1/2–28 UNEF	1 inch–14 UNEF	
10–24 UNC	9/16–12 UNC	1 inch–20 UNEF	
10–32 UNF	9/16–18 UNF	1-1/16–18 UNEF	
12–24 UNC	9/16–24 UNEF	1-1/8–7 UNC	
12–28 UNF		1-1/8–12 UNF	
12–32 UNEF		1-1/8–18 UNEF	

Table 15-2 **Metric thread sizes**

M1.6 × .35	M7 × 1	M18 × 2.5	M30 × 3.5
M1.8 × .35	M8 × 1.25	M18 × 1.5	M30 × 2
M2 × .4	M8 × 1	M20 × 2.5	M33 × 3.5
M2.2 × .45	M10 × 1.5	M20 × 1.5	M33 × 2
M2.5 × .45	M10 × 1.25	M22 × 2.5	M36 × 4
M3 × .5	M12 × 1.75	M22 × 1.5	M36 × 3
M3.5 × .6	M12 × 1.25	M24 × 3	M39 × 4
M4 × .7	M14 × 2	M24 × 2	M39 × 3
M4.5 × .75	M14 × 1.5	M27 × 3	
M5 × .8	M16 × 2	M27 × 2	
M6 × 1	M16 × 1.5		

Table 15-3 *Pipe thread sizes*

1/16–27	1/2–14	1-1/2–11-1/2	3-1/2–8
1/8–27	3/4–14	2–11-1/2	4–8
1/4–18	1–11-1/2	2-1/2–8	
3/8–18	1-1/4–11-1/2	3–8	

internal and external screw threads are assembled. They are also designed with straight threads, which are designated as NPS threads. They are also manufactured in various standard sizes and thread series (see Table 15-3).

Thread Features and Identification System

The UN screw thread features used for manufacturing purposes include the Major diameter, the Minor diameter, the Pitch diameter, the TPI, the Crest, the *Root,* the Pitch, the Lead, the thread depth, and the thread angle (see Figure 15-20).

The screw threads are basically identified by the Major diameter, the TPI, the thread form type, the thread series, the Class fit, and internal or external thread design (see Figure 15-21). It should be noted that the pipe threads use the nominal pipe size in place of the Major diameter and the Metric threads use pitch in place of TPI.

UN Thread and NPT Thread Description

The screw thread design typically requires external and internal threads as illustrated in Figures 15-18 and 15-19. The threads are machined to tolerances that will allow proper fit between the external and the internal thread.

UN Screw Thread Features

The UN screw thread described in Figure 15-20 illustrates and describes the main features associated with machining and identifying screw threads.

CNC LATHE ALARM CODES

The CNC lathe control continually diagnoses itself to determine if any error exists. When an error is detected, the machine will stop automatically and the CRT will display the word "Alarm." Table 15-4 describes a sample of typical lathe alarm codes. Each control manufacturer provides a specific alarm list for their control.

When an alarm error occurs, the CNC lathe will not operate until the error has been corrected and the control is reset. Additionally, there are other

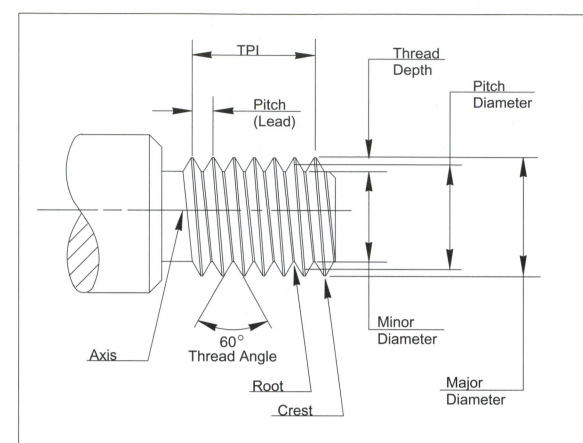

Thead Feature Descriptions:

1. Major Diameter: The large diameter of the screw thread, expressed as a nominal size not the actual machined size.
2. Minor Diameter: The small diameter of the screw thread.
3. Pitch Diameter: The theoretical contact diameter of the internal and the external screw, which is also used for measurement.
4. Thread Depth: The depth from the top to the bottom of a thread.
5. Pitch: The length of one complete revolution of the screw thread, which is also the Lead of the screw.
6. Thread Angle: The angle that is designed for the screw thread, which is machined or formed by the lathe tool.
7. TPI: The number of threads in a one inch length of the screw thread, expressed as course (C), fine (F), or extra-fine (EF).
8. Crest:The top of the thread.
9. Root: The bottom of the thread.
10. Axis: The center of the screw thread diameter.

Figure 15-20 *UN screw thread features and descriptions*

UN screw threads are identified by the system described below:

1. The Major Diameter (nominal OD size)
2. Threads Per Inch (TPI)
3. Thread Form—Unified National (UN)
4. Thread Series—Course (C), Fine (F), and Extra-Fine (EF)
5. Thread Class—1 = loose, 2 = normal, 3 = tight
6. Thread Design—A = external, B = internal

For screw thread: **1/4-20 UNC - 2A**

1. .250" nominal diameter
2. 20 TPI
3. UN 60-degree

6. A = External thread
5. 2 = Normal thread Class
4. C = Course thread series

NPT screw threads are identified by the system described below:

1. The Pipe Diameter (nominal OD pipe size)
2. Threads Per Inch (TPI)
3. Thread Form—American National (N)
4. Thread Series—Pipe (P)
5. Thread Design—T = taper, S = straight

For screw thread: **1/8-27 N P T**

1. 1/8" Pipe diameter
2. 27 TPI
3. American National

5. Taper Thread Design
4. Pipe Thread Series

Metric screw threads are identified by the system described below:

1. Metric thread
2. The Major Diameter (nominal OD millimeter size)
3. Thread Pitch (millimeter)
4. Thread Tolerance
5. Thread Design—H = internal, h = external

For screw thread: **M12 x 1.25 - 6H**

1. Metric thread
2. 12mm nominal diameter

5. H = Internal thread
4. 6 thread tolerance
3. 1.25mm Pitch

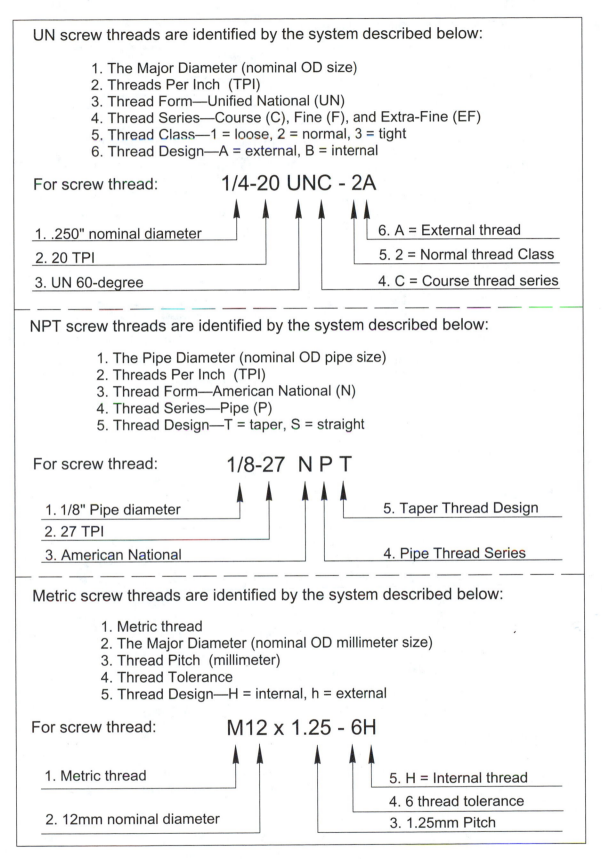

Figure 15-21 Screw thread identification system

■___Table 15-4 *Alarm code messages and descriptions*

ERROR CODE	ALARM TYPE	ALARM DESCRIPTION
1	Low lube oil	The lube oil reservoir is low, and must be refilled.
2	Thermal trip	A power interruption condition has caused the power panel to switch off, and must be reset.
3	Emergency stop	An emergency condition has caused the power panel to switch Off, and must be reset.
4	X-Axis limit travel exceeded	The X-Axis limit switch was reached, and the turret must be moved off the limit switch.
5	Z-Axis limit travel exceeded	The Z-Axis limit switch was reached, and the turret must be moved off the limit switch.
6	Safety door open	Safety door is open and must be closed before command can be executed.
7	Fuse Open	Fuse is blown and must be changed.
8	Servo Off	The servomotor is Off, and must be turned On.
9	Monitor Reset	The control monitor is in Feed Hold and must be Reset.
10	Hydraulic Off	The hydraulic motor is Off, and must be turned On.

situations that can cause the CNC lathe to stop operation. Some situations, which are unsafe, usually include the following:

- Safety doors are open when Cycle Start is pressed.
- Safety doors are open when Spindle Start is pressed.
- Chuck jaws are not "clamped."

When these types of situations exist, the CRT will usually display a "command rejected" message. In situations where the error is due to a machine mechanical or electrical malfunction, the CNC operator should contact the appropriate repair personnel. In situations where the error is due to a CNC program function, the CNC operator should contact the CNC programmer. The CNC operator should never attempt to correct any dangerous malfunctions that may cause injury.

CHAPTER SUMMARY

The key concepts presented in this chapter include the following:

- The names and descriptions of lathe machining calculations.
- The names and descriptions of lathe speed calculations.
- The names and descriptions of lathe feedrate selections.

- The names and descriptions of GD&T.
- The descriptions of GD&T applications.
- The characteristics and descriptions of surface finish.
- The descriptions of surface finish machining calculation.
- The characteristics and descriptions of screw threads.
- The types and descriptions of lathe alarms.

■ ■ REVIEW QUESTIONS

1. When calculating coordinates for a taper or radius, which tool feature is entered into the calculations?

2. What is the difference between SFM and RPM?

3. List the factors used to select a cutting speed (SFM) value.

4. List the factors used to select a feedrate (IPR) value.

5. List the cutting speed (SFM) range that is appropriate for rough turning (0.200 DOC) mild steel with a carbide insert tool.

6. List the feedrate (IPR) range for finish turning (0.060 inch DOC) aluminum with a carbide insert tool.

7. Calculate the spindle speed (RPM) for rough turning (0.250 inch DOC) a 3.5 inch cast iron diameter with a carbide insert tool.

8. Why is it important to optimize CNC lathe machining operations?

9. How is runout of a shaft diameter measured?

10. List the two main factors that control surface finishes on a CNC lathe.

11. List the four screw thread forms that are typically machined on CNC lathes.

12. List the features that are used for machining and identifying UN screw threads.

13. Calculate the feedrate (IPR) for a **3/4-10 UNC 2A** threaded diameter.

14. List the surface finish range produced by boring and turning.

15. What should the CNC operator do if the CRT displays an alarm message?

chapter 16

CNC LATHE RAPID AND FEED MOVES

Chapter Objectives

After studying and completing this chapter, the student should have knowledge of the following:

- *G00 and G01 codes*

- *G00 rapid movement in X- and Z-axes*

- *G01 feed movement in X- and Z-axes*

- *Calculating X-Z coordinates*

RAPID TRAVERSE AND FEED MOVES

The first set of illustrations in this chapter (Figures 16-1 through 16-3), describes the rapid movements for 2-axis CNC lathes. The second set of illustrations (Figures 16-4 through 16-6) describes the feed movements for 2-axis CNC lathes. These figures are designed to illustrate the differences between rapid moves and feed moves. Additionally, the chapter includes a coordinate calculation worksheet and a sample program to identify some basic CNC lathe program format.

G00 RAPID MOVEMENT

The G00 rapid code is primarily used to move the X-axis and the Z-axis as fast as possible from one point to another. Therefore, after the G00 code an axis or combination of axes will follow. The axis letter (X or Z) will include a positive or negative number, which represents the coordinate position of the machine. The positive sign is the machine default sign and therefore it is usually not programmed.

The G00 code is modal and will remain active for each subsequent line until it is canceled. Therefore, any axis letter present in a subsequent line of the CNC program will move in rapid traverse mode whether the G00 code is present or not.

When the G00 code along with an axis coordinate position is read (Cycle Start), the specified machine axis or axes will move automatically at a very rapid rate. This is only recommended when it is absolutely certain that a collision will not occur. Typically this is after the CNC program has been Dry Run or proved out. In situations where the rapid moves have not been proved out, the CNC operator should override the rapid traverse rate to a lower setting. This will allow the CNC operator time to react to a possible collision and stop the movement before the collision occurs.

G01 FEED MOVEMENT

The G01 feed code is primarily used to move the X-axis and Z-axis at a predetermined feedrate to cut the workpiece. Therefore, after the G01 code an axis or combination of axes will follow. The axis letter (X and Z) will include a positive or negative number, which represents the coordinate position of the machine. Additionally, for the G01 code a feedrate value is required. It should also be noted that cutting requires the spindle/chuck to be rotating.

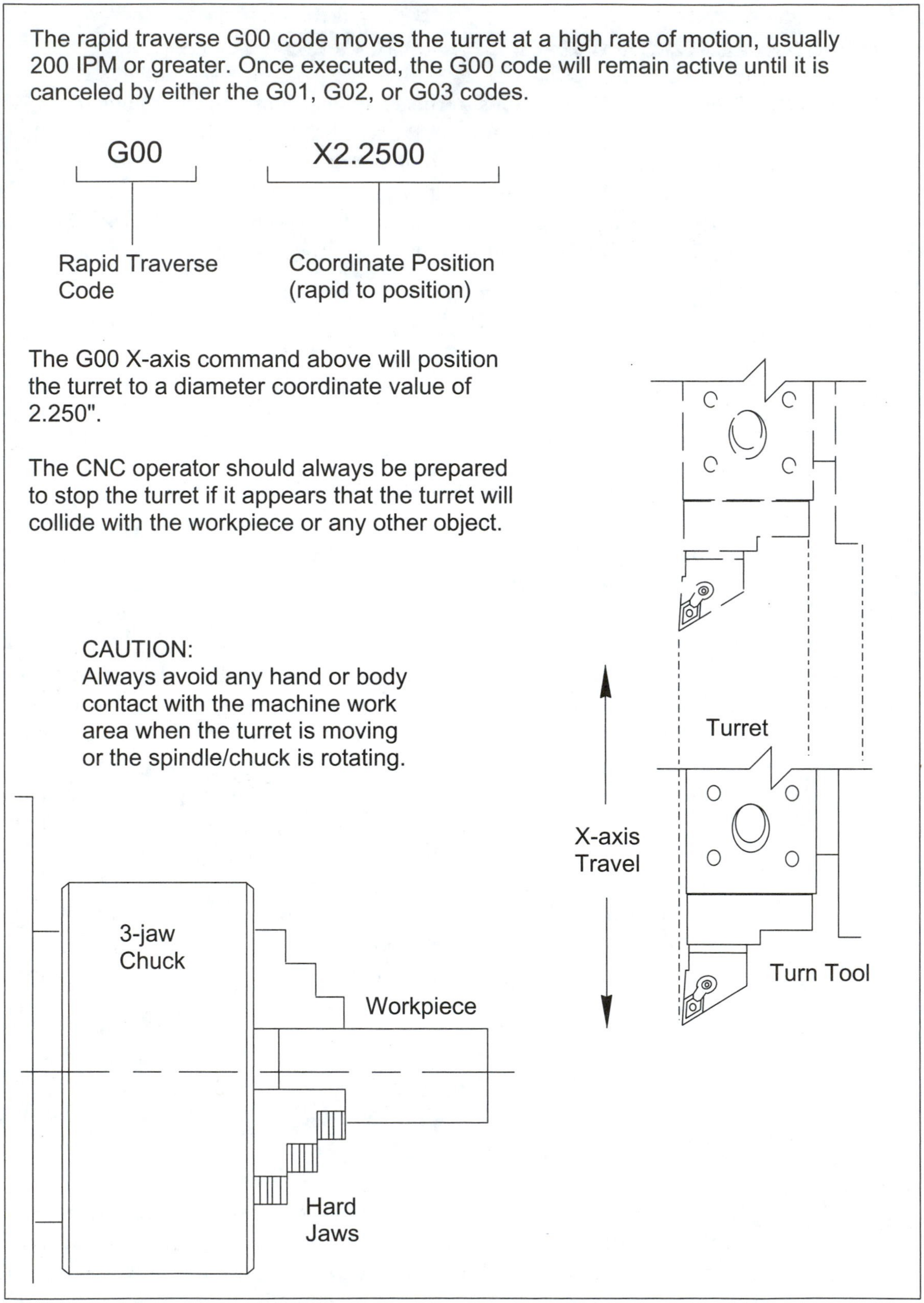

The rapid traverse G00 code moves the turret at a high rate of motion, usually 200 IPM or greater. Once executed, the G00 code will remain active until it is canceled by either the G01, G02, or G03 codes.

G00   X2.2500

Rapid Traverse   Coordinate Position
Code   (rapid to position)

The G00 X-axis command above will position the turret to a diameter coordinate value of 2.250".

The CNC operator should always be prepared to stop the turret if it appears that the turret will collide with the workpiece or any other object.

CAUTION:
Always avoid any hand or body contact with the machine work area when the turret is moving or the spindle/chuck is rotating.

Turret

X-axis
Travel

Turn Tool

3-jaw
Chuck

Workpiece

Hard
Jaws

Figure 16-1 **G00 rapid traverse in the X-axis**

The rapid traverse G00 code moves the turret at a high rate of motion, usually 200 IPM or greater. Once executed, the G00 code will remain active until it is canceled by either the G01, G02, or G03 codes.

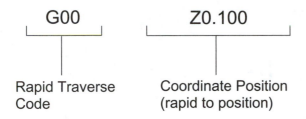

Rapid Traverse
Code

Coordinate Position
(rapid to position)

The G00 Z-axis command above will position the turret to a length-wise coordinate value of 0.100".

The CNC operator should always be prepared to stop the turret if it appears that the turret will collide with the workpiece or any other object.

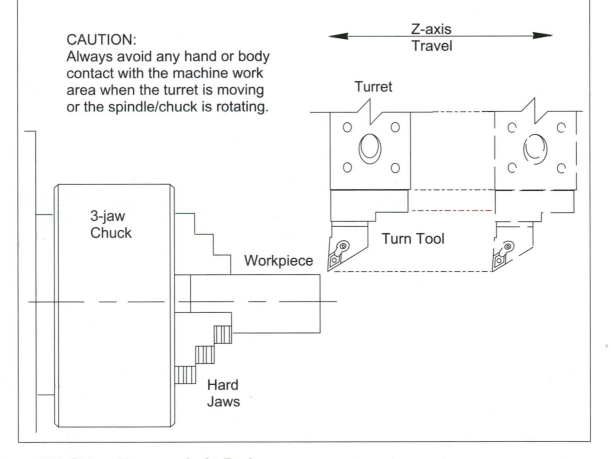

CAUTION:
Always avoid any hand or body contact with the machine work area when the turret is moving or the spindle/chuck is rotating.

Z-axis
Travel

Turret

Turn Tool

3-jaw
Chuck

Workpiece

Hard
Jaws

Figure 16-2 G00 rapid traverse in the Z-axis

The rapid traverse G00 code moves the turret at a high rate of motion, usually 200 IPM or greater. Once executed, the G00 code will remain active until it is canceled by either the G01, G02, or G03 codes.

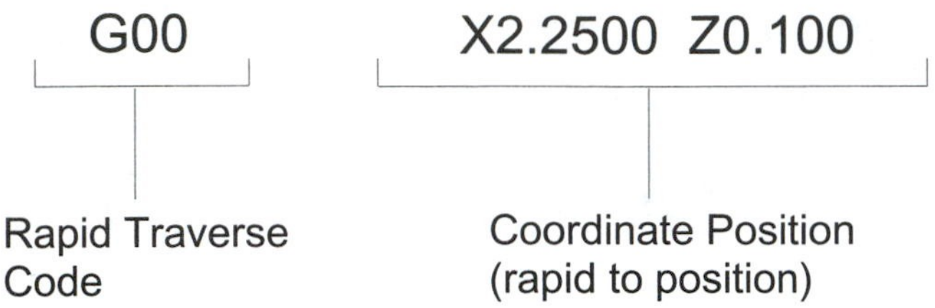

G00   X2.2500 Z0.100

Rapid Traverse   Coordinate Position
Code   (rapid to position)

The G00 X-axis and Z-axis command above will position the turret to a diameter coordinate value of 2.250" and a length-wise coordinate of 0.100". The X-Z axes will position simultaneously until either the X-axis or the Z-axis coordinate is reached, then the other axis will continue in a straight-line to the coordinate.

The CNC operator should always be prepared to stop the turret if it appears that the turret will collide with the workpiece or any other object.

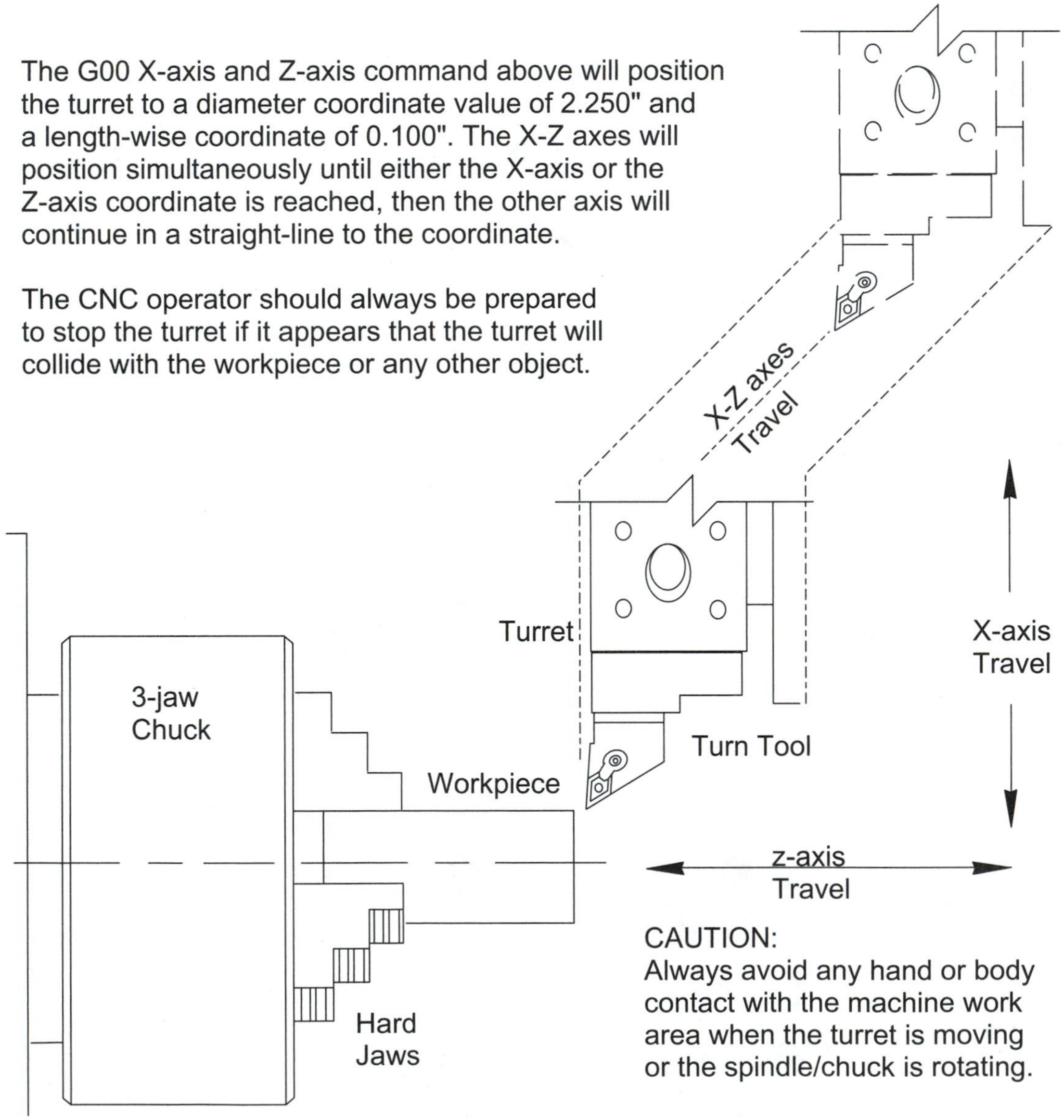

X-Z axes Travel

Turret

X-axis Travel

3-jaw Chuck

Turn Tool

Workpiece

z-axis Travel

Hard Jaws

CAUTION:
Always avoid any hand or body contact with the machine work area when the turret is moving or the spindle/chuck is rotating.

Figure 16-3 G00 rapid traverse in the X-axis and Z-axis

The feed G01 code moves the turret at a feedrate that is selected by the CNC programmer, usually between .001 to .030 IPR. Once executed, the G01 code will remain active until it is canceled by either the G00, G02, or G03 codes.

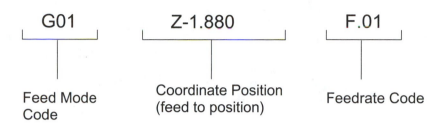

G01	Z-1.880	F.01
Feed Mode Code	Coordinate Position (feed to position)	Feedrate Code

The G01 Z-axis command above will position the turret/tool to the length-wise coordinate value of 1.880". This operation requires an F code, which is the feedrate that has been selected for the type of turn cut to be performed. It is important to note that the spindle must be rotating for the tool to cut.

The CNC operator should always be prepared to stop the turret/tool if it appears that the tool will collide with the chuck jaws or if the tool breaks.

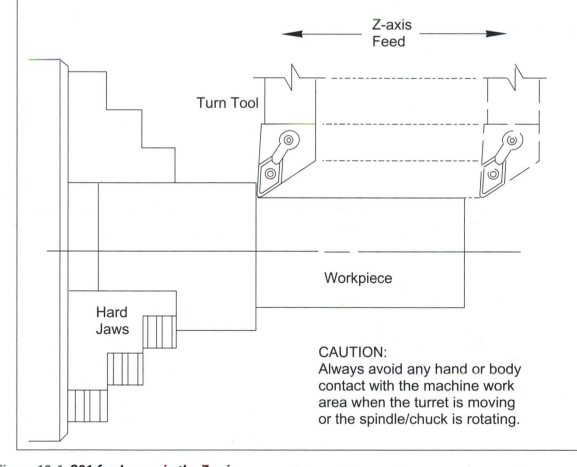

Z-axis Feed

Turn Tool

Workpiece

Hard Jaws

CAUTION:
Always avoid any hand or body contact with the machine work area when the turret is moving or the spindle/chuck is rotating.

*Figure 16-4 **G01 feed move in the Z-axis***

The feed G01 code moves the turret at a feedrate that is selected by the CNC programmer, usually between .001 to .030 IPR. Once executed, the G01 code will remain active until it is canceled by either the G00, G02, or G03 codes.

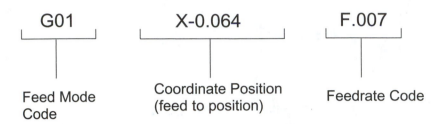

G01	X-0.064	F.007
Feed Mode Code	Coordinate Position (feed to position)	Feedrate Code

The G01 X-axis command above will position the turret/tool to the diameter coordinate value of -0.064". This operation requires an F code, which is the feedrate that has been selected for the type of face cut to be performed. It is important to note that the spindle must be rotating for the tool to cut.

The CNC operator should always be prepared to stop the turret/tool if it appears that the tool will collide with the chuck jaws or if the tool breaks.

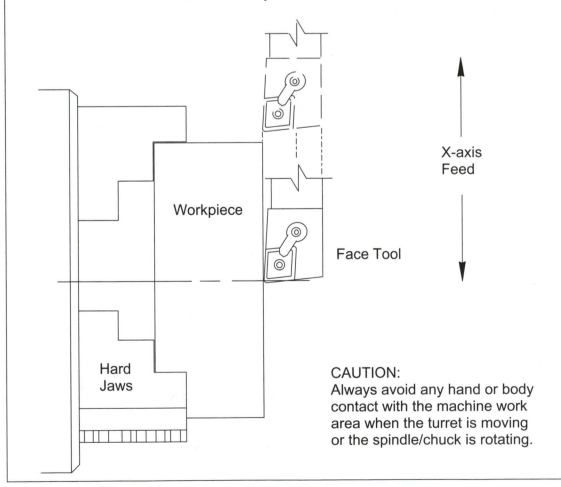

X-axis Feed

Workpiece

Face Tool

Hard Jaws

CAUTION:
Always avoid any hand or body contact with the machine work area when the turret is moving or the spindle/chuck is rotating.

Figure 16-5 **G01 feed move in the X-axis**

The feed G01 code moves the turret at a feedrate that is selected by the CNC programmer, usually between .001 to .030 IPR. Once executed, the G01 code will remain active until it is canceled by either the G00, G02, or G03 codes.

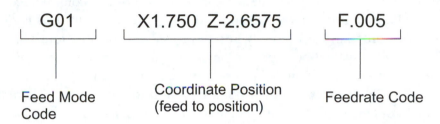

G01

Feed Mode Code

X1.750 Z-2.6575

Coordinate Position (feed to position)

F.005

Feedrate Code

The G01 X-axis and Z-axis command above will position the turret/tool to the diameter coordinate of 1.750" and to the length-wise coordinate value of 2.6575". This operation requires an F code, which is the feedrate that has been selected for the type of turn cut to be performed. It is important to note that the spindle must be rotating for the tool to cut.

The CNC operator should always be prepared to stop the turret/tool if it appears that the tool will collide with the chuck jaws or if the tool breaks.

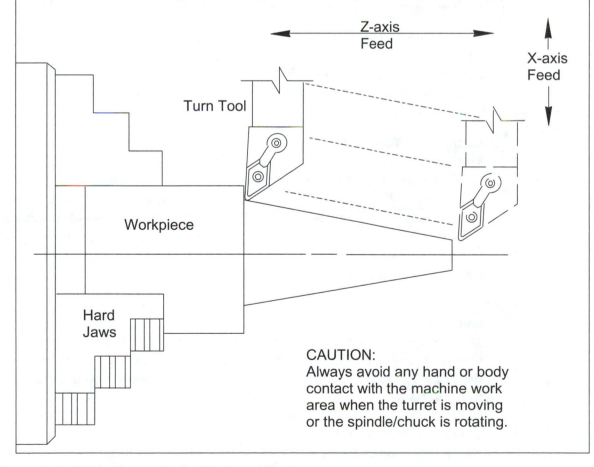

Z-axis Feed

X-axis Feed

Turn Tool

Workpiece

Hard Jaws

CAUTION:
Always avoid any hand or body contact with the machine work area when the turret is moving or the spindle/chuck is rotating.

Figure 16-6 ***G01 feed move in the X-axis and Z-axis***

The G01 code and the F code are also modal, and will remain active for each subsequent line until they are either changed or canceled.

When the G01 code along with an axis coordinate position and a feed rate is read (Cycle Start), the specified machine axis or axes will move at a relatively slow rate and cut the workpiece. This is only recommended when it is absolutely certain that a collision will not occur. Typically, this is also after the CNC program has been Dry Run or proved out. In situations where the feed moves have not been proved out, the CNC operator can override the feedrate to adjust or completely stop the feed movement.

The CNC program format that is common to most industrial applications is described in the CNC program exercises in the following chapters. The CNC programs in an industrial application should be created and maintained in a format that is consistent and as "operator-friendly" as possible. This will increase the efficiency of all related CNC operations throughout the plant.

INSTRUCTIONS FOR CNC PROGRAM EXAMPLE 6000

This section includes a coordinate worksheet, CNC documentation, and a CNC program for a CNC lathe. Each of these items is designed to illustrate the CNC setup and operation process. Therefore, it is important to understand the purpose and role of each of the following items:

Coordinate Worksheet

The coordinate worksheet is derived from the engineering drawing of this workpiece program example. It represents the CNC program coordinate requirements. Obtaining the dimensions from the engineering drawing will be required to complete it. The main purpose of this coordinate worksheet is to demonstrate how the X-Z coordinates of the CNC program are developed. It is only intended to be an instructional aid to help the student identify and read the coordinates in the CNC program.

Engineering Drawing

The engineering drawing is primarily used to determine that the workpiece is being manufactured to specifications. It is used to verify the part configuration, dimensions, tolerances, finish requirements, material, and datums locations on the workpiece.

A *process drawing*, which details only the specific process requirements, is sometimes substituted for the engineering drawing. It is used to define the workpiece requirements more clearly and eliminate items not included in that particular operation.

CNC Lathe Setup Plan

The setup plan is used to specify what device will be used to hold the workpiece, how the workpiece will be orientated, and where the Z-zero is located. The setup plan can be used to visualize and anticipate the machine moves from the CNC program.

CNC Lathe Tool List

The tool list is required to assemble and load the tools in the same order as the CNC program turret index commands. Therefore, the CNC operator can and should reference the tool list to verify the CNC program.

Instructions for CNC Program Example 6000

For this CNC program, the student is required to identify and note the main CNC program items listed below:

1. Program number

2. Information notes

3. Turret/tool indexes

4. Default code setting blocks

5. Rapid moves

6. Feed moves

The example illustrates how to identify and note the CNC program items above.

N130 T0100	*Turret/Tool Index to Station 1, and Offset Cancel*
N150 G0 X1.5 Z0.1 T0101	*Rapid Mode, to X1.5 and Z0.1, Activate Offset 1*
N160 G97 S1200 M3	*RPM Mode, 1200 RPM, Spindle On CW*
N170 G01 Z-1.88 F.012	*Feed Mode, Turn to Z-1.88, @ .012 IPR*

■ ■ ■ ■

COORDINATE WORKSHEET 6000

Instructions: Mark the X-Z zero origin, then calculate the coordinates for each point. Record the calculations in the chart. For the dimensions, use part drawing 6000 illustrated in Figure 16-7.

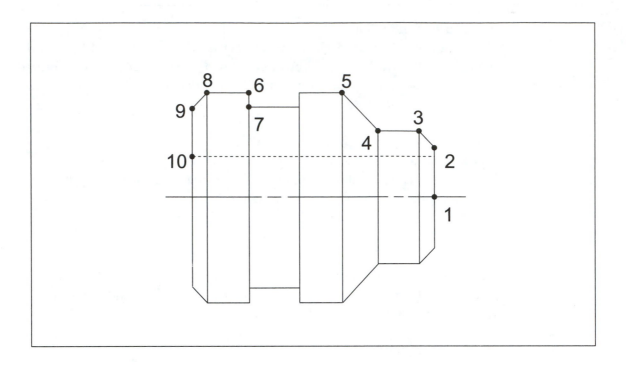

X- and Z-Coordinate Chart

PART NO.	X-COORDINATE	Z-COORDINATE
1		
2		
3		
4		
5		
6		
7		
8		
9		
10		

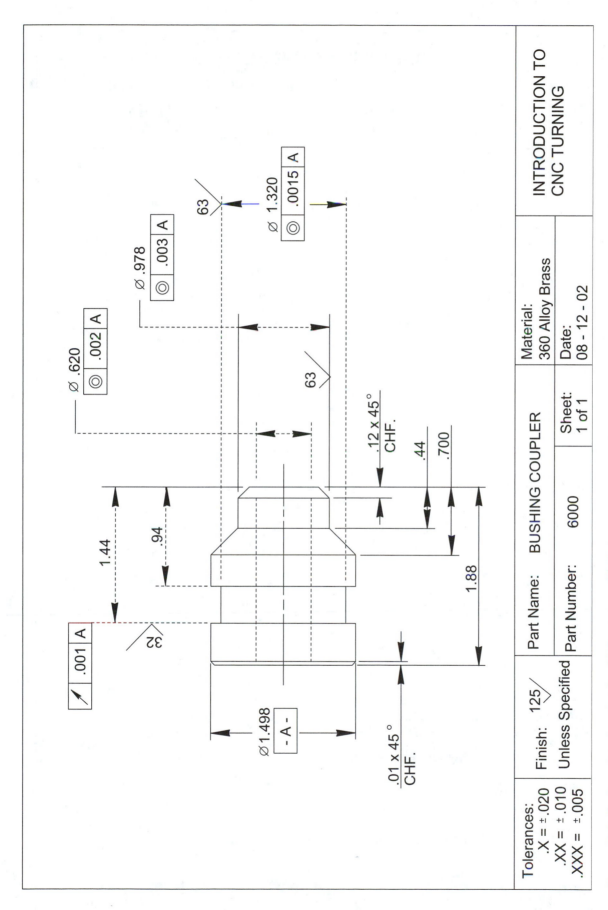

Figure 16-7 Engineering drawing for CNC program 6000

403

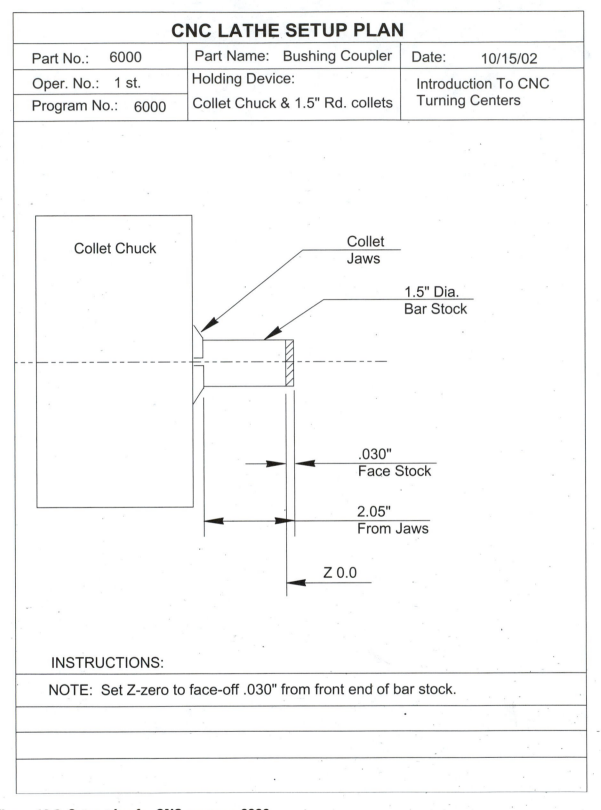

CNC LATHE SETUP PLAN

Part No.: 6000	Part Name: Bushing Coupler	Date: 10/15/02
Oper. No.: 1 st.	Holding Device:	Introduction To CNC Turning Centers
Program No.: 6000	Collet Chuck & 1.5" Rd. collets	

Collet Chuck

Collet Jaws

1.5" Dia. Bar Stock

.030" Face Stock

2.05" From Jaws

Z 0.0

INSTRUCTIONS:

NOTE: Set Z-zero to face-off .030" from front end of bar stock.

Figure 16-8 **Setup plan for CNC program 6000**

CNC TOOL LIST AND OPERATION

Part No.: 6000		Prog. No.: O6000	Oper. No.: 10
Part Name: Bushing Coupler		Material.: C.R. 1020	Date: 10/15/99

Station No.	Tool Offset	Tool No.	Tool Description and Operation
T01	01	8001	Holder No. DCLNR-164D Insert: CNMG-432 C2 Finish face & Rough turn .031 nose radius
T02	02	8022	Holder No. DDJNR-164D Insert: DNMG-432 C3 Finish turn .031 nose radius
T03	03	1015	.500" Dia. Drill - (carbide tip) Drill Thru.
T04	04	7002	Bar No. A06-STUNR2 Insert: TPG-221 C3 Finish bore .620 dia. .015 nose radius
T05	05	6007	Groove Holder No. NSR-163D Insert: NG-3R C2 Rgh. & Fin. Groove.
T05	05	8032	Cuttoff No. GSPR-125 Insert: GS-125N C2 Cutoff the part.

Figure 16-9 Tool list for CNC program 6000

■ ■ ■ ■

CNC PROGRAM EXAMPLE 6000

```
%
O6000
(BUSHING COUPLER - PROGRAM NO. 6000)
(DATE, 10/12/02)
(TOOL - 01 FINISH & ROUGH RIGHT - 80 DEG.)
(TOOL - 02 FINISH RIGHT - 55 DEG.)
(TOOL - 03 DRILL .5 DIA. CARBIDE)
(TOOL - 04 BORING BAR - .375 DIA.)
(TOOL - 05 OD GROOVE - .125 WIDE)
(TOOL - 06 CUTOFF - .125 WIDE)

N10G20G40G97
(TOOL - 01 RIGHT - 80 DEG.)
(FINISH FACE)
N15G0T0100
N20G97S1142M03
N25G0X2.3Z.1 T0101
N30G50S3600
N35G96S700
N40G1Z0.F.008M08
N45X-0.064
N50X.136Z.1
N55G0X2.0025

(ROUGH OD)
N60G96S600
N65G1Z-2.02F.012
N70X2.178
N75X2.3194Z-1.9493
N80G0Z.1
N85X1.827
N90G1Z-2.02
N95X2.0225
N100X2.1639Z-1.9493
N105G0Z.1
N110X1.6514
N115G1Z-2.02
N120X1.847
N125X1.9884Z-1.9493
N130G0Z.1
N135X1.4759
N140G1Z-.6731
N145X1.558Z-.7142
```

CNC PROGRAM EXAMPLE 6000 (*continued*)

N150Z-2.02
N155X1.6714
N160X1.8129Z-1.9493
N165G0Z.1
N170X1.3004
N175G1Z-.5854
N180X1.4959Z-.6831
N185X1.6373Z-.6124
N190G0Z.1
N195X1.1249
N200G1Z-.4976
N205X1.3204Z-.5954
N210X1.4618Z-.5247
N215G0Z.1
N220X.9494
N225G1Z-.0898
N230X1.038Z-.1342
N235Z-.4542
N240X1.1449Z-.5076
N245X1.2863Z-.4369
N250G0Z.1
N255X.7738
N260G1Z-.0021
N265X.9694Z-.0998
N270X1.1108Z-.0291
N275G0X2.2Z.1M09
N280G28U0.W0.M05
N285 T0100
N290M01

(TOOL - 2 OFFSET - 2)
(FINISH OD)
N295G0G20G40G97T0200
N300G97S3072M03
N305G0X.8703Z.0661T0202
N310G50S3600
N315G96S700
N320G1X.7288Z-.0046F.008M08
N325X.9659Z-.1231
N330G3X.978Z-.1377R.0206
N335G1Z-.4492
N340X1.4859Z-.7031
N345G3X1.498Z-.7177R.0206

■ ■ ■ ■
CNC PROGRAM EXAMPLE 6000 (*continued*)

N350G1Z-2.03
N355X2.178
N360G0X2.2Z.1M09
N365G28U0.W0.M05
N370T0200
N375M01

(TOOL - 3 OFFSET - 3)
(FINISH GROOVE)
N380G0 G20G40G97T0300
N385G97S700M03
N390G0X1.9Z.1 T0303
N395G50S3600
N400G96S300
N405 X1.638Z-1.065M08
N410G1X1.438F.003
N415Z-1.44
N420X1.638
N425G0Z-1.065
N430X1.578
N435G1X1.378
N440Z-1.44
N445X1.578
N450G0Z-1.065
N455X1.518
N460G1X1.32
N465Z-1.44
N470X1.518
N475G0X2.2Z.1M09
N480G28U0.W0.M05
N485T0300
N490M01

(TOOL - 4 OFFSET - 4)
(DRILL .5 DIA.)
N495G0G20G40G97T0400
N500G97S1500M03
N505G0X0.Z.25 T0404
N510Z.1M08
N515G1Z-2.1502F.011
N520G0Z.25M09
N525G28U0.W0.M05

CNC PROGRAM EXAMPLE 6000 (*continued*)

N530T0400
N535M01

(TOOL - 5 OFFSET - 5)
(FINISH BORING BAR .375 DIA.)
N540G0G20G40G97T0500
N545G97S1219M03
N550G0X.6267Z.1166T0505
N555G50S3600
N560G96S700
N565G1Z.0166F.007M08
N570X.5897Z-.0019
N575G2X.56Z-.0377R.0506
N580G1Z-1.98
N585X.5486Z-1.9093
N590G0Z.0954
N595X.6692
N600G1Z-.0046
N605X.6321Z-.0231
N610G2X.62Z-.0377R.0206
N615G1Z-1.98
N620X.5786Z-1.9093
N625G0Z.2M09
N630G28U0.W0.M05
N635T0500
N640M01

(TOOL - 6 OFFSET - 6)
(CUTOFF TOOL .125 WIDE)
N645G0G20G40G97T0600
N650G97S675M03
N655G0X1.698Z.1T0606
N660G50S3600
N665G96S300
N670Z-1.995M08
N675G1X1.498F.02
N680X1.478Z-2.005F.002
N685X.55
N690G0Z.1M09
N695G28U0.W0.M05
N700T0600
N705M30
%

CHAPTER SUMMARY

The key concepts presented in this chapter include the following:

- The G00 rapid traverse code and machine movement descriptions.

- The G01 feed (linear interpolation) code and machine movement descriptions.

- The descriptions of X- and Z-axis movements.

- The description of turret and tool movements.

- Calculations for CNC lathe coordinates.

- The description of an engineering drawing for CNC lathe machining.

- The description of a setup plan for CNC lathe machining.

- The description of a tool list for CNC lathe machining.

- The format description of a CNC program for CNC lathe machining.

- The basic codes required of a CNC program for CNC lathe machining.

REVIEW QUESTIONS

1. When moving in rapid G00 mode what must the CNC operator be careful to avoid?

2. Why is Dry Run and prove out required before running the CNC lathe in automatic mode?

3. Which lathe component moves when the control reads **G00 X 2.000**?

4. Which lathe component moves when the control reads **G00 Z .100**?

5. Which lathe component moves when the control reads **T0100**?

6. How are the following codes canceled: G00, G01?

7. What type of information can be derived from the CNC program manuscript?

8. List the two documents that are required to setup a CNC lathe.

9. Why is the engineering drawing required for a CNC lathe operation?

10. Discuss and identify the CNC program format structure and codes covered in this chapter. The items that should be identified include CNC program number, tool information notes, tool index blocks, default code setting blocks, spindle starts, coolant on, machine home command blocks, tool offset activation blocks.

CNC PROGRAM EXERCISE FOR CNC LATHES

Using the sketch in Figure 16-10, calculate the G00 rapid and G01 feed moves for the X- and Z-axes. Enter the codes and coordinates in the CNC program exercise on the following page.

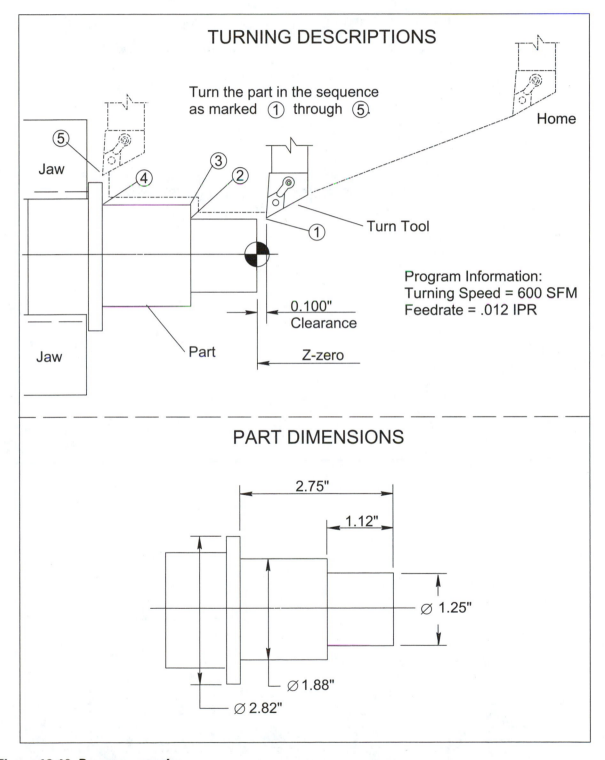

Figure 16-10 *Program exercise*

CNC PROGRAM EXERCISE 6001

Fill in the missing program information of CNC program 6001 as indicated below:

Block N130: enter the G code and X-Z coordinate values.

Block N150: enter the SFM cutting speed value.

Block N160: enter the G code, Z coordinate value, and IPR value.

Block N170: enter the X-coordinate value.

Block N180: enter the Z-coordinate value.

Block N190: enter the X-coordinate value.

```
%
O6001
(ROLL GUIDE - PROGRAM NO. 6001 DATE, 03/18/02)
(TOOL - 01 FINISH 55 - DEGREE)

N100 G20 G40 G97
(TOOL - 01 55 - DEG.)
N110 G0 T0100
N120 G97 S1900 M03
N130 G_____ X _____ Z _____ T0101
N140 G50 S5000
N150 G96 S_____
N160 G_____ Z _____ F_____ M08
N170 X _____
N180 Z _____
N190 X _____
N200 G97 G0 Z1.0 M09
N210 G28 U0.0 W0.0 M05
N220 T0100
N230 M30
%
```

chapter 17

CNC LATHE CIRCULAR INTERPOLATION

Chapter Objectives

After studying and completing this chapter, the student should have knowledge of the following:

- *G02 and G03 circular interpolation codes*

- *M03, M04, M05, and S spindle codes*

- *CW radius feed movement*

- *CCW radius feed movement*

- *Radius feed features*

- *Radius feed methods*

- *Calculating X-Z coordinates*

- *CNC program format descriptions*

CIRCULAR INTERPOLATION

This chapter introduces circular interpolation, which is the CNC method used to cut circular features on a workpiece. The performance of circular interpolation basically requires a code (G02 or G03), the arc radius size, the start point coordinates, and the end point coordinates of the arc. For CNC lathes, circular interpolation is typically performed using the insert indexable turn tool, which include 80°, 55°, and 35° toolholders. The examples in this chapter describe and illustrate circular interpolation as it is performed using similar insert toolholders.

The first illustration in this chapter (Figure 17-1) describes the radius turning features, codes, coordinate values, and movements associated with turning CW arcs. The illustration that follows (Figure 17-2) describes the codes, coordinate values, and movements associated with turning CCW arcs. These figures are designed to illustrate the differences between CW and CCW radius turning. Additionally, this chapter includes a coordinate calculation worksheet and a sample program to identify basic CNC program format.

As previously mentioned, an arc or radius cut is performed on CNC lathes by the process called "circular interpolation." The G02 and G03 codes are designated to perform this function. Both codes are modal and each will remain active for all subsequent lines until a G00 or G01 cancels either code. Therefore, any axis coordinate present in a subsequent line of the CNC program will move in radial feed mode whether the G02 or G03 code is present in the subsequent line or not.

G02 CIRCULAR INTERPOLATION CW

The G02 code is used to cut a radius in a clockwise direction (see Figure 17-1). It basically moves the X- and Z-axes simultaneously clockwise at a predetermined feedrate to cut a specific radius on the workpiece. Therefore, after the G02 code, a combination of two axis coordinates will follow.

The coordinate position of the turret/tool, which is represented by the axis letters (X and Z), will include a positive or negative valued number. Additionally, the G02 code requires a code for the radius value. This value, which is represented by the letter "R" or the letters "I" and "K," is typically derived from the engineering drawing specifications. Because the circular interpolation code is intended for cutting, the spindle/chuck must be rotating and a feedrate value (F.008, for example) is also required.

When cutting a radius in the X- and Z-axes a standard insert with a specified TNR is used. Therefore, when cutting a radius, it is necessary to

The feed G02 code moves the turret at a feedrate that is selected by the CNC programmer, usually between .001 to .030 IPR. Once executed, the G02 code will remain active until it is canceled by either the G00, G01, or G03 codes.

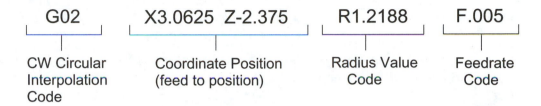

G02	X3.0625 Z-2.375	R1.2188	F.005
CW Circular Interpolation Code	Coordinate Position (feed to position)	Radius Value Code	Feedrate Code

The G02 CW circular interpolation command is designed to feed both the X-axis and Z-axis in a radial tool path. This command will position the turret/tool to the diameter coordinate of 3.0625 and to the length-wise coordinate value of −2.375 at a feedrate of .005 IPR. This radius, illustrated below is programmed as follows:

G00X0.625Z0.1   (position to turn the .625 diameter)
G01Z-1.1562F.005   (finish turn and position to start of radius)
G02X3.0625Z-2.375R1.2188 (finish radius)

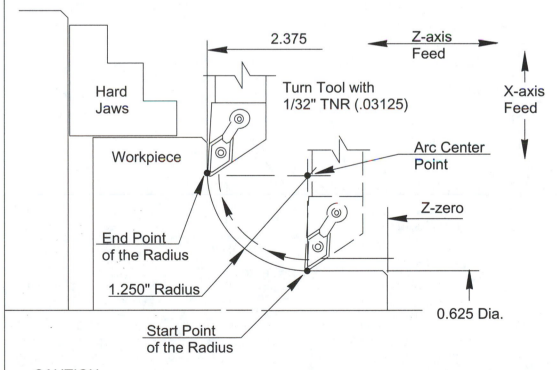

Figure 17-1 Circular interpolation G02 code

CAUTION:
Always avoid any hand or body contact with the machine work area when the turret is moving or the spindle/chuck is rotating.

The feed G03 code moves the turret at a feedrate that is selected by the CNC programmer, usually between .001 to .030 IPR. Once executed, the G03 code will remain active until it is canceled by either the G00, G01, or G02 codes.

G03	X2.875 Z-0.7812	R0.7812	F.005
CW Circular Interpolation Code	Coordinate Position (feed to position)	Radius Value Code	Feedrate Code

The G03 CCW circular interpolation command is designed to feed both the X-axis and Z-axis in a radial tool path. This command will position the turret/tool to the diameter coordinate of 2.875" and to the length-wise coordinate value of .7812" at a feedrate of .005 IPR. This radius, illustrated below is programmed as follows:

G00X1.3125Z0.1   (position to X-start diameter)
G01Z0.0F.005   (position to Z-start of radius)
G03X2.875Z-0.7812R0.7812   (finish radius)

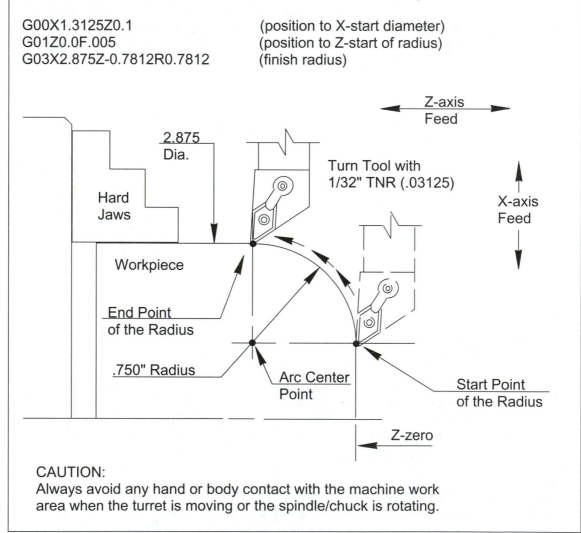

CAUTION:
Always avoid any hand or body contact with the machine work area when the turret is moving or the spindle/chuck is rotating.

Figure 17-2 Circular interpolation G03 code

compensate for the insert TNR. This is typically accomplished by either of two methods, which include calculating the compensation or programming TNR compensation codes into the CNC program. The TNR compensation codes commonly used to compensate for the insert TNR are the G41 and G42 codes. The examples in this section will be described and illustrated using the calculation method of determining the TNR compensations. The G41 and G42 TNR compensation codes are described and illustrated in the next chapter.

When the G02 code along with an axis coordinate position is read (Cycle Start), the specified machine axes will move automatically at a relatively slow feedrate and cut the radius on the workpiece. Initiating a cycle start is only recommended when it is absolutely certain that a collision will not occur. Typically this is after the CNC program has been Dry Run or proved out. In cases where the radius feed moves have not been proved out, the CNC operator can override the feedrate to adjust or completely stop the radius feed movement. This will allow the CNC operator time to react to a possible collision, heavy cuts, or an excessive feedrate and stop the movement before any damage occurs.

G03 CIRCULAR INTERPOLATION CCW

The G03 code is used to cut a radius in a CCW direction (see Figure 17-2). Similar to the G02 code, it basically moves the X- and Z-axes simultaneously CCW at a predetermined feedrate to cut a specific radius on the workpiece. The G03 and the G02 codes are similar in all respects except direction of cut.

SPINDLE FUNCTION CODES

As previously stated, metal cutting on a CNC machine requires the spindle/chuck to be rotating as the tool feeds into the workpiece. For this reason, the machine spindle is primarily used to rotate the spindle/chuck at a predetermined speed (RPM). The RPM is first calculated and then programmed by using the S code along with the numerical RPM value.

M03, M04, and S Codes
The M03 code is used to rotate the spindle in a CW direction. The M04 code is used to rotate the spindle in the CCW direction. To start the spindle rotation the CNC program must read the S code and RPM value (S2500, for

example) either before or at the same time that the M code is read. It is typically programmed as follows:

S2500 M03 *(both codes on the same block)*

These spindle rotation code descriptions are applied to all cutting situations. The M03, M04, and S codes are required for all types of machining operations. They are programmed and executed in the same manner for all machining situations.

M05 Spindle Stop Code

The M05 code is used in situations when the spindle rotation needs to be stopped. For instance, some situations when the spindle stop M05 code may be required include the following:

- To load and unload the workpiece

- To inspect a machined feature size

- To remove metal chips

- To reverse the spindle rotation

- To shift spindle gear range

The CNC lathe control can read the CNC program codes in various format configurations. However, the codes should be structured in a consistent format to increase the efficiency of reading CNC programs.

INSTRUCTIONS FOR CNC PROGRAM EXAMPLE 7000

This section includes a coordinate worksheet, CNC documentation, and a CNC program for a CNC lathe. As explained in chapter 16, each of these items is designed to illustrate the CNC setup and operation process. Therefore, it is important to review and understand the purpose and role of each CNC item. These items are listed below:

- Coordinate worksheet (for instructional purposes)

- Engineering drawing/process drawing

- CNC setup plan

- CNC tool list

This program is similar in format to the other programs in this manual. The similarities that should be noted include

- Program number

- Information notes

- Turret indexes

- Default code setting

- Spindle starts

- Coolant on

- Machine home command

- Geometry offset activation

- End of program

Instructions for CNC Program 7000

For this CNC program, the student is required to identify and note the main CNC program items that are listed below:

1. Program number

2. Information notes

3. Tool indexes

4. Default code setting blocks

5. Rapid moves

6. Feed moves

7. Angle feed moves

8. End of program

The example below illustrates how to identify and note the CNC program items above.

O7000	*Program Number*
N10 G20G40G97	*Default Code Setting Block*
(TOOL - 01 OFFSET - 01)	*Tool Information Note*
N20 T0500M41	*Turret Index 5, High Range*
N30 M01	*Optional Stop*
N40 G00X2.0Z. 1T0505	*Rapid X&Z, Activate Offset 5*
N50 S1142M03	*1142 RPM, Spindle On CW*
N60 G01Z-1.135F.012	*Feed to Z-1.135 at .012 IPR*

■ ■ ■ ■

COORDINATE WORKSHEET 7000

Instructions: Mark the X-Z zero origin, then calculate the coordinates for each point. Record the calculations in the chart. For the dimensions, use part drawing no. 7000 illustrated in Figure 17-3.

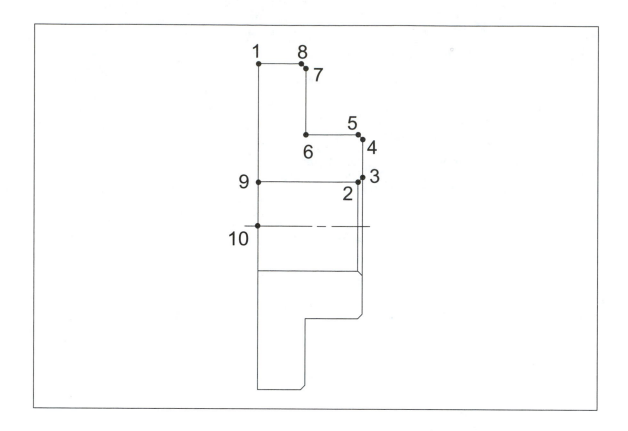

X- and Z-Coordinate Chart

PART NO.	X-COORDINATE	Z-COORDINATE
1		
2		
3		
4		
5		
6		
7		
8		
9		
10		

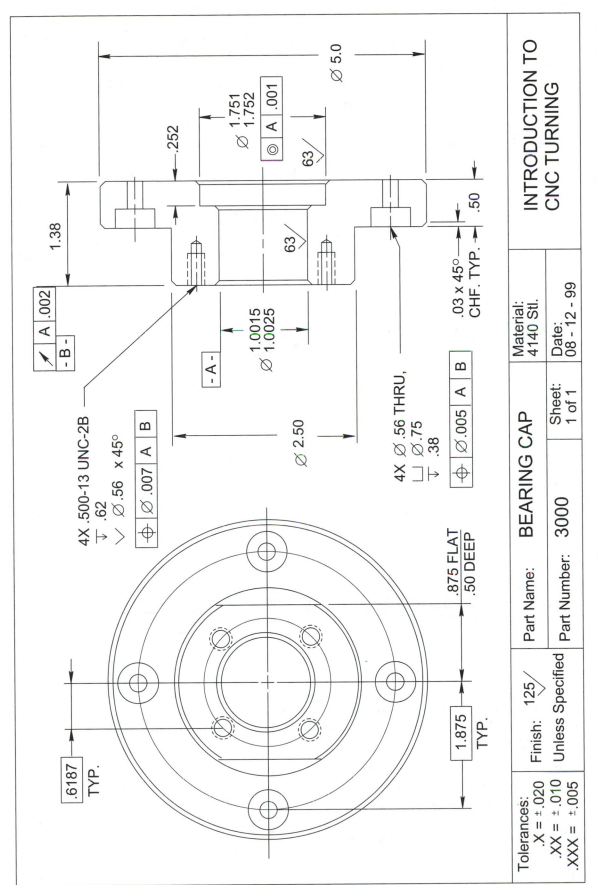

INTRODUCTION TO CNC TURNING

Part Name:	BEARING CAP	Material: 4140 Stl.
Part Number: 3000		Sheet: 1 of 1
		Date: 08 - 12 - 99

Finish: 125 / Unless Specified

Tolerances:
.X = ±.020
.XX = ±.010
.XXX = ±.005

Figure 17-3 Engineering drawing for CNC program 7000

421

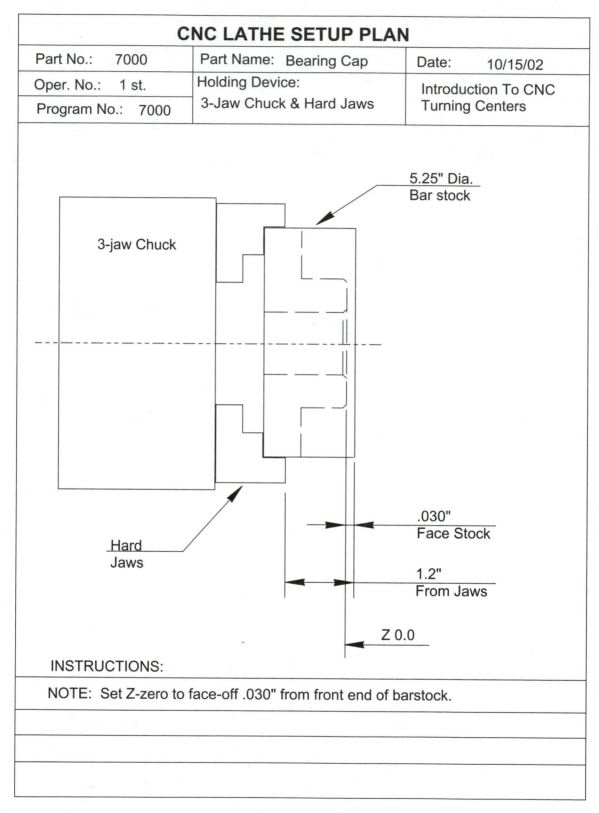

CNC LATHE SETUP PLAN

Part No.: 7000	Part Name: Bearing Cap	Date: 10/15/02
Oper. No.: 1 st.	Holding Device:	Introduction To CNC Turning Centers
Program No.: 7000	3-Jaw Chuck & Hard Jaws	

3-jaw Chuck

5.25" Dia. Bar stock

Hard Jaws

.030" Face Stock

1.2" From Jaws

Z 0.0

INSTRUCTIONS:

NOTE: Set Z-zero to face-off .030" from front end of barstock.

Figure 17-4 **Setup plan for CNC program 7000**

CNC TOOL LIST AND OPERATION

Part No.: 7000		Prog. No.: O7000	Oper. No.: 10
Part Name: Bearing Cap		Material.: C.R. 1020	Date: 10/15/99

Station No.	Tool Offset	Tool No.	Tool Description and Operation
T01	01	8001	Holder No. DCLNR-164D Insert: CNMG-432 C5 Rough face & turn .031 nose radius
T02	02	8022	Holder No. DDJNR-164D Insert: DNMG-432 C6 Finish face & turn .031 nose radius
T03	03	1015	.750" Dia. Drill - (carbide tip) Drill Thru.
T04	04	7002	Bar No. A06-STUNR2 Insert: TPG-221 C3 Rough bore .690 dia. .015 nose radius

Figure 17-5 Tool list for CNC program 7000

■ ■ ■ ■

CNC PROGRAM EXAMPLE 7000

```
%
O7000
(BEARING CAP - PROGRAM NO. 7000)
(DATE, 10/12/02)
(TOOL - 01  ROUGH RIGHT - 80 DEG.)
(TOOL - 02  FINISH RIGHT - 55 DEG.)
(TOOL - 03  DRILL .75 DIA.  CARBIDE)
(TOOL - 04  BORING BAR - .500 DIA.)

N10G20G40G97
(TOOL - 1 OFFSET - 1)
(FINISH FACE & ROUGH TURN)
N15G28U0.W0.
N20G0T0100M41
N25G50S5000
N30G97S924M3
N35 X5.4Z.1 T0101M8
N40G96S600
N45 G1Z0.F.02
N50X-.064F.012
N55G0Z.1
N60 X5.0066
N65G1X4.9066F.02
N70Z-.87F.012
N75X4.9516
N80X5.1066Z-.9474
N85X5.1266
N90G0 Z.1
N95X4.8066
N100G1X4.7066F.02
N105Z-.87F.012
N110X4.9266
N115G0 Z.1
N120X4.6066
N125G1X4.5066F.02
N130Z-.87F.012
N135X4.7266
N140G0 Z.1
N145X4.4066
N150G1X4.3066F.02
N155Z-.87F.012
N160X4.5266
N165G0 Z.1
```

■ ■ ■ ■
CNC PROGRAM EXAMPLE 7000 (*continued*)

N170X4.2066
N175G1X4.1066F.02
N180Z-.87F.012
N185X4.3266
N190G0 Z.1
N195X4.0066
N200G1X3.9066F.02
N205Z-.87F.012
N210X4.1266
N215G0 Z.1
N220X3.8066
N225G1X3.7066F.02
N230Z-.87F.012
N235X3.9266
N240G0 Z.1
N245X3.6066
N250G1X3.5066F.02
N255Z-.87F.012
N260X3.7266
N265G0 Z.1
N270X3.4066
N275G1X3.3066F.02
N280Z-.87F.012
N285X3.5266
N290G0 Z.1
N295X3.2066
N300G1X3.1066F.02
N305Z-.87F.012
N310X3.3266
N315G0 Z.1
N320X3.0066
N325G1X2.9066F.02
N330Z-.87F.012
N335X3.1266
N340G0 Z.1
N345X2.8066
N350G1X2.7066F.02
N355Z-.87F.012
N360X2.9266
N365G0 Z.1
N370X2.66
N375G1X2.56F.01

CNC PROGRAM EXAMPLE 7000 (*continued*)

N380Z-.87F.012
N385X2.7266
N390G0X5.1616
N395X5.4Z.1
N400G28U0.W0.M5
N405T0100
N410M01

(TOOL - 2 OFFSET - 2)
(FINISH TURN - 55 DEG.)
N415G28U0.W0.
N420G0T0200M41
N425G50S5000
N430G97S1104M3
N435X2.4217Z.1T0202M8
N440G96S700
N445G1X2.4216Z-.0092F.01
N450X2.5Z-.0483F.007
N455Z-.88
N460X4.9034
N465X5.0524Z-.9545
N470G0Z.1
N475M9
N480G28U0.W0.M5
N485T0200
N490M01

(TOOL - 3 OFFSET - 3)
(DRILL .75 DIA.)
N495G28U0.W0.
N500G0T0300M41
N505G97S1500M3
N510X0.Z.25 T0303M8
N515Z.1
N520G1Z-1.6502F.009
N525G0Z.1
N530Z1.0
N535M9
N540G28U0.W0.M5
N545T0300
N550M01

CNC PROGRAM EXAMPLE 7000 (*continued*)

```
(TOOL - 4 OFFSET - 4)
(FINISH BORE, MIN. .5 DIA.)
N555G28U0.W0.
N560G0T0400M41
N565G50S5000
N570G97S2121M3
N575 X1.0803Z.25T0404M8
N580Z.1
N585G96S600
N590G1X1.0804Z-.0092F.01
N595X.882Z-.1083F.007
N600Z-1.5
N605X.842Z-1.48
N610G0Z1.0
N615M9
N620G28U0.W0.M5
N625T0400
N630M30
%
```

CHAPTER SUMMARY

The following key concepts that were presented in this chapter include:

- The G02 CW circular interpolation code and turret/tool movement descriptions.

- The G03 CCW circular interpolation code and turret/tool movement descriptions.

- The M03 spindle/chuck CW rotation code description.

- The M04 spindle/chuck CCW rotation code description.

- The S spindle RPM speed code description.

- The M05 spindle stop code description.

- The description of radius turning on CNC lathes.

- Calculations for CNC lathe coordinates.

- The format description of a CNC program for CNC lathes.

- The basic codes required of a CNC program for CNC lathes.

REVIEW QUESTIONS

1. Which type of tool is typically used for cutting a radius (G02 or G03) on a CNC lathe?

2. What control feature/s can the operator use to verify that the programmed circular interpolation feed moves are correct and at optimum level?

3. Which CNC lathe components move when the control reads **G02 X2.0 Z-1.5 R.5**?

4. When cutting a G02 or G03 circular interpolation, which two codes specify the programmed radius?

5. What other codes are required for the G02/G03 to function properly?

6. Which CNC lathe components are activated when the control reads an M03 or M04?

7. What other code is required for the M03/M04 to function properly?

8. Which codes cancel the G02 or G03 in a CNC lathe program?

9. When G02 or G03 circular interpolation is programmed, what item is used to compensate the X-Z axes?

10. Discuss and identify the CNC program format structure and codes covered in this chapter. The items that should be identified include CNC program number, tool information notes, tool index blocks, default code setting blocks, spindle starts, coolant on and off, machine home command blocks, tool offset activation blocks.

CNC PROGRAM EXERCISE FOR CNC LATHES

Using the sketch in Figure 17-6 calculate the G02 and G03 circular feed moves for the X- and Z-axes. Enter the codes and coordinates in the CNC program on the following page. Note: Compensate for TNR when calculating G02/G03 coordinates.

CNC PROGRAM EXERCISE 7001

Fill in the missing program information of CNC program 7001 as indicated below:

Block N130: enter the G code and X-Z coordinate values.

Block N150: enter the SFM cutting speed value.

Block N160: enter the G code, Z-coordinate value, and IPR value.

Block N170: enter the G code, X-Z coordinates, and R value.

Block N180: enter the G code and Z-coordinate value.

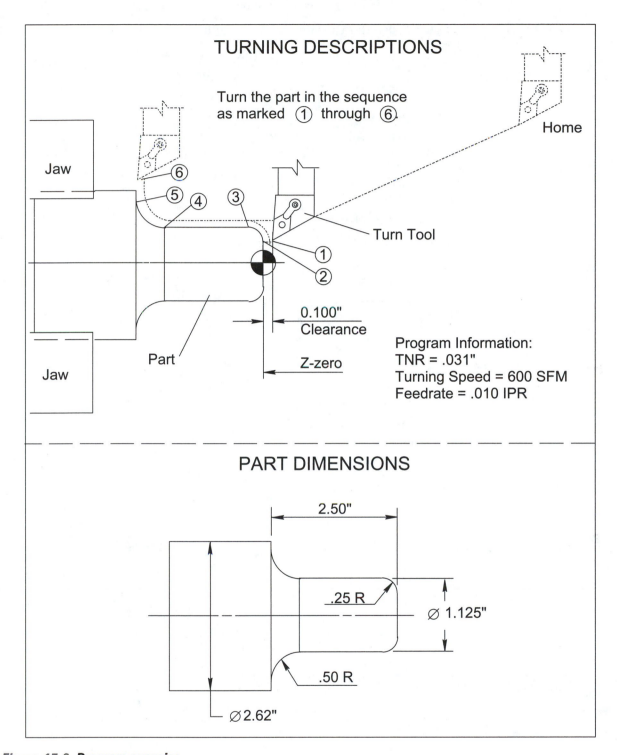

Figure 17-6 **Program exercise**

Block N190: enter the G code, X-Z coordinates, and R value.

Block N200: enter the G code and X-coordinate value.

```
%
O7001
(SHAFT PIN - PROGRAM NO. 7001 DATE, 11/23/01)
(TOOL - 01  FINISH 55 - DEGREE)

N100 G20 G40 G97
(TOOL - 01  55 - DEG.)
N110 G0 T0100
N120 G97 S1900 M03
N130 G _____ X _____ Z _____ T0101
N140 G50 S5000
N150 G96 S _____
N160 G _____ Z _____ F _____ M08
N170 G _____ X _____ Z _____ R _____
N180 G _____ Z _____
N190 G _____ X _____ Z _____ R _____
N200 G _____ X _____
N210 G97 G0 Z1.0 M09
N220 G28 U0.0 W0.0 M05
N230 T0100
N240 M30
%
```

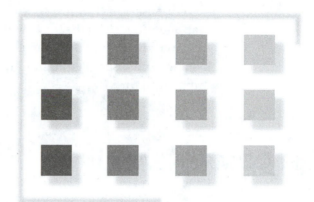

chapter 18

TOOL NOSE RADIUS COMPENSATION

Chapter Objectives

After studying and completing this chapter, the student should have knowledge of the following:

- **G41 tool nose radius compensation left**

- **G42 tool nose radius compensation right**

- **G40 tool nose radius cancel**

- **T-code offset activate and cancel**

- **G28 machine home return**

- **G50, G96, and G97 spindle mode codes**

- **Calculating X-Z Coordinates**

- **CNC program format descriptions**

The set of illustrations in this chapter (Figures 18-1 to 18-4) describe the TNR compensation features, codes, and methods. TNR compensation is typically applied to CNC lathe parts with taper and radius features. Additionally, this chapter includes other miscellaneous codes, a coordinate calculation worksheet, and a sample program to familiarize the basic CNC lathe program format structure.

TNR COMPENSATION

The typical application for TNR compensation is CNC lathe parts with taper and/or radius features. The TNR compensation is either calculated by the programmer, or calculated by the CNC control when it reads the programmed G41 or G42 codes from the CNC program. These codes, once activated, will command the CNC lathe control to calculate and then compensate the appropriate axis accordingly.

The G41/G42 method also requires data input into the CNC control offset memory by the CNC operator or CNC setup person. The required data includes the insert TNR size, which is specified by the insert identification number, and the direction that the tool is oriented (see Figure 18-1). The offset number on the CNC program that activates TNR compensation must be matched to the tool specified on the tool list and to the offset memory number on the CNC control. When the control reads the G41 or G42 code, the appropriate axis letters (X or Z) will shift by the amount that is stored in the memory (see Figure 18-2). Likewise, when the TNR compensation is canceled, it will shift the axis away by the same amount. It is important to note that, while TNR compensation is active, the tool can change direction with another axis (X to Z, or Z to X), but if the same axis (X to X, or Z to Z) changes direction 180° an alarm will occur. Additionally, because this code is intended for cutting applications, the spindle/chuck must be rotating and a feedrate value is required.

When the G41 or G42 code along with an axis coordinate position is read (Cycle Start), the specified machine axes will shift and move automatically to cut the workpiece. This is only recommended when it is absolutely certain that collisions will not occur. Typically this is after the CNC program has been Dry Run or proved out. In cases where the compensation moves have not been proved out, the CNC operator can use the override features to slow the axis movement or completely stop the axis movement. This will allow the CNC operator time to react to a possible collision and stop the movement before any damage occurs.

G41 TNR Compensation Left

The G41 TNR compensation code is used to shift the axes (X, Z) by a specified radius amount to the left of the workpiece diameter or face surface (see Figure 18-3). The TNR compensation (left or right) is determined by viewing the back of the tool as it feeds away. If the tool is to the left of the part surface, the G41 code is programmed. The position (X-Z) prior to TNR activation must be away from the part surface by an amount greater than twice the TNR compensation amount.

G42 TNR Compensation Right

The G42 TNR compensation code (see Figure 18-4) performs the same function as the G41 TNR compensation code. The only difference is that G42 is used to shift the insert tool to the right. All the items described for the G41 code also apply to the G42 code.

TNR and Vector Orientation

The TNR size for inserts, as previously described, is available in standard sizes. The common sizes include 1/64 inch (0.0156) and 1/32 inch (0.0312). However, the TNR size specified must be setup accordingly for each tool. This is important because the CNC control accesses the TNR size input by the setup person to calculate the amount of compensation. This calculation is performed each time that the CNC program executes a programmed X-Z axes simultaneous move with G41 or G42 active.

The theoretical tool tip vector number is used to specify the direction and orientation of tool tip. The machine controller will use this input number to determine the direction of the tool movement when a compensation command (G41 or G42) is executed.

G40 TNR Compensation Cancel

The G41 and G42 TNR compensation codes are both modal, and therefore require a cancel code when the tool cutting is finished or when a change of direction is required. The G40 is the code selected for this purpose. When this code is read by the control, it will shift the turret axes back by the amount of TNR compensation.

Geometry Offset Activation

The tool geometry offset activation is used to compensate each tool in both the X- and Z-axes by a specified amount. The amount is the actual tool tip difference from one tool to another in both the X- and Z-axes. The tool tip of each tool, as previously discussed, must be measured, recorded, and stored in the CNC control memory. The CNC operator or setup person is responsible for entering the *geometry offsets* into the control memory in the specified offset number.

The offset number is specified on the tool list and in the CNC program. When the control reads the T0101 code, it will shift the turret in X and Z by the amount stored in the memory. The third and fourth digits of the T01**01** code specify the offset number to be activated. The first and second digits of the T**01**01 code specify the turret index position. Usually, the first and second numbers of the "T" code are the same as the third and fourth numbers

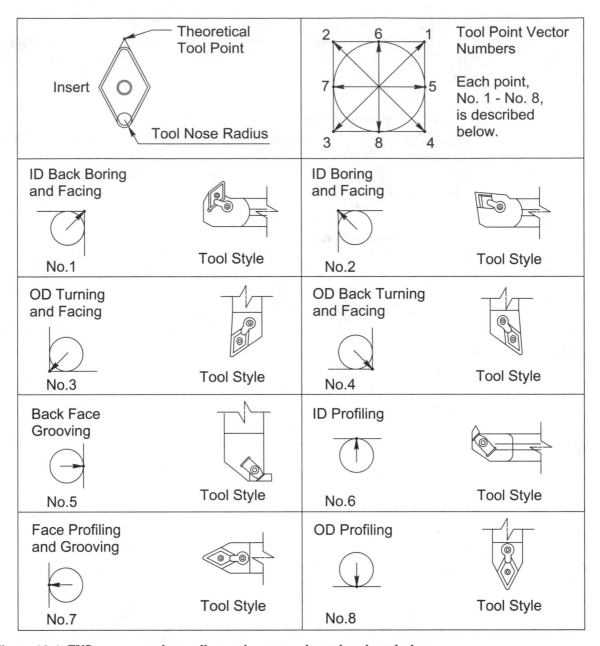

Figure 18-1 **TNR compensation radius and vector orientation descriptions**

for each subsequent tool. For example, the subsequent tools would be programmed T0202, T0303, T0404, and so on.

When the T0101 code along with an X-axis and/or Z-axis coordinate position is read (Cycle Start), the turret will shift and move automatically to the workpiece coordinates. This is also only recommended when it is absolutely certain that a collision will not occur. Again, the CNC operator can use the override features to slow the axis movement or completely stop the axis movement and allow the CNC operator time to react to a possible collision and stop the movement before any damage occurs. This is required for each tool geometry offset in the CNC program.

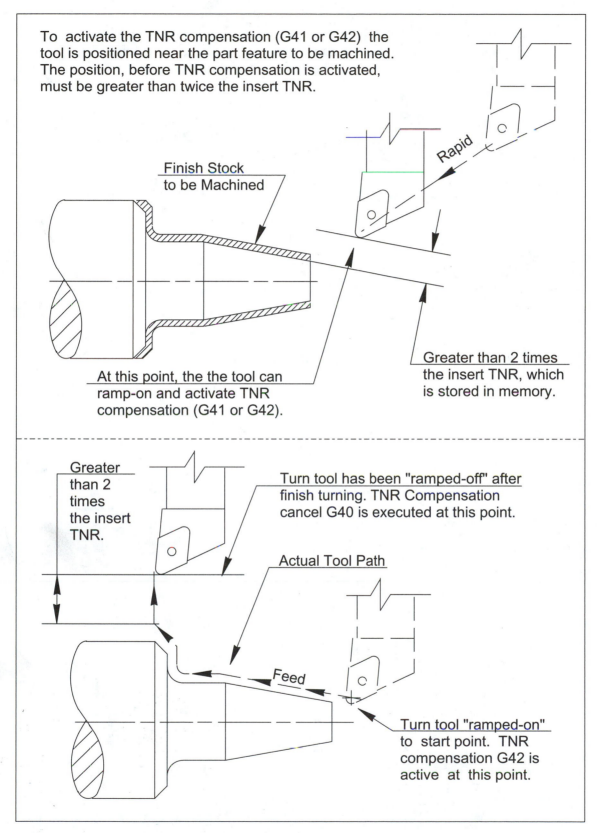

To activate the TNR compensation (G41 or G42) the tool is positioned near the part feature to be machined. The position, before TNR compensation is activated, must be greater than twice the insert TNR.

Finish Stock to be Machined

Rapid

At this point, the the tool can ramp-on and activate TNR compensation (G41 or G42).

Greater than 2 times the insert TNR, which is stored in memory.

Greater than 2 times the insert TNR.

Turn tool has been "ramped-off" after finish turning. TNR Compensation cancel G40 is executed at this point.

Actual Tool Path

Feed

Turn tool "ramped-on" to start point. TNR compensation G42 is active at this point.

Figure 18-2 *TNR code activation and cancel descriptions*

The G41 TNR left compensation is illustrated for internal (taper bore) and external (chamfer) applications. Note that the tool is viewed from behind, as it feeds away, to determine if the TNR compensation is left or right.

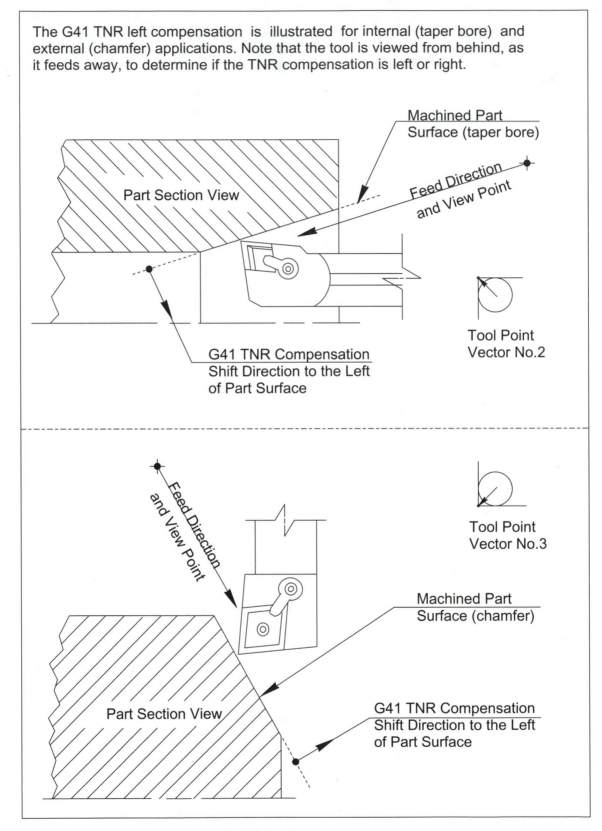

Figure 18-3 **G41 TNR compensation left description**

The G42 TNR right compensation is illustrated for internal (c'bore face) and external (OD taper) applications. Note that the tool is viewed from behind, as it feeds away, to determine if the TNR compensation is left or right.

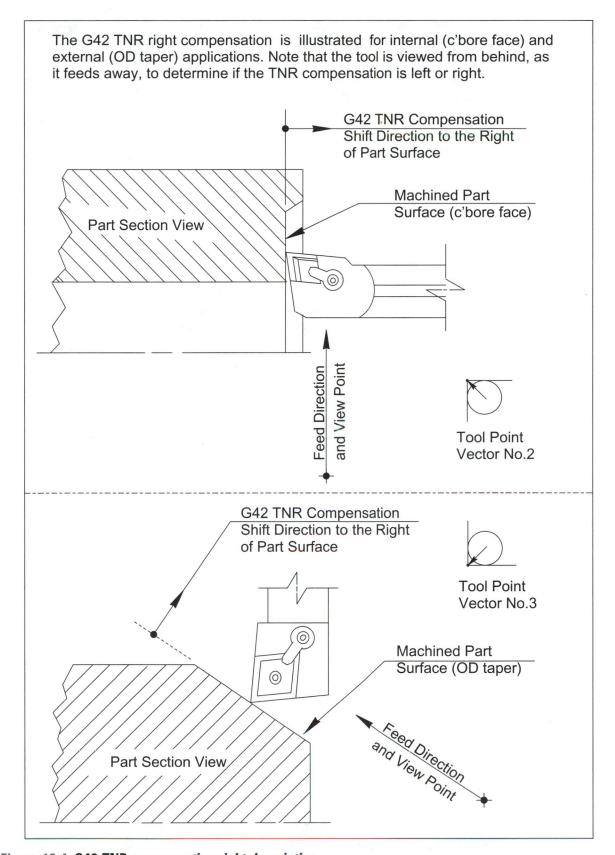

G42 TNR Compensation
Shift Direction to the Right
of Part Surface

Machined Part
Surface (c'bore face)

Part Section View

Feed Direction
and View Point

Tool Point
Vector No.2

G42 TNR Compensation
Shift Direction to the Right
of Part Surface

Tool Point
Vector No.3

Machined Part
Surface (OD taper)

Feed Direction
and View Point

Part Section View

Figure 18-4 **G42 TNR compensation right description**

The cutting tool machines and controls the size in diameter (X-axis) and in length (Z-axis) of the part features. Some sizes may require a higher degree of accuracy. This is determined by the tolerance on the engineering drawing. In these situations, the CNC operator or setup person must first measure the diameter or length of the part feature, determine if an adjustment is required, then adjust the tool by adding or subtracting the required amount from the geometry offset. It is important to note that if the adjustment amount exceeds 0.100 inch the CNC program or tool length offset should be checked for a possible error.

Geometry Offset Cancel

The geometry offset compensation, once activated, remains modal, and therefore it must be canceled when the tool is finished machining. This is required for every tool that is specified in the CNC program. The geometry offset is also canceled with the T code. To cancel a geometry offset, the third and forth digits of the T01**00** code are programmed as zeros.

The geometry offset cancel is typically programmed at the end of each tool before the next tool is programmed to index. In addition, it can be programmed at the start of a new tool as a precaution to avoid accidentally omitting the cancel code or axis overtravels.

G28 Machine Home Return

The CNC lathe must be homed in all axes (X and Z) before the CNC program can be cycled. The CNC lathe axes can be homed manually with the control, or automatically with the G28 code. The G28 code can be cycled in MDI mode or Memory mode. Some CNC lathes substitute the "U" letter code for the "X" letter code and the "W" letter code for the "Z" letter code. Both command methods to home the CNC lathe X- and Z-axes are described below:

G28X0.0Z0.0

or

G28U0.0W0.0

The CNC operator can also use the control rapid overrides to slow the axes movements or completely stop them to visually verify that a collision will not occur when homing the machine axes.

▪ ▪ ▪ ▪
SPINDLE SPEED CONTROL CODES

G50 Maximum RPM and Coordinate System Setting

There are situations when the spindle, chuck, jaws, or workpiece will limit how fast (RPM) the spindle can rotate. For example, the machine spindle and the chuck are typically designed to operate within a specified RPM range that cannot to be exceeded. Also, there are situations when the jaws or the workpiece are not symmetrically balanced and can cause excessive machine vibration. In either case, the spindle RPM must be limited so that the CNC lathe operates safely and effectively. The command to set the spindle limit—for example, to 1200 RPM—is programmed to read as follows:

> G50S1200

The G50 code can also be used to set the coordinates of the tool tip to the workpiece zero point (origin). To set the tool tip coordinates, each tool is accurately fastened to a specified distance in X and Z. For example, if the X-distance = 15.0 and Z-distance = 20.0, the command to set the tool tip coordinates is programmed to read as follows:

> G50X15.0Z20.0

G97 Constant Spindle RPM Mode

The G97 code is programmed for situations when the spindle rotates at a constant RPM. A typical situation includes when the tool is positioned at X-zero for drilling, tapping, and reaming. Another situation is when the RPM would increase and decrease needlessly, such as when single-point threading, causing unnecessary machine wear. Therefore, drilling, tapping, reaming, and single-point threading typically use the G97 constant RPM mode. When G97 is active, the S code specifies RPM (S2200).

G96 Constant Spindle SFM Mode

There are situations when it is beneficial for the spindle to rotate at a constant SFM cutting speed. For these situations, the G96 code is programmed. Typical situations include when the tool X-diameter position changes, such as for facing, grooving, taper turning, and turning multiple diameters. Therefore, facing, grooving, taper, and multiple-diameter turning typically use the G96 constant SFM mode. When G96 is active, the S code specifies SFM (S700).

INSTRUCTIONS FOR CNC PROGRAM EXAMPLE 8000

This section includes a coordinate worksheet, CNC documentation, and a CNC program for a CNC lathe. As explained previously, each of these items is designed to illustrate the CNC setup and operation process. Therefore, it is important to review and understand the purpose and role of each CNC item. These items are listed below:

- Coordinate worksheet (for instructional purposes)
- Engineering drawing/process drawing
- CNC lathe setup plan
- CNC tool list

This program is similar in format to the other programs in this manual. The similarities that should be noted include

- Program number
- Information notes
- Turret indexes
- Default code setting
- Spindle starts
- Coolant on
- Machine home command
- Geometry offset activation
- End of program

Instructions for CNC Program Example 8000

For this CNC program, the student is required to identify and note the main CNC program items that are listed below:

1. Program number
2. Information notes
3. Turret indexes
4. Default code setting blocks
5. Rapid moves
6. Feed moves
7. RPM codes
8. SFM codes
9. Maximum spindle RPM code

The example below illustrates how to identify and note the CNC program items above.

N20 T0400M41	*Turret Index 4, High Range*
N30 G97G50S2200	*RPM Mode, Max. RPM - 2200*
N40 G00X5.0Z.1T0404	*Rapid X&Z, Activate Offset No. 4*
N50 S1200M03	*1200 RPM, Spindle On CW*
N55 G96S750	*SFM Mode - 750 SFM*
N60 G01X.532F.005	*Feed to X.532 at 0.005 IPR*

COORDINATE WORKSHEET 8000

Instructions: Mark the X-Z zero origin, then calculate the coordinates for each point. Record the calculations in the chart. For the dimensions use part drawing no. 8000 illustrated in Figure 18-5.

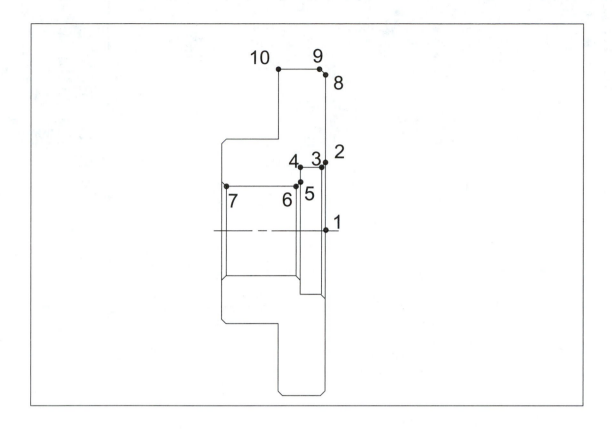

X- and Z-Coordinate Chart

PART NO.	X-COORDINATE	Z-COORDINATE
1.		
2.		
3.		
4.		
5.		
6.		
7.		
8.		
9.		
10.		

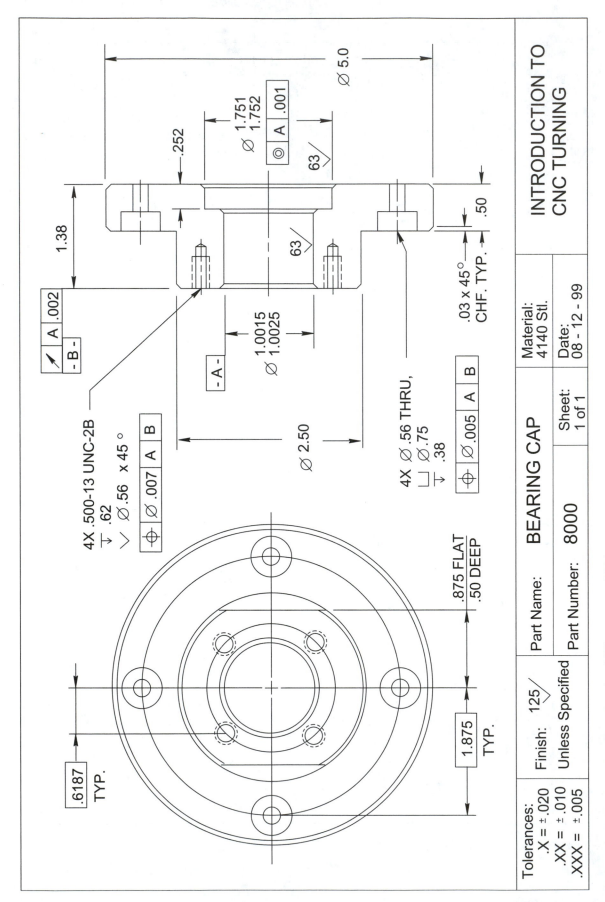

Figure 18-5 Engineering drawing for CNC program 8000

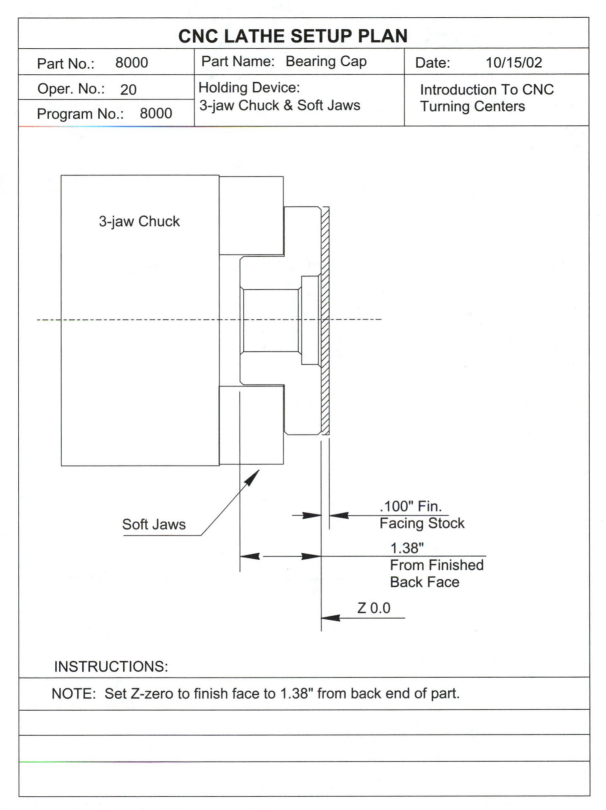

CNC LATHE SETUP PLAN		
Part No.: 8000	Part Name: Bearing Cap	Date: 10/15/02
Oper. No.: 20	Holding Device: 3-jaw Chuck & Soft Jaws	Introduction To CNC Turning Centers
Program No.: 8000		

3-jaw Chuck

Soft Jaws

.100" Fin. Facing Stock

1.38" From Finished Back Face

Z 0.0

INSTRUCTIONS:

NOTE: Set Z-zero to finish face to 1.38" from back end of part.

Figure 18-6 Setup plan for CNC program 8000

CNC TOOL LIST AND OPERATION

Part No.: 8000			Prog. No.: O8000	Oper. No.: 20
Part Name: Bearing Cap			Material.: C.R. 1020	Date: 10/15/99

Station No.	Tool Offset	Tool No.	Tool Description and Operation
T01	01	8001	Holder No. DCLNR-164D Insert: CNMG-432 C5 Rough face & turn .031 nose radius
T02	02	8022	Holder No. DDJNR-164D Insert: DNMG-432 C6 Finish face & turn .031 nose radius
T03	03	7002	Bar No. A06-STUNR2 Insert: TPG-221 C3 Rough bore .690 dia. .015 nose radius

Figure 18-7 Tool list for CNC program 8000

■ ■ ■ ■

CNC PROGRAM EXAMPLE 8000

```
%
O8000
(BEARING CAP – PROGRAM NO. 8000)
(DATE, 11/05/02)
(TOOL – 1 FINISH FACE & TURN OD)
(TOOL – 2 ROUGH BORE & C'BORE)
(TOOL – 3 FINISH BORE & C'BORE)

N10G20G40G97
(TOOL - 1 OFFSET - 1)
(FINISH FACE)
N20G0T0100M41
N30G97S495M3
N35X5.4Z.1 T0101M8
N40G50S3600
N45G96S700
N50G1Z0.F.01
N55X.8F.007
N60G0Z.1
N65X5.1
N70Z-.49F.012
N75X5.3
N80G0Z.1
N85X5.06
N90G1Z-.49
N95X5.12
N100G0Z.1
N105X4.9217Z.0908
N110G96S700
N115G1X4.9216Z-.0092F.01
N120X5.Z-.0483F.007
N125Z-.5
N130X5.3
N135G0Z.1M9
N140G28U0.W0.M5
N145T0100
N150M01

(TOOL - 2 OFFSET - 2)
(ROUGH ID AND C'BORE)
N155G0T0200
N160G50S5000
N165G97S1845M3
```

■ ■ ■ ■

CNC PROGRAM EXAMPLE 8000 (*continued*)

```
N170 X1.142Z.1 T0202M8
N175G96S600
N180G1Z-.242F.012
N185X1.0504
N190X.942Z-.2962
N200Z-1.4
N205X.922
N210G0Z.1
N215X1.342
N220G1Z-.242F.012
N225X1.122
N230G0 Z.1
N235X1.542
N240Z-.242F.012
N245X1.322
N250G0 Z.1
N255X1.742
N260Z-.0189F.012
N265X1.6916Z-.0442
N270Z-.242
N275X1.522
N280G0 Z.1
N285X1.7756
N290X1.722Z-.0289F.012
N300G0Z.1M9
N305G28U0.W0.M5
N310T0200
N315M01

(TOOL - 3 OFFSET - 3)
(FINISH ID AND C'BORE)
N320G0T0303
N325G50S5000
N330G97S1461M3
N335X1.8298 Z.2M8
N340Z.0908
N345G96S700
N350G1Z-.0092F.01
N355X1.7516Z-.0483F.007
N360Z-.252
N365X1.0986
N370X1.002Z-.3003
N375Z-1.4
```

CNC PROGRAM EXAMPLE 8000 (*continued*)

```
N380X1.7516Z-.0483
N385G0Z.0908
N390X1.8298
N395M9
N400G28U0.W0.M5
N405T0300
N410M30
%
```

CHAPTER SUMMARY

The key concepts presented in this chapter include the following:

- TNR compensation description.
- The G41 TNR compensation left code description.
- The G42 TNR compensation right code description.
- The G40 TNR compensation cancel code description.
- The T code turret index and offset compensation descriptions.
- The G50, G96, and G97 spindle code descriptions.
- The G28 machine home return code description.
- Calculations for CNC lathe coordinates.
- The format description of a CNC program for CNC lathes.
- The basic codes required of a CNC program for CNC lathes.

REVIEW QUESTIONS

1. How does TNR size affect a radius that is programmed for a CNC lathe?

2. Why do tools require tool geometry compensation (T0101)?

3. How is a tool geometry offset activated, and canceled?

4. When the G41 or G42 codes are programmed, what are the responsibilities of the CNC operator or setup person?

5. How is a maximum spindle 2500 RPM set on a CNC lathe?

6. Which codes are used to maintain the spindle at a constant SFM—650, for example?

7. List some situations when it is necessary to set a maximum spindle RPM.

8. What control features can the CNC setup person use to avoid machine collisions during a G28 machine home rapid move?

9. When is it appropriate to use spindle mode (G97) and when is it appropriate to use spindle mode (G96)?

10. Discuss and identify the CNC program format structure and codes covered in this chapter. The items that should be identified include CNC program number, tool information notes, Turret/Tool index blocks, tool geometry offset activation blocks, default code setting blocks, spindle modes, maximum setting, start and stop, coolant on and off, machine home command blocks.

■ ■ ■ CNC PROGRAM EXERCISE FOR CNC LATHES

Using the sketch in Figure 18-8 calculate the G02 and G03 circular feed moves for the X- and Z-axes using G41/G42 and G40 TNR compensation codes. Enter the codes and coordinates in the CNC program exercise that is provided.

■ ■ ■ CNC PROGRAM EXERCISE 8001

Fill in the missing program information of CNC program 8001 as indicated below:

Block N130: enter the G code and X-Z coordinate values.

Block N150: enter the SFM cutting speed value.

Block N160: enter the G codes, Z-coordinate value and IPR value.

Block N170: enter the G code, X-Z coordinates and R value.

Block N180: enter the G code and Z-coordinate value.

Block N190: enter the G code, X-Z coordinates and R value.

Block N200: enter the G code and X-coordinate value.

Block N210: enter the G code.

```
%
O8001
(SHAFT PIN - PROGRAM NO. 8001 DATE, 08/12/01)
(TOOL - 01  FINISH 55 - DEGREE)

N100 G20 G40 G97
(TOOL - 01 55 - DEG.)
N110 G0 T0100
N120 G97 S1900 M03
N130 G _____ X _____ Z _____ T0101
```

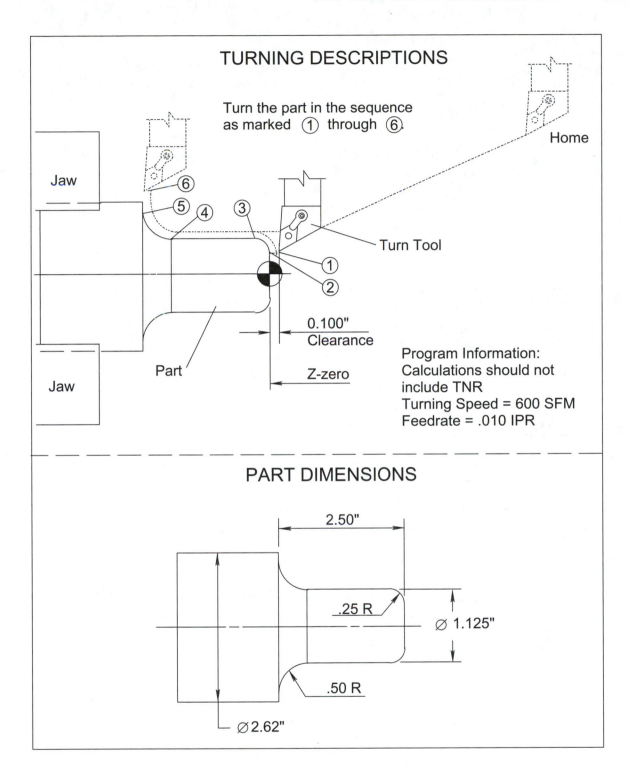

TURNING DESCRIPTIONS

Turn the part in the sequence as marked ① through ⑥.

Home

Jaw

⑥
⑤ ④ ③

Turn Tool

①
②

0.100"
Clearance

Part

Z-zero

Jaw

Program Information:
Calculations should not
include TNR
Turning Speed = 600 SFM
Feedrate = .010 IPR

PART DIMENSIONS

2.50"

.25 R

Ø 1.125"

.50 R

Ø 2.62"

*Figure 18-8 **Program exercise***

N140 G50 S5000
N150 G96 S _____
N160 G _____ G _____ Z _____ F _____ M08
N170 G _____ X _____ Z _____ R _____
N180 G _____ Z _____
N190 G _____ X _____ Z _____ R _____
N200 G _____ X _____
N210 G _____
N220 G97 G0 Z1.0 M09
N230 G28 U0.0 W0.0 M05
N240 T0100
N250 M30
%

chapter 19

CNC LATHE FIXED CYCLES G70–G74

Chapter Objectives

After studying and completing this chapter, the student should have knowledge of the following:

- *G70 finishing cycle*
- *G71 rough turning/boring cycle*
- *G72 multiple repetitive facing cycle*
- *G73 pattern repeat cycle*
- *G74 face grooving cycle*
- *G74 peck drilling cycle*
- *Calculating X-Z coordinates*
- *CNC program format descriptions*

CNC lathe fixed cycle codes were basically developed to increase programming efficiency. These fixed cycles are typically used for CNC lathe applications that require contouring and repetitive patterns. The applications include turning, facing, boring, grooving, drilling, and single-point threading. The common CNC lathe fixed cycle codes are identified and described in Table 19-1. Additionally, this chapter includes a coordinate calculation worksheet designed to illustrate how the drilling coordinates are determined for the CNC program.

CNC LATHE FIXED CYCLES

These fixed cycle codes are designed to perform machining calculations and execute a combination of axis moves automatically. The axis moves usually include the X- and Z-axes in rapid mode and feed mode. The machining calculations, which are derived from the programmed code values, typically include calculating the number of turning, boring, facing, grooving, and threading passes.

The combination of calculations and axis moves depends on the type of machining operation to be completed. Therefore, the fixed cycle code selected and programmed must be appropriately matched to the type of CNC lathe operation to be performed. Additionally, the data programmed with the fixed cycle must be accurate so that the calculations and moves are executed correctly.

Table 19-1 *Fixed cycle application and description*

CODE	CNC LATHE APPLICATION	CODE DESCRIPTION
G70	Finishing cycle	Used for finish turning, boring, and facing after roughing operations.
G71	Rough turn/bore cycle	Used for rough turning or boring of complex contours.
G72	Rough face cycle	Used for rough OD or ID facing of complex contours.
G73	Pattern repeat cycle	Used for rough OD or ID turning of castings, forgings, and roughed parts.
G74	Multiple face grooving cycle	Used for multiple-face grooving cuts.
G74	Peck drilling cycle	Used for drilling deep holes.
G75	Multiple diameter grooving cycle	Used for multiple OD grooving cuts.
G76	Multiple thread cutting cycle	Used for finish threading by multiple-pass cuts.
G90	Straight turn/bore cycle	Used for finish straight turning and boring by multiple-pass cuts.
G90	Taper turn/bore cycle	Used for finish taper turning and boring by multiple-pass cuts.
G92	Multiple thread cutting cycle	Used for finish threading by multiple pass cuts.
G94	End face turning cycle	Used for finish straight facing by multiple-pass cuts.

CNC FIXED CYCLES G70–G74

The CNC operator and CNC setup person are responsible for verifying that every fixed cycle tool path move (feed and rapid) in the CNC program is safe and correct. When the control reads, "Cycle Start," a fixed cycle code, it automatically executes the rapid and feed moves required by the fixed cycle. Therefore, when the fixed cycle codes have not been proved out, the CNC operator can use the control override features to slow the axis movement or completely stop the axis movement. This will allow the CNC operator time to react to a possible collision before any damage occurs. This is required for all fixed cycles and moves in the CNC program. The fixed cycle G70–G74 are described and illustrated in this section.

G70 Finishing Cycle

The G70 finishing cycle is designed to finish a contour that was previously roughed by the G71, G72, or G73 roughing cycle codes. Therefore, the G70 code always follows the roughing codes. The G70 code uses the same programmed blocks of the roughing code that it follows (see Figure 19-1). The T-tool code and the S-spindle RPM code are not allowed on the same block as the G70 code. Therefore, they must be programmed on the block or blocks before the G70 code block.

G71 Rough Turn/Bore Cycle

The G71 cycle is designed to rough cut an OD or ID contour, which can consist of any combination of turn cuts, face cuts, and radius cuts (see Figure 19-2). The G70 code usually follows the G71 to finish machine the contour. The T-tool code, the S-spindle RPM code, and F-feed code are allowed on the same block as the G71 code.

G72 Rough Face Cycle

The G72 cycle is designed to rough face cut an OD or ID contour, which can consist of any combination of turn cuts, face cuts, and radius cuts (see Figure 19-3). The G70 code usually follows the G72 to finish machine the contour. The T-tool code, the S-spindle RPM code, and F-feed code are allowed on the same block as the G72 code block.

G73 Pattern Repeat Cycle

The G73 cycle is designed to rough cut an OD or ID pattern, which can consist of any combination of turn cuts, face cuts, and radius cuts (see Figure 19-4). This code is typically used for castings and forgings where the rough pattern is defined. The G70 code usually follows the G73 to finish machine the contour. The T-tool code, the S-spindle RPM code, and F-feed code are ignored on the same block as the G73 code block.

The G70 code example below will execute a finish cut after the contour is rough machined as programmed between blocks N130 and N155 at a feedrate of .007 IPR. The CNC program procedure is the same for both OD and ID contours as illustrated below.

G70	P0130	Q0155
Finishing Fixed Cycle Code	Start Block of the Finish Contour	End Block of the Finish Contour

N130 G00 X.5352 Z.1
N135 G01 X1.125 Z-1.2332 F.007
N140 G01 Z-1.7156
N145 G02 X1.5626 Z-1.9344 R.2188
N150 G01 X3.1872
N155 G01 X3.750 Z-2.1226
N160 G70 P0130 Q0155

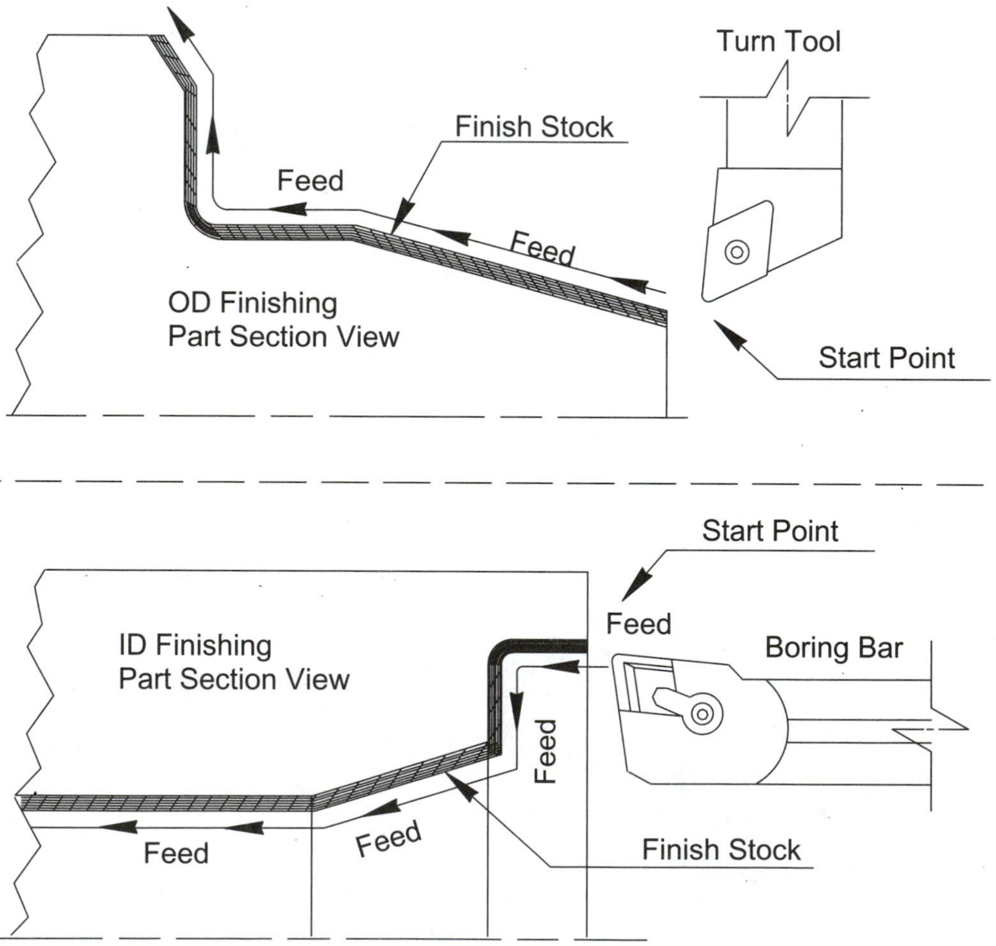

Turn Tool

Finish Stock

Feed

Feed

OD Finishing
Part Section View

Start Point

Start Point

Feed

Boring Bar

ID Finishing
Part Section View

Feed

Feed

Feed

Feed

Finish Stock

Figure 19-1 G70 finishing fixed cycle code

The G71 code example below will execute a series of rough cuts as programmed between blocks N130 and N155. The control will automatically calculate the number of passes required based on the large OD, small OD, and .1500 DOC. The CNC program procedure is the same for both OD and ID contours as described below.

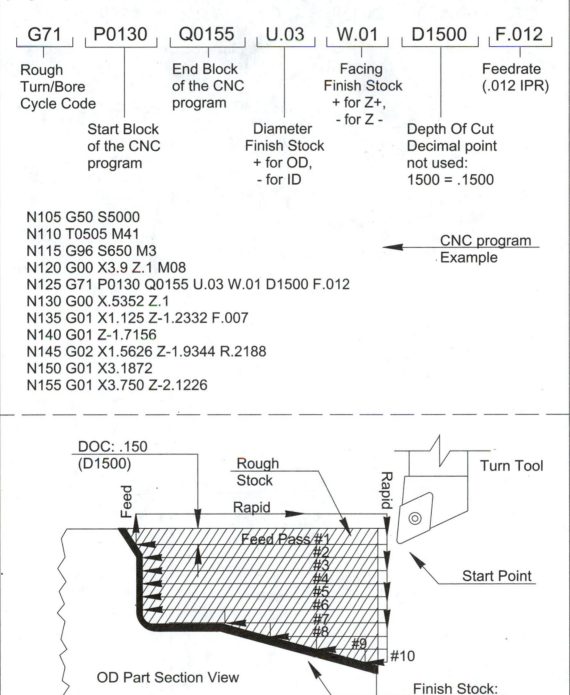

| G71 | P0130 | Q0155 | U.03 | W.01 | D1500 | F.012 |

Rough Turn/Bore Cycle Code

End Block of the CNC program

Facing Finish Stock + for Z+, - for Z -

Feedrate (.012 IPR)

Start Block of the CNC program

Diameter Finish Stock + for OD, - for ID

Depth Of Cut Decimal point not used: 1500 = .1500

N105 G50 S5000
N110 T0505 M41
N115 G96 S650 M3
N120 G00 X3.9 Z.1 M08
N125 G71 P0130 Q0155 U.03 W.01 D1500 F.012
N130 G00 X.5352 Z.1
N135 G01 X1.125 Z-1.2332 F.007
N140 G01 Z-1.7156
N145 G02 X1.5626 Z-1.9344 R.2188
N150 G01 X3.1872
N155 G01 X3.750 Z-2.1226

← CNC program Example

DOC: .150 (D1500)

Rough Stock

Rapid

Feed

Rapid

Turn Tool

Feed Pass #1
#2
#3
#4
#5
#6
#7
#8
#9
#10

Start Point

OD Part Section View

Finish Stock: U.03 and W.01

Figure 19-2 G71 rough turn/bore fixed cycle code

The G72 code example below will execute a series of rough facing cuts parallel to the X-axis. The other items of this code are otherwise similar to the G71 code as described and illustrated below.

G72	P0130	Q0160	U.03	W.01	D1200	F.012

Rough
Facing
Cycle Code

End Block
of the CNC
program

Finish Stock
+ for Z+,
- for Z -

Feedrate
(.012 IPR)

Start Block
of the CNC
program

Finish Stock
+ for OD,
- for ID

Depth Of Cut
Decimal point
not used:
1200 = .1200

```
N110 T0303 M41
N115 G96 S650 M3
N120 G00 X3.9 Z.1 M08
N125 G72 P0130 Q0160 U.03 W.01 D1200 F.012
N130 G00 Z-2.1226
N135 G01 X3.1872 Z-1.9344 F.007
N140 G01 X1.5626
N145 G03 X1.125 Z-1.2332 R.2188
N150 G01 Z-.2823
N155 G01 X.6107 Z0.0
N160 G01 X-.064
```

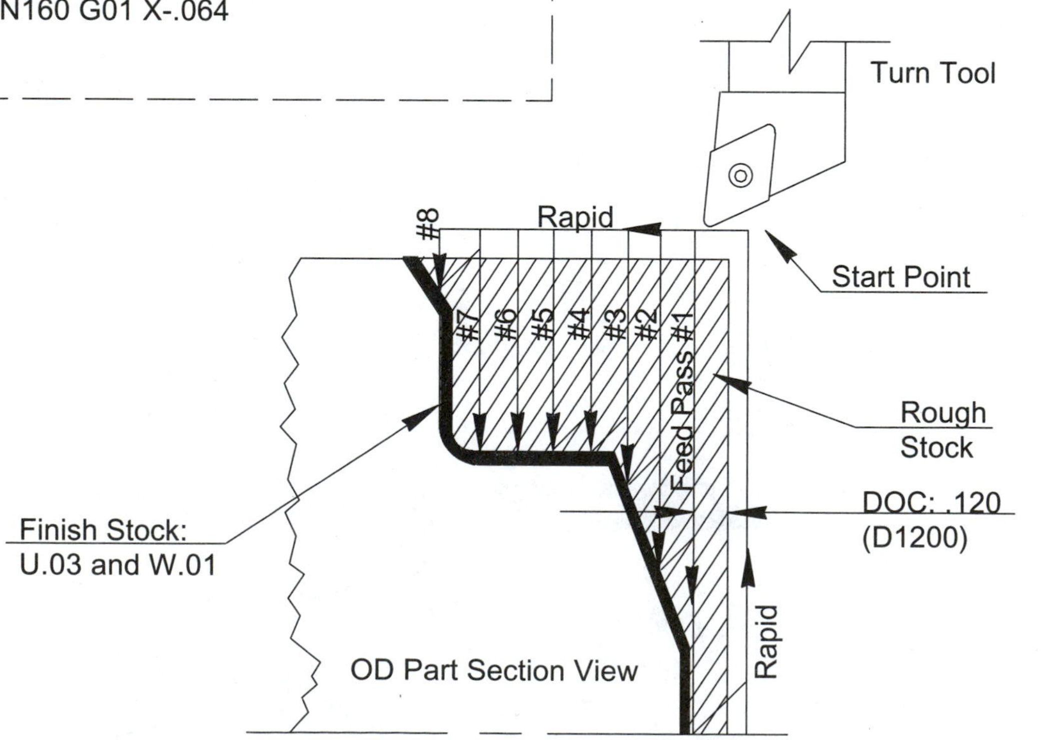

CNC program
Example

Turn Tool

#8   Rapid

Start Point

#7 #6 #5 #4 #3 #2 #1

Feed Pass #1

Rough
Stock

DOC: .120
(D1200)

Finish Stock:
U.03 and W.01

Rapid

OD Part Section View

Figure 19-3 **G72 rough facing fixed cycle code**

The G73 code example below will execute a series of rough cuts that will repeat the finished part contour as programmed between blocks N215 and N240. The control will automatically repeat the rough passes based on the "D" number. Note that the G70 finish cycle is also in the program example (N245).

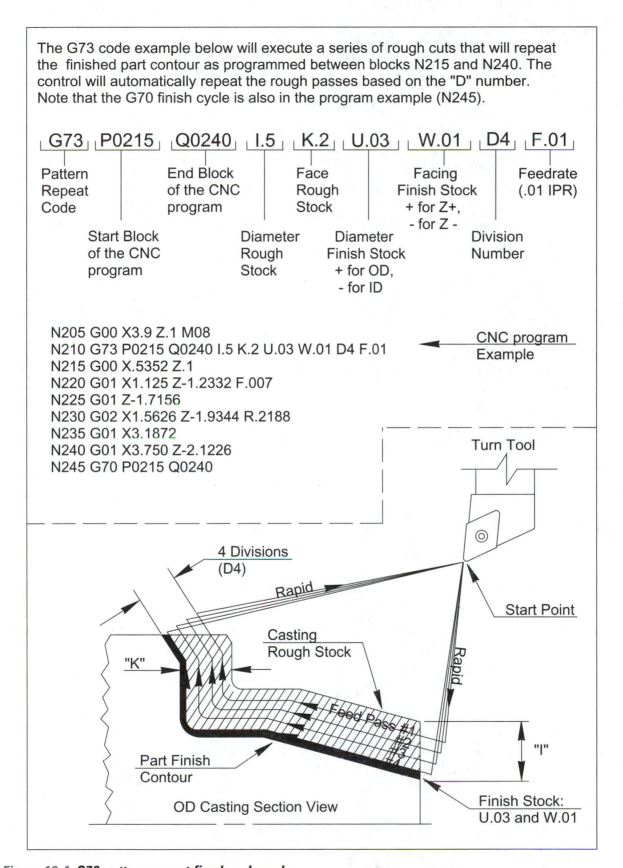

| G73 | P0215 | Q0240 | I.5 | K.2 | U.03 | W.01 | D4 | F.01 |

Pattern Repeat Code

Start Block of the CNC program

End Block of the CNC program

Face Rough Stock

Diameter Rough Stock

Facing Finish Stock + for Z+, - for Z -

Diameter Finish Stock + for OD, - for ID

Feedrate (.01 IPR)

Division Number

```
N205 G00 X3.9 Z.1 M08
N210 G73 P0215 Q0240 I.5 K.2 U.03 W.01 D4 F.01
N215 G00 X.5352 Z.1
N220 G01 X1.125 Z-1.2332 F.007
N225 G01 Z-1.7156
N230 G02 X1.5626 Z-1.9344 R.2188
N235 G01 X3.1872
N240 G01 X3.750 Z-2.1226
N245 G70 P0215 Q0240
```

CNC program Example

4 Divisions (D4)

Rapid

Turn Tool

Start Point

Rapid

Casting Rough Stock

"K"

Feed Pass #1 #2 #3 #4

"I"

Part Finish Contour

Finish Stock: U.03 and W.01

OD Casting Section View

Figure 19-4 **G73 pattern repeat fixed cycle code**

G74 Multiple-Face Grooving Cycle

The G74 cycle is designed to perform multiple-face groove cuts (see Figure 19-5). The G74 code interrupts the grooving feed move to break up the occurrence of a continuous chip. It automatically retracts and reenters to resume grooving until the depth is reached.

G74 Peck Drilling Cycle

The G74 cycle is also designed for peck drilling (see Figure 19-5). The G74 code interrupts the drilling feed move to break up the occurrence of a continuous chip. It automatically retracts and reenters to resume drilling until the depth is reached.

INSTRUCTIONS FOR CNC PROGRAM EXAMPLE 9000

This section includes a coordinate worksheet, CNC documentation, and a CNC program for a CNC lathe. As explained previously, each of these items is designed to illustrate the CNC setup and operation process. It is important to review and understand the purpose and role of each CNC item. These items are listed below:

- Coordinate worksheet (for instructional purposes)
- Engineering drawing
- CNC lathe setup plan
- CNC tool list

This program is similar in format to the other sample programs. The similarities that should be noted include

- Program number
- Information notes
- Turret indexes
- Default code setting
- Spindle starts
- Coolant on
- Machine home command

The G74 code is used for two different lathe operations. It is used for multiple-face grooves and for peck drilling deep holes. Both of these code applications are described and illustrated below.

G74 Multiple Face Grooving Code Description:

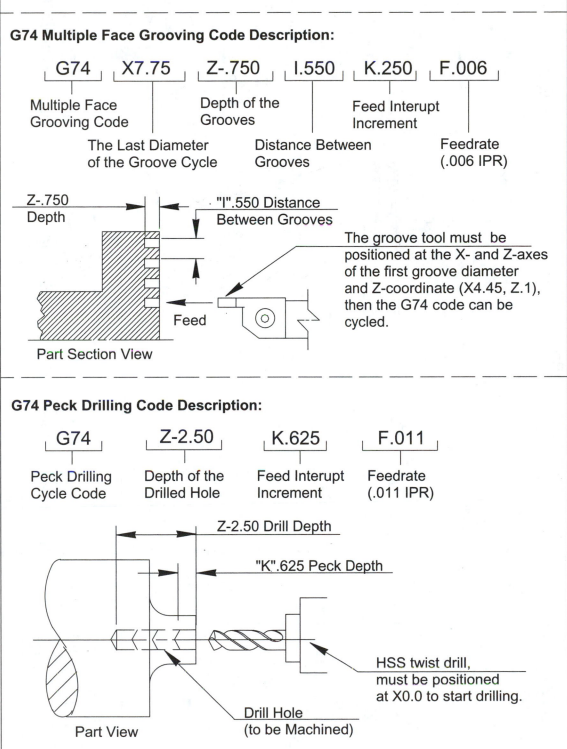

| G74 | X7.75 | Z-.750 | I.550 | K.250 | F.006 |

Multiple Face Grooving Code

The Last Diameter of the Groove Cycle

Depth of the Grooves

Distance Between Grooves

Feed Interupt Increment

Feedrate (.006 IPR)

Z-.750 Depth

"I".550 Distance Between Grooves

The groove tool must be positioned at the X- and Z-axes of the first groove diameter and Z-coordinate (X4.45, Z.1), then the G74 code can be cycled.

Feed

Part Section View

G74 Peck Drilling Code Description:

| G74 | Z-2.50 | K.625 | F.011 |

Peck Drilling Cycle Code

Depth of the Drilled Hole

Feed Interupt Increment

Feedrate (.011 IPR)

Z-2.50 Drill Depth

"K".625 Peck Depth

HSS twist drill, must be positioned at X0.0 to start drilling.

Drill Hole (to be Machined)

Part View

Figure 19-5 *G74 face grooving and peck drilling fixed cycle code*

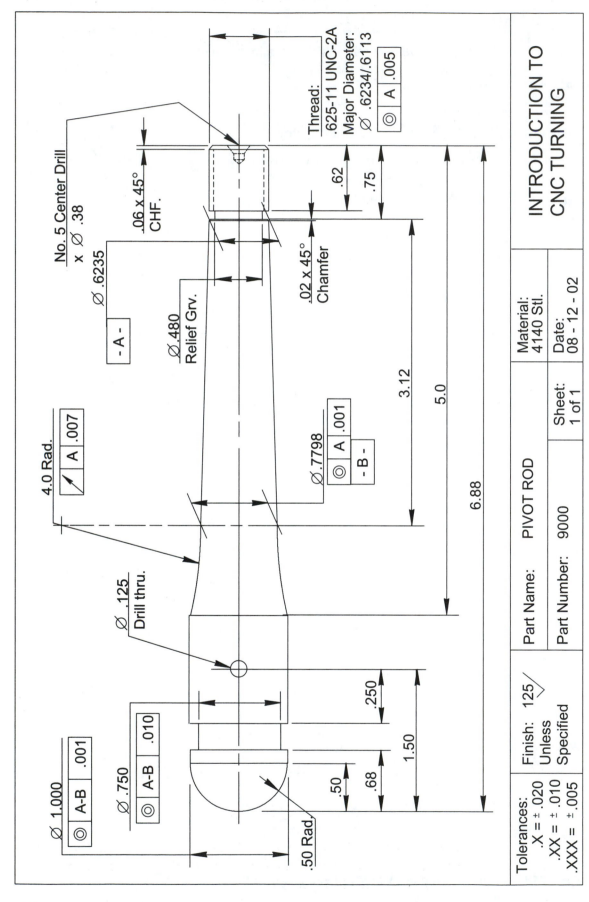

Figure 19-6 Engineering drawing for CNC program 9000

- Geometry offset activation

- End of program

Instructions for CNC Program Example 9000

For this CNC program, the student is required to identify and note the main CNC program items that are listed below:

1. Program number

2. Activate geometry offset

3. Cancel geometry offset

4. Turret indexes

5. Default code setting blocks

6. Rapid moves

7. Feed moves

8. RPM codes

9. SFM codes

10. Maximum spindle RPM code

The example below illustrates how to identify and note the CNC program items above.

(Turn Tool 80-Deg.)	*Information Note- tool description*
N20 T0200M41	*Turret Index 2, Cancel Offset, High Range*
N30 G97G50S2200	*RPM Mode, Max. RPM - 2200*
N40 G00X5.0Z.1T0202	*Rapid X&Z, Activate Offset 2*
N50 S2000M03	*2000 RPM, Spindle On CW*
N55 G96S550	*SFM Mode - 550 SFM*
N60 G01X-.064F.007	*Feed to X-.064 at .007 IPR*

COORDINATE WORKSHEET 9000

Instructions: Mark the X-Z zero origin, then calculate the coordinates for each point. Record the calculations in the chart. For the dimensions, use part drawing 9000 illustrated in Figure 19-6.

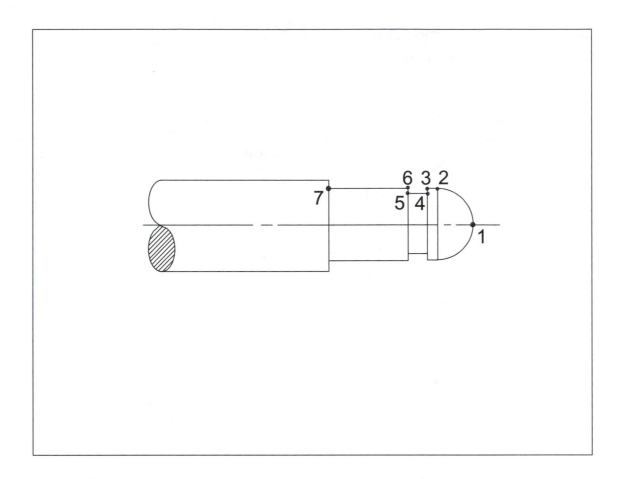

X- and Z-Coordinate Chart

PART NO.	X-COORDINATE	Z-COORDINATE
1		
2		
3		
4		
5		
6		
7		

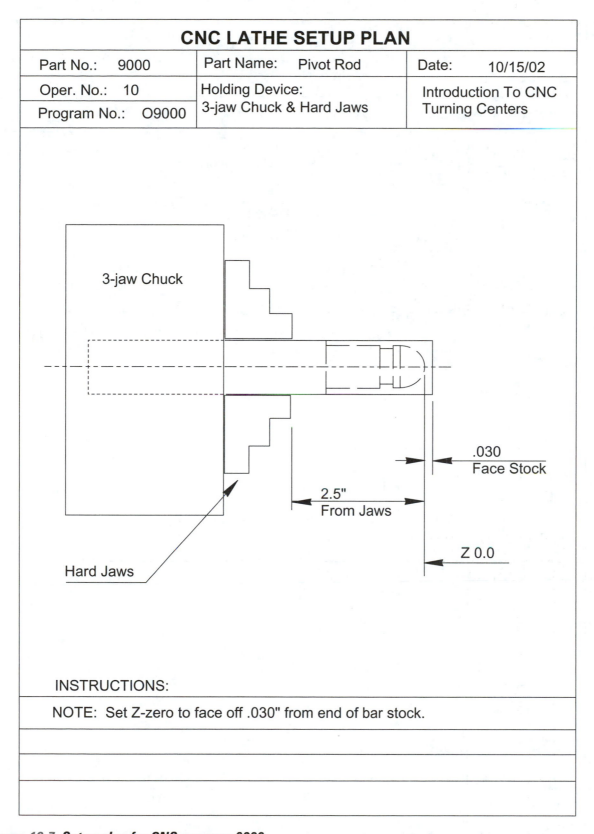

CNC LATHE SETUP PLAN

Part No.: 9000	Part Name: Pivot Rod	Date: 10/15/02
Oper. No.: 10	Holding Device:	Introduction To CNC
Program No.: O9000	3-jaw Chuck & Hard Jaws	Turning Centers

3-jaw Chuck

Hard Jaws

.030
Face Stock

2.5"
From Jaws

Z 0.0

INSTRUCTIONS:

NOTE: Set Z-zero to face off .030" from end of bar stock.

Figure 19-7 Setup plan for CNC program 9000

CNC TOOL LIST AND OPERATION

Part No.: 9000	Prog. No.: O9000	Oper. No.: 10
Part Name: Pivot Rod	Material.: C.R. 4140	Date: 10/15/02

Station No.	Tool Offset	Tool No.	Tool Description and Operation
T01	01	8001	Holder No. DCLNR-164D Finish face & Rough turn Insert: CNMG-432 C5 .031 nose radius
T02	02	8022	Holder No. DDJNR-164D Finish face & turn Insert: DNMG-432 C6 .031 nose radius
T03	03	6007	Groove Holder No. NSR-163D Rgh. & Fin. Groove. Insert: NG-3R C5

Figure 19-8 Tool list for CNC program 9000

■ ■ ■ ■
CNC PROGRAM EXAMPLE 9000

```
%
O9000
(PIVOT ROD – 4140 STEEL)
(PROGRAM NO. 8000)
(DATE, 11/05/02)
(TOOL – 1 ROUGH TURN OD)
(TOOL – 2 FINISH TURN OD)
(TOOL – 3 FINISH GROOVE)
N10G20G40G97

(TOOL - 1 OFFSET - 1)
(ROUGH TURN)
N15G0T0100
N20G97S1875M3
N25G0X1.22Z.1 T0101M8
N30G50S3600
N35G96S600
N40G71P45Q75U.03W.01D1200F.011
N45G0X-.0313Z.1
N50G1Z0.F.007
N55G3X1.Z-.5156R.5156
N60G1Z-1.88
N65X1.4
N70X1.5414Z-1.8093
N75G0Z.1
N80G28U0.W0.M05
N85T0100M9
N90M01

(TOOL - 2 OFFSET - 2)
(FINISH TURN OD)
N95G0T0200
N100G97S3600M3
N105G0X-.0313Z.1T0202M8
N110G50S3600
N115G96S700
N120G70P45Q75
N125G0Z1.0M9
N130G28U0.W0.M05
N135T0200
N140M01

(TOOL - 3 OFFSET - 3)
(FINISH OD GROOVE)
```

CNC PROGRAM EXAMPLE 9000 (*continued*)

```
N145G0T0300
N150G97S1042M3
N155G0X1.1Z-.805 T0303M8
N160G50S3600
N165G96S300
N170G1X.9F.003
N175Z-.93
N180X1.1
N185G0Z-.805
N190X1.
N195G1X.8
N200Z-.93
N205X1.
N210G28U0.W0.M05
N215T0300M9
N220M30
%
```

CHAPTER SUMMARY

The key concepts presented in this chapter include the following:

- Fixed cycle code descriptions and applications.
- The G70 finishing code descriptions.
- The G71 rough turn/bore code descriptions.
- The G72 rough facing code descriptions.
- The G73 pattern repeat code description.
- The G74 multiple-face grooving code description.
- The G74 peck drilling code description.
- Calculations for CNC lathe coordinates.
- The format description of a CNC program for CNC lathes.
- The basic codes required of a CNC program for CNC lathes.

REVIEW QUESTIONS

1. Which types of tools are used for G70 finishing application on OD contours?

2. Which types of tools are used for G70 finishing application on ID contours?

3. When the G71 rough turn/bore code is executed, which axis (X or Z) feeds to cut the stock?

4. When the G72 rough facing code is executed, which axis (X or Z) feeds to cut the stock?

5. Which types of parts are suitable for the G73 pattern repeat roughing code?

6. List the applications and tools that are associated with the G74 peck drilling code.

7. List the other machine functions/codes that are required before a fixed cycle code (G70–G74) is executed.

8. Which letter in a fixed cycle code specifies the amount of stock left for finishing on the diameter?

9. Which letter in a fixed cycle code specifies the amount of stock left for finishing on the face?

10. Which letter in a fixed cycle code specifies the DOC machined by each pass?

11. Discuss and identify the CNC program format structure and codes covered in this chapter. The items that should be identified include CNC program number, tool information notes, tool index blocks, tool offset activation blocks, tool offset cancel blocks, default code setting blocks, spindle RPM or SFM mode, maximum RPM, spindle starts and stops, fixed cycle codes, machine home command blocks and end of program block.

CNC PROGRAM EXERCISE FOR CNC LATHES

Using the sketch in Figure 19-9, determine the G70/G71 fixed codes and values to turn the part as shown. Enter the codes and values in the CNC program exercise that is provided.

CNC PROGRAM EXERCISE 9001

Fill in the missing program information of CNC program 9001 as indicated below:

Block N130: enter the G code and X-Z coordinate values.

Block N160: enter the G code, P-Q-U-W and D values and IPR value.

Block N170: enter the G code, the X-Z coordinates.

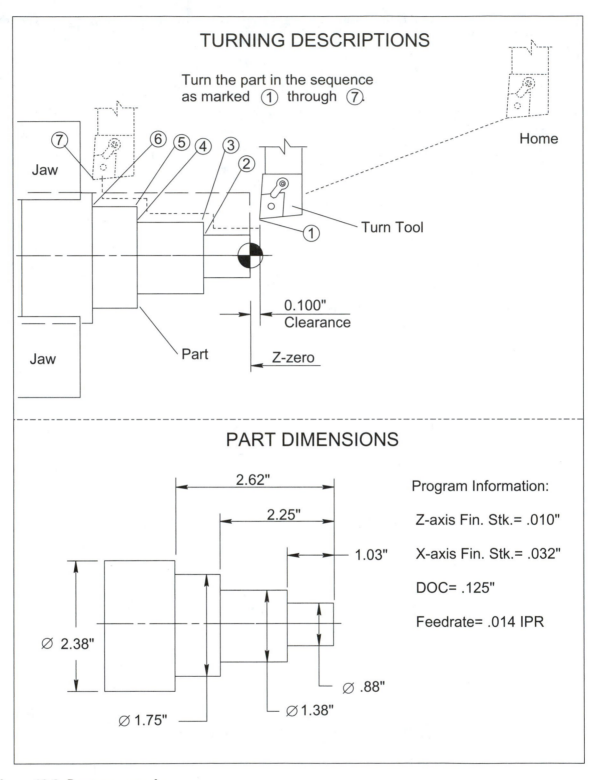

TURNING DESCRIPTIONS

Turn the part in the sequence as marked ① through ⑦.

Home

Jaw

⑦ ⑥ ⑤ ④ ③ ②

Turn Tool

①

0.100" Clearance

Z-zero

Jaw

Part

PART DIMENSIONS

2.62"

2.25"

1.03"

Ø 2.38"

Ø .88"

Ø 1.38"

Ø 1.75"

Program Information:

Z-axis Fin. Stk.= .010"

X-axis Fin. Stk.= .032"

DOC= .125"

Feedrate= .014 IPR

*Figure 19-9 **Program exercise***

Block N180: enter the G code, and the Z-coordinate.

Block N190: enter the X-coordinate value.

Block N200: enter the Z-coordinate value.

Block N210: enter the X-coordinate value.

Block N220: enter the Z-coordinate value.

Block N230: enter the X-coordinate value.

Block N240: enter the G code, P and Q values.

```
%
O9001
(SHAFT - PROGRAM NO. 9001 DATE, 04/22/02)
(TOOL - 01 RGH. & FINISH 80 - DEGREE)
N100 G20 G40 G97
(TOOL - 01 80 - DEG.)
N110 G0 T0100
N120 G97 S1500 M03

N130 G0 X _____ Z _____ T0101
N140 G50 S5000
N150 G96 S700 M08

N160 G_____ P _____ Q _____ U _____ W _____
        D_____ F_____

N170 G_____ X _____ Z _____
N180 G_____ Z _____ F.007
N190 X _____
N200 Z _____
N210 X _____
N220 Z _____
N230 X _____
N240 G_____ P _____ Q _____

N250 G97 G0 Z1.0 M09
N260 G28 U0.0 W0.0 M05
N270 T0100
N280 M30
%
```

chapter 20

CNC LATHE FIXED CYCLES G75–G94

Chapter Objectives

After studying and completing this chapter, the student should have knowledge of the following:

- **G75 multiple diameter grooving cycle**
- **G76 multiple thread cutting cycle**
- **G90 OD and ID cutting cycle**
- **G90 OD and ID taper cutting cycle**
- **G92 multiple thread cutting cycle**
- **G94 end face cutting cycle**
- **G04 dwell code**
- **Calculating X-Z coordinates**
- **CNC program format description**

The fixed cycle codes described in this chapter, G75 diameter grooving, G76 multiple thread cutting cycle, G90 OD and ID cutting cycle, G92 multiple thread cutting cycle, and G94 end face cutting cycle are also listed in Table 19-1, previously discussed. Additionally, this chapter describes the G04 dwell code, which is typically used with grooving operations. It also includes a coordinate calculation worksheet for X-Z program coordinate analysis and introduces a fixed cycle code into the CNC program.

CNC FIXED CYCLES G75–G94

As previously mentioned, these fixed cycle codes are typically available on most CNC lathes used in industry. It is also important to restate that the CNC operator and setup person are responsible for verifying that all the fixed cycle tool path moves (feed and rapid) in the CNC program are safe and correct. When the control reads, "Cycle Start," a fixed cycle code, it will automatically execute the programmed command. Therefore, with these fixed cycle codes it is also a recommended practice to Dry Run or proveout the CNC program to ensure that collisions will not occur.

The CNC operator can use the override features to slow the axis movement or completely stop any axis movement. The Feed Hold button is also a means of controlling the axis movement during proveout. This should also allow the CNC operator time to react to a possible collision and stop the movement before any damage occurs. However, it is important to note that the Feed Hold button and the Feedrate Override control will not stop the G76 and G92 multiple threading fixed cycles during the Z-axis cutting move. Consequently, care should be taken with these and the other fixed cycles when they are executed with the CNC program. If a collision occurs, the Emergency Stop button should be pressed.

G75 Multiple Diameter Grooving Cycle

The G75 cycle is designed to perform multiple diameter groove cuts (see Figure 20-1). The G75 code also interrupts the grooving feed move to break up the occurrence of a continuous chip or chatter. It automatically retracts and reenters to resume grooving until the specified depth is reached. The T-tool code, the S-spindle RPM code, and the F-feed code are required and typically programmed on the block or blocks before the G75 code block.

G76 Multiple Thread Cutting Cycle

The G76 cycle code is used for CNC lathe single-point threading applications (see Figure 20-2). This code allows the CNC programmer to enter the data

The G75 code is used for two different types of groove operations. It is used for a multiple series of grooves and also for a wide groove. Both of these groove types along with the code information are described and illustrated below.

G75 Multiple Diameter Grooving Code Description:

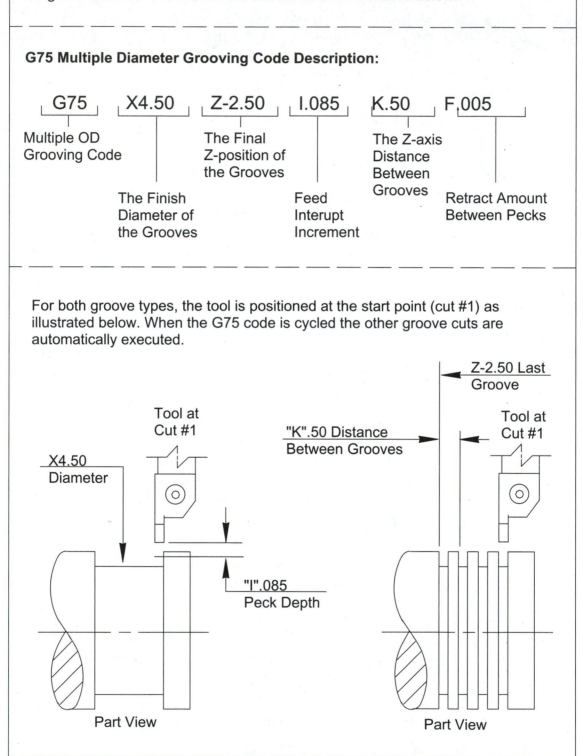

| G75 | X4.50 | Z-2.50 | I.085 | K.50 | F.005 |

Multiple OD Grooving Code

The Finish Diameter of the Grooves

The Final Z-position of the Grooves

Feed Interupt Increment

The Z-axis Distance Between Grooves

Retract Amount Between Pecks

For both groove types, the tool is positioned at the start point (cut #1) as illustrated below. When the G75 code is cycled the other groove cuts are automatically executed.

Tool at Cut #1

X4.50 Diameter

"I".085 Peck Depth

Part View

Z-2.50 Last Groove

Tool at Cut #1

"K".50 Distance Between Grooves

Part View

Figure 20-1 G75 multiple OD grooving fixed cycle code

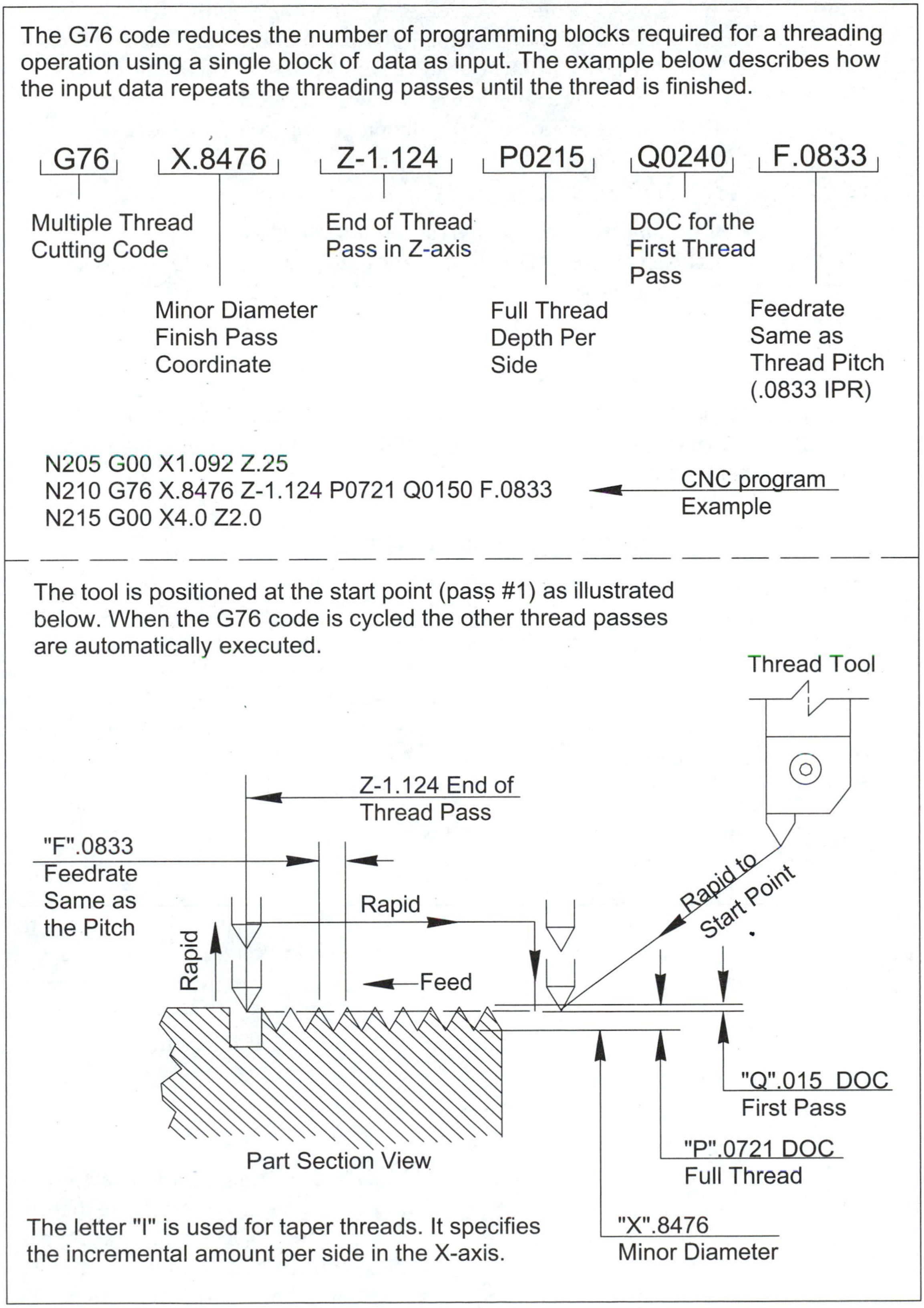
The G76 code reduces the number of programming blocks required for a threading operation using a single block of data as input. The example below describes how the input data repeats the threading passes until the thread is finished.
G76
X.8476
Z-1.124
P0215
Q0240
F.0833
Multiple Thread Cutting Code
Minor Diameter Finish Pass Coordinate
End of Thread Pass in Z-axis
Full Thread Depth Per Side
DOC for the First Thread Pass
Feedrate Same as Thread Pitch (.0833 IPR)
N205 G00 X1.092 Z.25
N210 G76 X.8476 Z-1.124 P0721 Q0150 F.0833
N215 G00 X4.0 Z2.0
CNC program Example
The tool is positioned at the start point (pass #1) as illustrated below. When the G76 code is cycled the other thread passes are automatically executed.
Thread Tool
Z-1.124 End of Thread Pass
"F".0833 Feedrate Same as the Pitch
Rapid
Rapid
Feed
Rapid to Start Point
Part Section View
"Q".015 DOC First Pass
"P".0721 DOC Full Thread
"X".8476 Minor Diameter
The letter "I" is used for taper threads. It specifies the incremental amount per side in the X-axis.

Figure 20-2 ***G76 multiple thread cutting fixed cycle code***

required for a specified thread size and the control will determine the number of thread passes and automatically position the tool. The code can be applied for both OD and ID threads. The F-feed code is typically programmed on the same block as the G76 code. The T-tool code and the S-spindle RPM code are also required and typically programmed on the block or blocks before the G76 code block.

G90 OD and ID Cutting Cycle (Straight Cutting)

The G90 cycle code is used to define a series of straight OD or ID cuts. The difference between the G90 and the G71 fixed cycle codes is that the G90 allows the CNC programmer to specify the diameter and length of each pass (see Figure 20-3).

Also, the G90 code, which is modal, must be canceled with a G00 code at the end of the operation. The T-tool code, the S-spindle RPM code, and the F-feed code are required and typically programmed on the block or blocks before the G90 code block.

G90 OD and ID Cutting Cycle (Taper Cutting)

The G90 cycle code can also be used to define a series of taper OD or ID cuts. This requires that the incremental amount between the small and large diameters be specified with the G90 code (see Figure 20-4). Also, the G90 taper cutting code is modal and must be canceled with a G00 code at the end of the operation. The T-tool code, the S-spindle RPM code, and the F-feed code are required and typically programmed on the block or blocks before the G90 code block.

G92 Multiple Thread Cutting Fixed Cycle

The G92 cycle code is also used for CNC lathe single-point threading applications (see Figure 20-5). The difference between the G92 and the G76 threading cycle codes is that the G92 allows the CNC programmer to specify the DOC for each thread pass. This code will basically position the tool retract moves automatically. The code can be applied for both OD and ID threads. The F-feed code is typically programmed on the same block as the G92 code. The T-tool code and the S-spindle RPM code are also required and typically programmed on the block or blocks before the G92 code block.

G94 End Face Cutting Cycle

The G94 cycle code is used to define a series of straight or tapered face cuts. The difference between the G94 and the G72 fixed cycle codes is that the G94 allows the CNC programmer to specify the DOC for each facing pass (see Figure 20-6). Also, the G94 end face cutting code is modal and must be canceled with a G00 code at the end of the operation. The T-tool code, the S-spindle RPM code, and the F-feed code are required and typically programmed on the block or blocks before the G94 code block.

The G90 code reduces the number of programming blocks required for OD and ID operations. The example and illustration below describes how the input data repeats the turn or bore passes until the part is finished. When finished, the G00 code cancels the G90 code. The process is the same for both OD and ID cutting.

G90 Code Description:

| G90 | X3.50 | Z-1.125 | F.012 |

OD and ID
Cutting Code

X-diameter
Position

Turn to
Z-position

Feedrate
(.012 IPR)

N205 G00 X3.85 Z.1
N210 G90 X3.5 Z-1.125 F.012
N215 X3.250 Z-1.125
N220 X3.0 Z-.625
N225 X2.75 Z-.625
N230 G00 X4.0 Z5.0

← CNC program
Example

The tool is positioned at the start point as illustrated below. When the G90 code is cycled the turn cuts are executed. The G00 code is used to cancel the 90 code.

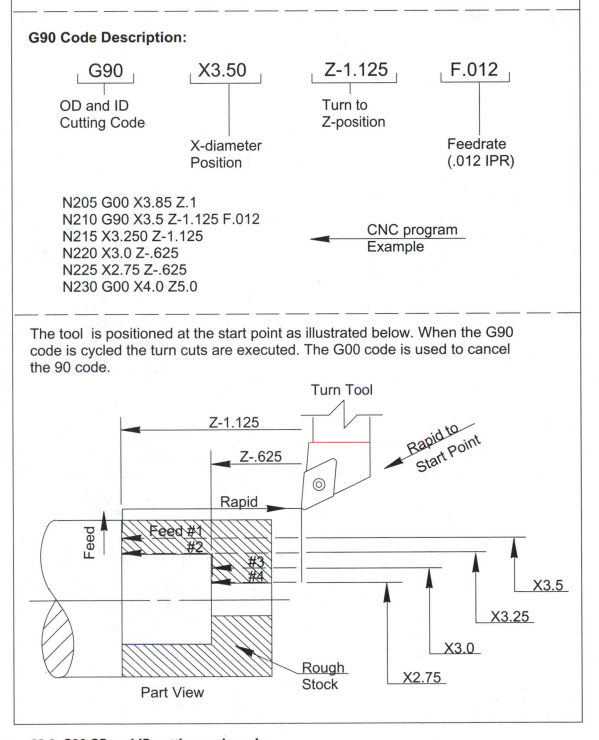

Figure 20-3 G90 OD and ID cutting cycle code

The example and illustration below describes how the input data repeats taper turn or taper bore passes until the part is finished. The process is the same for both OD and ID cutting.

G90 Code Description:

| G90 | X2.5 | Z-1.50 | I.125 | F.012 |

OD and ID
Cutting Code

X-diameter
Position

Turn to
Z-position

Taper Cut
Increment
Per Side

Feedrate
(.012 IPR)

```
N205 G00 X3.0 Z.1
N210 G90 X2.5 Z-1.5 I.125 F.012
N215 X2.25 Z-1.5
N220 X2.0 Z-1.5
N225 X1.75 Z-1.5
N230 G00 X4.0 Z5.0
```

← CNC program
Example

The tool is positioned at the start point as illustrated below. When the G90 code is cycled the taper turn cuts are executed. The G00 code is used to cancel the 90 code.

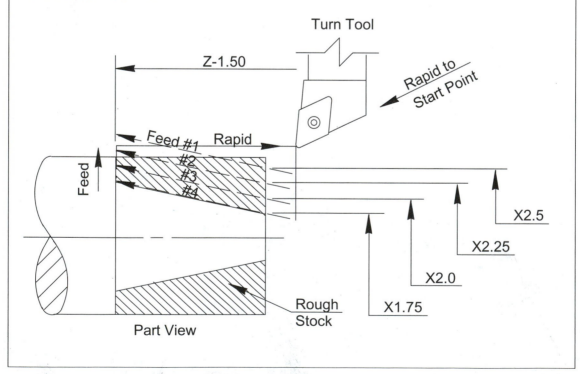

Figure 20-4 G90 OD and ID taper cutting cycle code

The G92 code reduces the number of programming blocks required for a threading operation using a single block of data as input. The example below describes how the input data repeats the threading passes until the thread is finished.

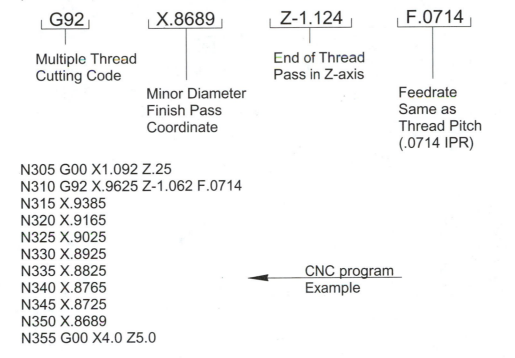

| G92 | X.8689 | Z-1.124 | F.0714 |

Multiple Thread Cutting Code

Minor Diameter Finish Pass Coordinate

End of Thread Pass in Z-axis

Feedrate Same as Thread Pitch (.0714 IPR)

N305 G00 X1.092 Z.25
N310 G92 X.9625 Z-1.062 F.0714
N315 X.9385
N320 X.9165
N325 X.9025
N330 X.8925
N335 X.8825
N340 X.8765
N345 X.8725
N350 X.8689
N355 G00 X4.0 Z5.0

← CNC program Example

The tool is positioned at the start point as illustrated below. When the G92 code is cycled, the thread passes are executed. The G00 code is used to cancel the G92 code.

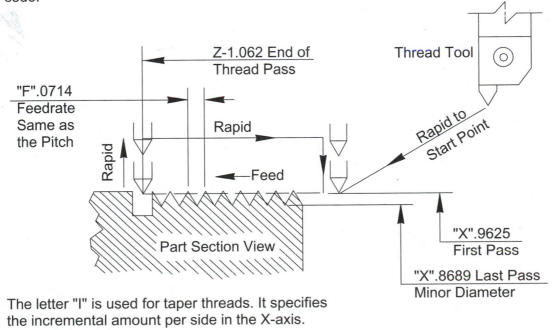

The letter "I" is used for taper threads. It specifies the incremental amount per side in the X-axis.

Figure 20-5 **G92 multiple thread cutting fixed cycle code**

The G94 code reduces the number of programming blocks required for end facing operations. The example and illustration below describes how the input data repeats the facing passes until the part is finished. When finished, the G00 code cancels the G94 code.

G94 Code Description:

G94	X.53	Z-.125	F.008
End Face Cutting Code	Face to X-diameter	Z-position to Start Facing	Feedrate (.008 IPR)

```
N205 G00 X1.7 Z.1
N210 G94 X.53 Z-.125 F.008
N215 Z-.25
N220 Z-.375                    ◄─── CNC program
N225 Z-.5                            Example
N230 G00 X4.0 Z5.0
```

The tool is positioned at the start point as illustrated below. When the G94 code is cycled, the other facing cuts are executed. The G00 code is used to cancel the G94 code.

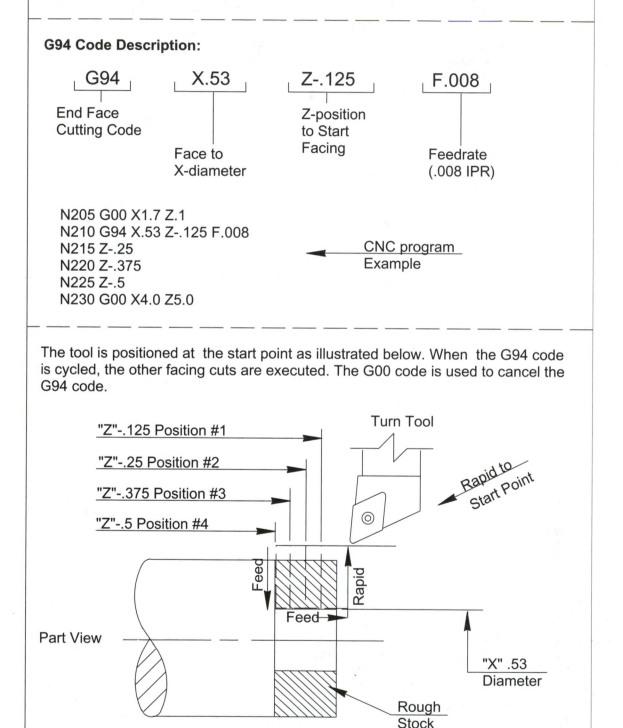

Part View

"Z"-.125 Position #1
"Z"-.25 Position #2
"Z"-.375 Position #3
"Z"-.5 Position #4

Turn Tool

Rapid to Start Point

Feed

Rapid

Feed ►

"X" .53 Diameter

Rough Stock

Figure 20-6 G94 end face cutting cycle code

G04 Dwell Code

There are various situations that require a dwell in time before the next CNC program command should be executed. There are dwell time situations that are related to actual machining of the workpiece and there are situations that are related to CNC machine operation.

For instance, when OD grooving the tool should remain at the finish position for approximately 1–2 seconds. This dwell allows the workpiece to rotate sufficiently to machine a groove diameter that is round. Another instance is when the spindle is started or changed in RPM or speed range (low or high). This dwell allows the CNC machine to reach the specified RPM or speed range shift. The CNC machine operation related dwells are typically designed permanently into the control. For the machining related dwell situations, the G04 code is programmed. The letter "U" is used to specify the time in seconds. For example, a dwell of one second is programmed as follows:

G04U1000

When a dwell code is required, the CNC programmer determines where in the machine cycle the dwell code is required and the amount in time that will be sufficient.

INSTRUCTIONS FOR CNC PROGRAM EXAMPLE 9500

The example in this section illustrates how to identify various codes including a fixed cycle code in the CNC program. This section also includes a coordinate worksheet, CNC documentation, and a CNC program for a CNC lathe. As explained previously, each of these items is designed to illustrate the CNC setup and operation process. It is important to review and understand the purpose and role of each CNC item. These items are listed below:

- Coordinate worksheet (for instructional purposes)
- Engineering drawing
- CNC lathe setup plan
- CNC tool list

This program is similar in format to the other sample programs. The similarities that should be noted include

- Program number
- Information notes
- Turret indexes
- Default code setting

- Spindle starts

- Coolant on

- Machine home command

- Geometry offset activation

- End of program

Instructions for CNC Program Example 9500

For this CNC program, the student is required to identify and note the main CNC program items that are listed below:

1. Program number

2. Activate geometry offset

3. Cancel geometry offset

4. Turret indexes

5. Fixed cycle codes

6. Rapid moves

7. Feed moves

8. RPM codes

9. SFM codes

10. Maximum spindle RPM code

The example below illustrates how to identify and note the CNC program items above.

(Turn Tool 55-Deg.)	*Information Note- tool description*
N20 T0700M41	*Turret Index 7, Cancel Offset, High Range*
N30 G97G50S1500	*RPM Mode, Max. RPM - 1500*
N40 G00X5.0Z.1T0707	*Rapid X&Z, Activate Offset No. 7*
N50 S1000M03	*1000 RPM, Spindle On CW*
N55 G00 X2.2 Z.1 M08	*Rapid X&Z - Coolant On*
N60 G90 X1.75 Z-1.3 F.015	*Fixed Cycle, Start at X1.75 To Z-1.3 @ .015 IPR*

■ ■ ■ ■

COORDINATE WORKSHEET 9500

Instructions: Mark the X-Z zero origin, then calculate the coordinates for each point. Record the calculations in the chart. For the dimensions, use part drawing no. 9500 illustrated in Figure 20-7.

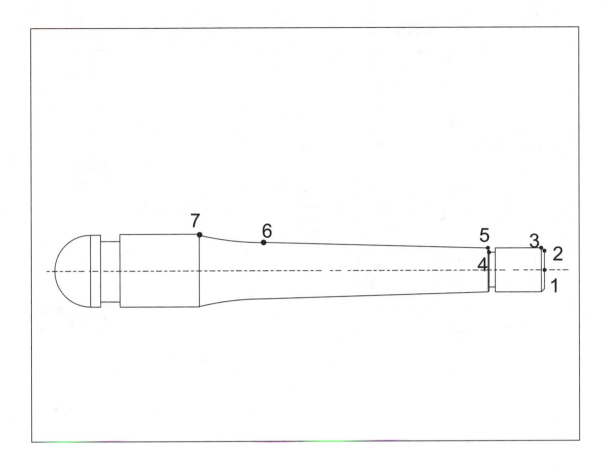

X- and Z-Coordinate Chart

PART NO.	X-COORDINATE	Z-COORDINATE
1		
2		
3		
4		
5		
6		
7		

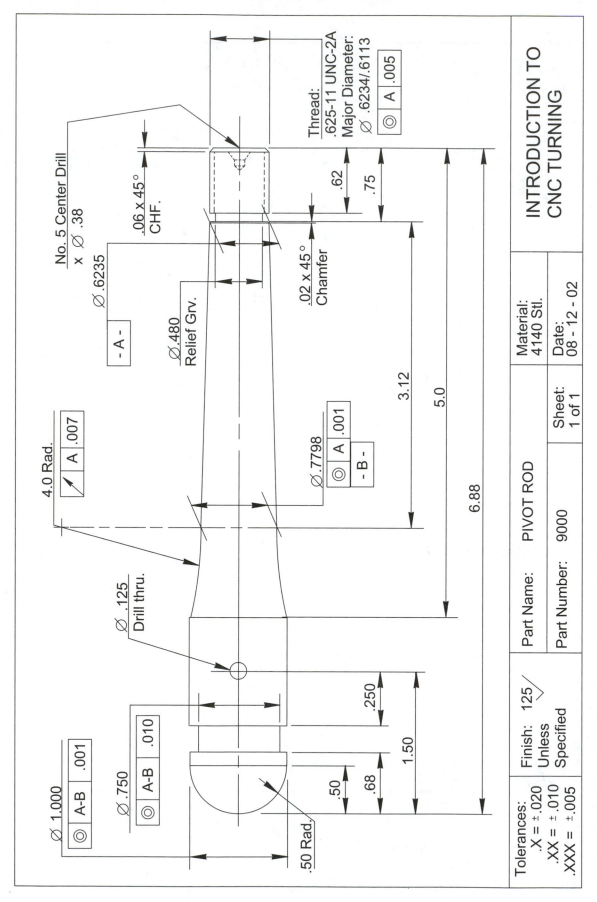

Figure 20-7 *Engineering drawing for CNC program 9500*

CNC LATHE SETUP PLAN

Part No.: 9500	Part Name: Pivot Rod	Date: 10/15/02
Oper. No.: 20	Holding Device:	Introduction To CNC
Program No.: O9500	3-jaw Chuck and Soft Jaws	Turning Centers

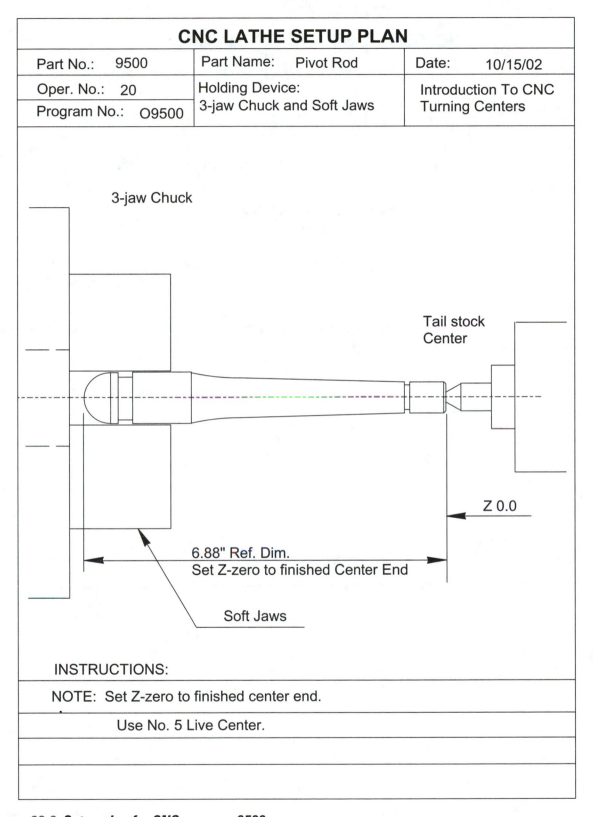

3-jaw Chuck

Tail stock
Center

Z 0.0

6.88" Ref. Dim.
Set Z-zero to finished Center End

Soft Jaws

INSTRUCTIONS:

NOTE: Set Z-zero to finished center end.

Use No. 5 Live Center.

Figure 20-8 Setup plan for CNC program 9500

			CNC TOOL LIST AND OPERATION		
Part No.: 9000			**Prog. No.:** O9500		**Oper. No.:** 20
Part Name: Pivot Rod			**Material.:** C.R. 4140		**Date:** 10/15/02
Station No.	Tool Offset	Tool No.	Tool Description and Operation		
T01	01	8001	Holder No. DCLNR-164D Finish face & Rough turn	Insert: CNMG-432 C5 .031 nose radius	
T02	02	8022	Holder No. DDJNR-164D Finish face & turn	Insert: DNMG-432 C6 .031 nose radius	
T03	03	6007	Groove Holder No. NSR-163D Rgh. & Fin. Groove	Insert: NG-3R C5	
T04	04	8502	Threading Holder No. NSR-164D Fin. Thread	Insert: NG-4R C5	

Figure 20-9 Tool list for CNC program 9500

■ ■ ■ ■
CNC PROGRAM EXAMPLE 9500

```
%
O9500
(PIVOT ROD – PROGRAM NO. 9500)
(DATE, 11/05/02)
(TOOL – 1 ROUGH TURN OD)
(TOOL – 2 FINISH TURN OD)
(TOOL – 3 FINISH GROOVE)
(TOOL – 4 FINISH OD THREAD)

N10G20G40G97
(TOOL - 1 OFFSET - 1)
(ROUGH TURN)
N15G0T0100
N20G97S2405M03
N25G0X.9528Z.1 T0101 M8
N30G50S3600
N35G96S600
N40G1Z-4.7722F.01
N45G2X1.0576Z-5.0215R3.9587
N50G1X1.1991Z-4.9508
N55G0Z.1
N60X.8479
N65G1Z-4.3358
N70G2X.9728Z-4.8265R3.9587
N75G1X1.1142Z-4.7557
N80G0Z.1
N85X.743
N90G1Z-1.9678
N95X.8398Z-3.9007
N100Z-4.2624
N105G2X.8679Z-4.4607R3.9587
N110G1X1.0093Z-4.39
N115G0Z.1
N120X.6381
N125G1Z-.0545
N130X.6774Z-.0742
N135Z-.658
N140X.763Z-2.3671
N145X.9044Z-2.2964
N150G0Z.1
N155X.5332
N160G1Z-.0021
N165X.6581Z-.0645
N170X.7995Z.0062
```

CNC PROGRAM EXAMPLE 9500 (*continued*)

```
N175G0X1.0876 M9
N180G28U0.W0.M05
N185T0100
N190M01

(TOOL - 2 OFFSET - 2)
(FINISH TURN)
N195G0T0200
N200G97S3600M03
N205G0X.4882Z.0954T0202M8
N210G50S3600
N215G96S650
N220G1Z-.0046F.007
N225X.6173Z-.0692
N230Z-.6427
N235X.7798Z-3.8854
N240Z-4.2473
N245G2X.9991Z-5.0119R3.9844
N250G1X1.1405Z-4.9412
N255G28U0.W0.M05
N260T0200 M9
N265M01

(TOOL - 3 OFFSET - 3)
(OD GROOVE )
N270G0T0300
N275G97S1006M03
N280G0X.7597Z-.702 T0303 M8
N285G50S3600
N290G96S200
N295G1X.6183Z-.7727F.002
N300X.5705Z-.75
N305X.48
N310X.68 M9
N315G28U0.W0.M05
N320T0300
N325M01

(TOOL - 4 OFFSET - 4)
(OD THREAD RIGHT)
N330G0T0400
N335G97S1600M03
N340G0X.7588Z.2 T0404 M8
N345G76X.4626Z-.65P0150Q0787F.0909
N350G0X1.5M9
N355Z1.5
N360G28U0.W0.M05
```

CNC PROGRAM EXAMPLE 9500 (*continued*)

```
N365T0400
N370M30
%
```

CHAPTER SUMMARY

The key concepts presented in this chapter include the following:

- The G75 grooving code function and descriptions.
- The G76 threading code function and descriptions.
- The G90 OD and ID cutting code function and descriptions.
- The G90 OD and ID taper cutting code function and descriptions.
- The G92 threading code function and descriptions.
- The G94 facing code function and descriptions.
- The G04 dwell code applications and descriptions.
- Calculations for CNC lathe machining coordinates.
- The format description of a CNC program for CNC lathes.
- The basic codes required of a CNC program for CNC lathe.

REVIEW QUESTIONS

1. Which types of tools are used with the G75 fixed cycle applications?
2. Which types of tools are used with the G76 fixed cycle applications?
3. Which types of tools are used with the G90 fixed cycle applications?
4. How is the feedrate determined and programmed for the G92 and the G76 threading fixed cycle?
5. What will occur when the Feed Hold is pressed during automatic cycle of the G92 or the G76 threading fixed codes?
6. Explain how the threading passes amounts and number of passes are determined for the G92 and G76 threading cycles.
7. What are the differences between the G92 and the G76 codes?
8. Explain how the G90, G92, and G94 codes are cancelled.
9. List some CNC lathe applications where the G04 dwell code is used.
10. Discuss and identify the CNC program format structure and codes covered in this chapter. The items that should be identified include: CNC program number, tool information notes, tool index blocks, tool offset activation blocks, tool offset cancellation blocks, default code setting blocks, rapid and feed moves, circular interpolation blocks, fixed cycle codes, spindle starts and stops, maximum RPM, spindle RPM or SFM mode, coolant on and off, machine home command blocks, and end of program block.

■ ■ CNC PROGRAM EXERCISE FOR CNC LATHES

Using the sketch in Figure 20-10, determine the fixed cycle G code and values required to cut the thread for the part described. Apply the single-line fixed cycle code and enter the code, values, and coordinates in the CNC program on the following page.

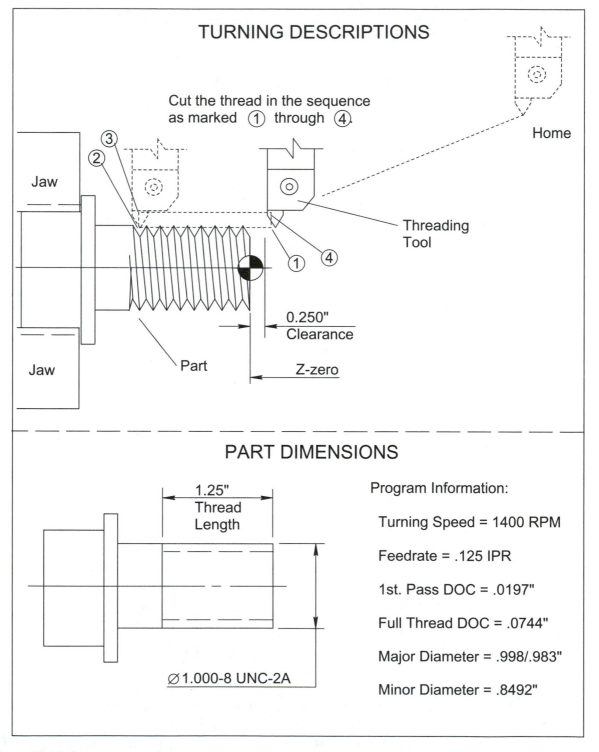

TURNING DESCRIPTIONS

Cut the thread in the sequence as marked ① through ④.

Home

Jaw

③
②

Jaw

Part

Threading Tool

0.250" Clearance

Z-zero

PART DIMENSIONS

1.25" Thread Length

Ø1.000-8 UNC-2A

Program Information:

Turning Speed = 1400 RPM

Feedrate = .125 IPR

1st. Pass DOC = .0197"

Full Thread DOC = .0744"

Major Diameter = .998/.983"

Minor Diameter = .8492"

*Figure 20-10 **Program exercise***

■ ■ CNC PROGRAM EXERCISE 9501

Fill in the missing program information of CNC program 9501 as indicated below:

Block N260: enter the G code and X-Z coordinate values.

Block N280: enter the S RPM value.

Block N240: enter the G code, the X-Z coordinates, P–Q and F values.

```
%
O9501
(SCREW SHAFT - PROGRAM NO. 9501 DATE, 09/27/02)
(TOOL - 01 55-DEG. TURN TOOL)
(TOOL - 02 60-DEG. THREADING TOOL)

N100 G20 G40 G97
(TOOL - 01 55-DEG. TURN TOOL)
N110 G0 T0100
N120 G97 S1500 M03
N130 G0 X.7165 Z.1T0101
N140 G50 S5000
N150 G96 S700 M08
N160 G1 Z0.0 F.025
N170 X.9905 Z-.136 F.007
N180 Z-1.75
N190 X2.3
N200 G97 G0 Z1.0 M09
N210 G28 U0.0 W0.0 M05
N220 T0100

N230 G20 G40 G97
(TOOL - 02 60-DEG. THREAD)
N240 G0 T0200
N250 G97 S1500 M03
N260 G_____ X _____ Z _____ T0202
N270 G50 S5000
N280 G97 S_____ M08
N290 G_____ X _____ Z _____ P _____ Q _____ F_____
N300 G1 X1.2 Z-1.15 F.025
N310 G0 Z1.0 M09
N320 G28 U0.0 W0.0 M05
N330 T0100
N340 M30
%
```

CNC TOOL LIBRARY EXAMPLE

Note: The following CNC tool library example includes the typical types of tools and related features that are used for CNC machining centers. This list is not all-inclusive or complete—it is only intended to be an example.

Drills

Tool No.	Tool Dia.	Flute Length	Gage Length	Description
1000	0.125	2.0	5.500	Solid carbide
1001	0.187	2.0	5.500	Solid carbide
1002	0.125	2.0	5.500	HSS tin coated
1003	0.187	2.0	5.500	HSS tin coated
1004	0.250	3.0	6.500	HSS tin coated
1005	0.312	4.0	6.500	HSS tin coated
1006	0.375	4.0	7.250	HSS tin coated
1007	0.437	4.5	7.250	HSS tin coated
1008	0.500	3.0	5.500	HSS tin coated
1009	0.500	4.75	7.500	HSS tin coated
1010	0.531	4.75	7.500	HSS tin coated
1011	0.562	4.75	7.625	HSS tin coated
1012	0.625	5.0	7.750	HSS tin coated

Drills

Tool No.	Tool Dia.	Flute Length	Gage Length	Description
1013	0.625	6.5	9.250	HSS tin coated
1014	0.687	5.0	7.750	HSS tin coated
1015	0.750	5.5	8.00	HSS tin coated
1016	0.875	5.75	8.250	HSS tin coated
1017	0.937	6.0	9.500	HSS tin coated
1018	1.000	6.5	9.500	HSS tin coated
1019	1.250	6.5	9.750	HSS tin coated
1020	1.375	7.0	10.625	HSS tin coated
1021	1.500	7.0	10.750	HSS tin coated

Insert Drills

Tool No.	Tool Dia.	Flute Length	Gage Length	Description
2011	0.750	5.5	8.00	Tin-coated carbide insert
2012	0.875	5.75	8.250	Tin-coated carbide insert
2013	0.937	6.0	9.500	Tin-coated carbide insert
2014	1.000	6.5	9.500	Tin-coated carbide insert
2015	1.250	6.5	9.750	Tin-coated carbide insert
2016	1.375	7.0	10.625	Tin-coated carbide insert
2017	1.500	7.0	10.625	Tin-coated carbide insert
2018	1.625	7.0	10.625	Tin-coated carbide insert
2019	1.750	7.0	10.750	Tin-coated carbide insert
2020	1.875	7.0	10.750	Tin-coated carbide insert
2021	2.00	7.0	10.750	Tin-coated carbide insert

Spot Drills

Tool No.	Tool Dia.	Flute Length	Gage Length	Description
3000	0.250	1.0	2.500	Solid carbide
3001	0.250	1.0	2.500	HSS tin coated
3002	0.500	1.0	3.500	Solid carbide
3003	0.500	1.0	3.500	HSS tin coated
3004	0.750	1.0	4.500	HSS tin coated
3005	1.00	1.0	4.500	HSS tin coated
3006	1.50	1.0	4.500	HSS tin coated
3007	1.50	1.0	4.500	Solid carbide

Reamers

Tool No.	Tool Dia.	Flute Length	Gage Length	Description
4000	0.125	2.0	5.500	Solid carbide
4001	0.125	2.0	5.500	HSS tin coated
4002	0.250	2.0	5.500	Solid carbide
4003	0.250	2.0	5.500	HSS tin coated
4004	0.375	2.5	6.000	HSS tin coated
4005	0.500	4.0	6.500	HSS tin coated
4006	0.625	4.5	7.000	HSS tin coated
4007	0.750	5.0	7.500	HSS tin coated
4008	0.875	5.0	7.500	HSS tin coated
4009	1.000	5.5	8.000	HSS tin coated

Taps

Tool No.	Dia.-Thrd.	Flute Length	Gage Length	Description
5000	10–32	0.500	4.750	HSS tin coated
5001	1/4–20	0.750	6.125	HSS tin coated
5002	1/4–28	0.750	6.125	HSS tin coated
5003	5/16–18	0.750	6.125	HSS tin coated
5004	3/8–16	1.125	6.750	HSS tin coated
5005	3/8–24	1.125	6.750	HSS tin coated
5006	7/16–14	1.500	7.000	HSS tin coated
5007	1/2–13	1.500	7.000	HSS tin coated
5008	1/2–20	1.500	7.000	HSS tin coated
5009	9/16–12	1.625	7.000	HSS tin coated
5010	5/8–11	1.750	7.125	HSS tin coated
5011	5/8–18	1.750	7.125	HSS tin coated
5012	3/4–10	1.750	7.125	HSS tin coated
5013	3/4–16	1.750	7.125	HSS tin coated
5014	7/8–9	1.750	7.250	HSS tin coated
5015	1.0–8	2.000	7.250	HSS tin coated
5016	1.0–12	2.000	7.250	HSS tin coated

Boring Bars

Tool No.	Tool Dia.	Flute Length	Gage Length	Description
6000	0.35–0.50	2.0	5.500	Carbide cartridge
6001	0.50–0.68	2.0	5.500	Carbide cartridge
6002	0.68–0.85	2.0	5.500	Tin-coated carbide insert
6003	0.85–0.98	2.0	5.500	Tin-coated carbide insert
6004	0.98–1.3	2.5	5.500	Tin-coated carbide insert
6005	1.1–1.6	3.0	6.500	Tin-coated carbide insert
6006	1.5–1.9	3.0	6.500	Tin-coated carbide insert
6007	1.8–2.3	3.0	6.500	Tin-coated carbide insert

Boring Bars

Tool No.	Tool Dia.	Flute Length	Gage Length	Description
6008	2.2–2.6	4.0	7.500	Tin-coated carbide insert
6009	2.5–3.0	4.0	7.500	Tin-coated carbide insert
6010	2.9–3.5	4.0	7.500	Tin-coated carbide insert
6011	3.4–4.0	4.0	7.500	Tin-coated carbide insert
6012	3.9–4.5	4.0	7.500	Tin-coated carbide insert

End Mills

Tool No.	Tool Dia.	Flute Length	Gage Length	Description
7000	0.125	0.50	4.500	2 Flt. solid carbide
7000	0.125	0.50	4.500	4 Flt. HSS tin coated
7001	0.250	0.50	4.500	2 Flt. solid carbide
7001	0.250	0.50	4.500	2 Flt. HSS tin coated
7002	0.375	0.63	4.500	2 Flt. HSS tin coated
7002	0.375	0.63	4.500	4 Flt. HSS tin coated
7003	0.500	1.0	5.000	2 Flt. HSS tin coated
7004	0.500	1.0	5.000	4 Flt. HSS tin coated
7004	0.500	1.0	5.000	4 Flt. solid carbide
7005	0.625	1.2	5.500	2 Flt. HSS tin coated
7006	0.625	1.2	5.500	4 Flt. HSS tin coated
7006	0.625	3.2	7.000	4 Flt. HSS tin coated
7007	0.750	1.5	5.500	2 Flt. HSS tin coated
7007	0.750	4.0	7.500	4 Flt. HSS tin coated
7008	0.750	1.5	5.500	2 Flt. carbide insert
7007	0.750	4.0	7.500	4 Flt. roughing HSS
7009	1.000	2.0	6.000	2 Flt. HSS tin coated
7010	1.000	2.0	6.000	2 Flt. carbide insert
7011	1.000	2.0	6.000	2 Flt. HSS tin coated
7011	1.000	2.0	6.000	4 Flt. roughing HSS

End Mills

Tool No.	Tool Dia.	Flute Length	Gage Length	Description
7012	1.250	1.5	5.000	3 Flt. carbide insert
7013	1.250	2.0	6.000	4 Flt. single end HSS
7013	1.250	2.0	6.000	4 Flt. single end HSS
7014	1.500	1.50	5.500	4 Flt. roughing HSS
7014	1.500	2.50	6.500	4 Flt. single end HSS
7015	1.500	2.50	6.500	4 Flt. carbide insert
7016	2.000	2.50	6.500	4 Flt. single end HSS
7017	2.000	2.50	6.500	4 Flt. carbide insert
7018	2.000	2.50	6.500	6 Flt. carbide insert

Face Mills

Tool No.	Tool Dia.	Gauge Length	Insert No.	Description
8000	2.000	4.0	TPG-322	Pos/Pos, 45 Lead
8001	2.000	4.0	TPG-322	Pos/Pos, 0 Lead
8002	2.500	4.0	TPG-321	Pos/Pos, 15 Lead
8003	2.500	4.0	TPG-322	Neg/Neg, 45 Lead
8004	3.000	4.5	SPG-322	Pos/Pos, 0 Lead
8005	3.000	5.0	SPG-323	Pos/Pos, 45 Lead
8006	4.000	5.0	SPG-322	Pos/Pos, 0 Lead
8007	4.000	5.0	SPG-322	Neg/Neg, 45 Lead
8008	5.000	5.250	SPG-432	Pos/Pos, 0 Lead
8009	5.000	5.250	SPG-433	Pos/Pos, 45 Lead
8010	6.000	5.500	SPG-432	Pos/Pos, 0 Lead
8011	6.000	5.500	SPG-433	Neg/Neg, 45 Lead
8012	8.000	5.500	SPG-532	Pos/Pos, 0 Lead
8013	8.000	5.500	SPG-533	Pos/Pos, 45 Lead
8014	10.000	6.250	SPG-633	Pos/Pos, 45 Lead

FORMULAS FOR MACHINING CENTERS

Note: The following formulas are typically used for CNC machining center calculations. This only includes spindle speed and feedrate related formulas.

SPINDLE SPEED IN REVOLUTIONS PER MINUTE (RPM)

Required information:

CS is the cutting speed expressed in surface feet per minute (SFM)—this value is selected from the appropriate chart.

D is the diameter of the cutting tool.

$$RPM = \frac{4 \times CS}{D}$$

CUTTING SPEED IN SURFACE FEET PER MINUTE (SFM)

Required information:

RPM is the spindle speed expressed in revolutions per minute (RPM).

D is the diameter of the cutting tool.

$$CS = \frac{RPM \times D}{4}$$

■ ■ ■ ■

FEEDRATE IPR AND IPM

Feedrates can be expressed in either of two ways, inches per revolution (IPR) or, inches per minute (IPM).

IPR can typically range from 0.001 inch/revolution to 0.035 inch/revolution, and is determined by selecting an appropriate value from the feed charts.

IPM = IPR × RPM **(for drilling, tapping, boring, reaming)**

Required information:

T is the chip load per tooth or insert.

N is the number of teeth or inserts on the milling cutter.

RPM is the revolutions per minute.

IPM = RPM × T × N **(for milling)**

Tapping feed rate

Required information:

T is the number of threads per inch.

$$IPR \text{ (tapping)} = \frac{1}{T}$$

MILLING CUTTING SPEED (SFM)

The cutting speeds listed below are recommended starting values for end milling and face milling operations. Selection is based on the rigidity of the setup, the machine, the workpiece, and the tool.

Workpiece Material		Cutting Tool Materials			
	HSS	Uncoated Carbide	Coated Carbide	Ceramic	Diamond
Aluminum					
Low silicon	300–800	700–1400			1000–5000
High silicon					500–2500
Brass and bronze	65–130	500–700			1000–3000
Cast iron gray	50–80	250–450	350–500	700–2000	
Cast iron chilled	50–80	250–450	350–500	250–600	
Mild steel, free-machining	60–100	250–350	500–900	1000–2500	
Alloy steel	40–70		350–600	500–1500	
Tool steel	40–70		250–500	500–1200	
Stainless steel 300 series	30–80	100–250	400–650	300–1100	
Stainless steel 400 series	30–80	100–250	250–350	400–1200	
Superalloys		70–100	90–150	500–1000	

DRILLING, REAMING, AND TAPPING CUTTING SPEED (SFM) (HSS TOOLS)

The cutting speeds listed below are recommended starting values. Selection is based on the rigidity of the setup, the machine, the workpiece, and the tool.

Workpiece Material	Drilling Cutting Speed (SFM)	Reaming Cutting Speed (SFM)	Tapping Cutting Speed (SFM)	Cutting Lubricant
Aluminum	300–800	130–200	90–100	Kerosene Light base oil
Brass	65–130	130–200	90–100	Soluble oil Light base oil
Bronze	65–130	50–70	30–60	Mineral oil Light base oil
Cast iron gray	50–80	70–100	70–80	Dry Soluble oil
Cast iron chilled	50–80	50–70	35–60	Dry Soluble oil

Workpiece Material	Drilling Cutting Speed (SFM)	Reaming Cutting Speed (SFM)	Tapping Cutting Speed (SFM)	Cutting Lubricant
Mild steel, free-machining	60–100	35–70	35–80	Soluble oil
Alloy steel	40–70	35–40	10–35	Sulfur-base oil
Stainless steel 300 series	30–80	15–30	10–35	Sulfur-base oil
Stainless steel 400 series	30–80	20–50	10–35	Sulfur-base oil
Superalloys	Use carbide	10–35	10–35	Sulfur-base oil

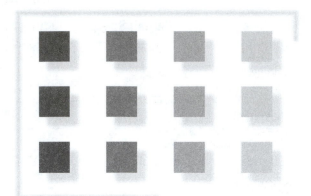

MILLING FEEDS (FPT)

The feedrate is determined by considering various factors, such as

1. Rigidity of the setup.

2. Rigidity of the workpiece structure.

3. Rigidity and extention of the spindle and tool.

4. Cutter diameter—use lower feed rates for smaller diameter cutters.

Workpiece Material	Type of Milling Cut and Tool Material							
	Face Milling Feed/Tooth		Side Milling Feed/Tooth		End Milling Feed/Tooth		Saw Milling Feed/Tooth	
	HSS	Carb.	HSS	Carb.	HSS	Carb.	HSS	Carb.
Aluminum	0.007 to 0.020	0.007 to 0.020	0.004 to 0.013	0.004 to 0.012	0.002 to 0.011	0.002 to 0.010	0.001 to 0.005	0.001 to 0.005
Brass and bronze	0.005 to 0.014	0.004 to 0.012	0.003 to 0.008	0.002 to 0.007	0.002 to 0.007	0.002 to 0.006	0.001 to 0.003	0.001 to 0.003
Cast iron	0.004 to 0.016	0.006 to 0.020	0.002 to 0.009	0.003 to 0.012	0.001 to 0.020	0.007 to 0.020	0.001 to 0.005	0.002 to 0.006

	Type of Milling Cut and Tool Material							
Workpiece Material	**Face Milling Feed/Tooth**		**Side Milling Feed/Tooth**		**End Milling Feed/Tooth**		**Saw Milling Feed/Tooth**	
	HSS	Carb.	HSS	Carb.	HSS	Carb.	HSS	Carb.
Mild steel, Free- machining	0.003 to 0.012	0.004 to 0.016	0.002 to 0.008	0.003 to 0.009	0.001 to 0.007	0.0015 to 0.008	0.001 to 0.003	0.001 to 0.004
Alloy steel	0.001 to 0.010	0.001 to 0.012	0.001 to 0.005	0.001 to 0.008	0.001 to 0.004	0.002 to 0.006	0.0005 to 0.002	0.001 to 0.004
Stainless steel 300 series 400 series	0.001 to 0.011	0.0015 to 0.015	0.001 to 0.008	0.0015 to 0.010	0.0005 to 0.005	0.0005 to 0.005	0.0005 to 0.004	0.0005 to 0.005

DRILLING FEEDRATES (IPR) (HSS DRILLS)

The feedrate is determined by considering various factors, such as

1. Rigidity of the setup.
2. Rigidity of the workpiece structure.
3. Rigidity and extention of the spindle and tool.
4. Rigidity of the machine.

Use lower feedrates for less rigid situations.

Workpiece Material	Diameter of the Drill (inch)							
	0.125 1/8	0.250 1/4	0.375 3/8	0.500 1/2	0.750 3/4	1.0	1.250 1-1/4	1.500 1-1/2
Aluminum	0.003	0.005	0.006	0.008	0.010	0.013	0.015	0.016
Brass and bronze	0.004	0.007	0.010	0.014	0.018	0.022	0.026	0.030
Cast iron	0.004	0.006	0.009	0.012	0.016	0.020	0.025	0.028
Mild steel free-machining	0.003	0.005	0.007	0.010	0.014	0.016	0.022	0.025
Alloy steel	0.003	0.004	0.005	0.006	0.010	0.012	0.014	0.016
Stainless steel 300 series 400 series	0.002	0.004	0.006	0.008	0.014	0.016	0.016	0.020

CARBIDE INSERT DRILLING CUTTING SPEEDS (SFM) AND FEEDRATES (IPR)

The cutting speed and feedrate values below assume optimum conditions, such as

1. Rigidity of the setup.

2. Rigidity of the workpiece structure.

3. Rigidity and extention of the spindle and tool.

4. Thru-the-tool coolant pressure.

Use lower cutting speeds and feedrates for less rigid situations.

Workpiece Material	Cutting Speed (SFM)	Feedrate (IPR) by Drill Diameter (inch) Range		
		0.500 Diameter to 1.00 Diameter	1.00 Diameter to 1.50 Diameter	1.50 Diameter and Over
Aluminum	500–2000	0.004–0.006	0.005–0.012	0.008–0.015
Brass and bronze	200–400	0.003–0.005	0.005–0.008	0.008–0.012
Cast iron	200–400	0.005–0.007	0.006–0.009	0.008–0.012
Mild steel free-machining	200–600	0.003–0.005	0.004–0.008	0.007–0.010
Alloy steel	200–500	0.003–0.005	0.004–0.008	0.007–0.015
Stainless steel 300 series 400 series	200–450	0.003–0.005	0.004–0.006	0.005–0.012

TAP DRILL SIZE CHART (INCH AND METRIC)

Tap Size	Tap Drill	Tap Size	Tap Drill	Tap Size	Tap Drill	Tap Size	Tap Drill
0–80	56 3/64	6–40	34 33 32	5/16–18	F G 17/64 H	7/8–9	49/64 25/32
1–64	54 53	8–32	29 28	5/16–24	H I J	7/8–14	51/64 13/16
1–72	53 1/16	8–36	29 28 9/64	3/8–16	5/16 O P	1 inch to 8	55/64 7/8 57/64 29/32
2–56	51 50 49	10–24	27 26 25 24 23 5/32 22	3/8–24	21/64 Q R	1 inch to 12	29/32 59/64 15/16
2–64	50 49	10–32	5/32 22 21 20 19	7/16–14	T 23/64 U 3/8 V	1 inch to 14	59/64 15/16
3–48	48 5/64 47 46 45	12–24	11/64 17 16 15 14	7/16–20	W 25/64 X	1-1/8–7	31/32 63/64 1 inch
3–56	46 45 44	12–28	16 15 14 13 3/16	1/2–13	27/64 7/16	1-1/8–12	1-3/64

Metric Tap Sizes

Tap Size	Tap Drill
M1.6 x .35	1.25
M1.8 x .35	1.45
M2 x .4	1.60
M2.2 x .45	1.75
M2.5 x .45	2.05
M3 x .5	2.50
M3.5 x .6	2.90
M4 x .7	3.30
M4.5 x .75	3.70
M5 x .8	4.20
M6 x 1	5.00
M7 x 1	6.00
M8 x 1.25	6.70
M8 x 1	7.00
M10 x 1.5	8.50
M10 x 1.25	8.70
M12 x 1.75	10.20
M12 x 1.25	10.80
M14 x 2	12.00
M14 x 1.5	12.50
M16 x 2	14.00
M16 x 1.5	14.50
M18 x 2.5	15.50
M18 x 1.5	16.50
M20 x 2.5	17.50
M20 x 1.5	18.50
M22 x 2.5	19.50
M22 x 1.5	20.50
M24 x 3	21.00
M24 x 2	22.00
M27 x 3	24.00
M27 x 2	25.00

								Metric Tap Sizes	
Tap Size	Tap Drill	Tap Size	Tap Drill	Tap Size	Tap Drill	Tap Size	Tap Drill	Tap Size	Tap Drill
4–40	44	1/4–20	9	1/2–20	29/64	1-1/4–7	1-1/8	M30 x 3.5	26.50
	43		8					M30 x 2	28.00
	42		7					M33 x 3.5	29.50
	3/32		13/64					M33 x 2	31.00
			6					M36 x 4	32.00
			5					M36 x 3	33.00
			4					M39 x 4	35.00
4–48	42	1/4–28	3	9/16–12	15/32	1-1/4–12	1-11/16	M39 x 3	36.00
	3/32		7/32		31/64				
	41		2						
5–40	40			9/16–18	1/2	1-3/8–6	1-15/64		
	39				33/64				
	38								
	37								
5–44	38			5/8–11	17/32	1-3/8–12	1-19/64		
	37				35/64				
	36								
6–32	37			5/8–18	9/16	1-1/2–6	1-23/64		
	36				37/64				
	7/64								
	35								
	34								
	33								
				3/4–10	41/64	1-1/2–12	1-27/64		
					21/32				
				3/4–16	11/16				

NPT PIPE TAP DRILL SIZE CHART

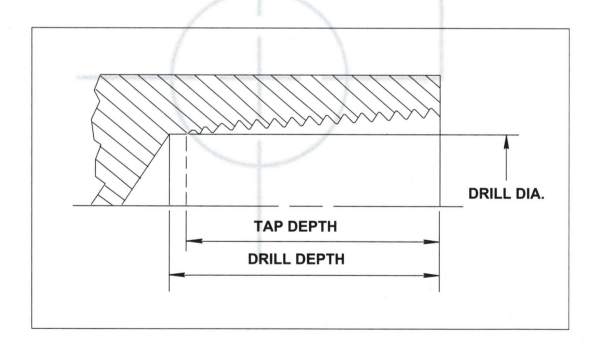

DRILL DIA.

TAP DEPTH

DRILL DEPTH

NPT Size	Drill Size	Drill Depth	Tap Depth
1/16–27	D	19/32 min.	0.555 max.
1/8–27	R	19/32 min.	0.555 max.
1/4–18	7/16	25/32 min.	0.722 max.
3/8–18	37/64	13/16 min.	0.757 max.
1/2–14	45/64	1-1/32 min	0.962 max.
3/4–14	59/64	1-1/32 min.	0.967 max.
1 inch to 11-1/2	1-5/32	1-1/4 min.	1.172 max.
1-1/4–11-1/2	1-1/2	1-9/32 min.	1.200 max.
1-1/2–11-1/2	1-47/64	1-5/16 min.	1.213 max.
2 inches to 11-1/2	2-7/32	1-9/32 min.	1.197 max.
2-1/2–8	2-5/8	1-27/32 min.	1.701 max.
3 inches to 8	3-1/4	1-29/32 min.	1.785 max.
3-1/2–8	3-3/4	2 inches min.	1.884 max.
4 inches to 8	4-1/4	2-1/16 min.	1.919 max.

MACHINING CENTER PROGRAM CODE DESCRIPTIONS

Note: The following is a typical CNC program used for CNC machining center applications. This is only intended as a guide for interpreting CNC programs.

CNC Program Codes	Code Description
%	*Program file start code*
O1000	*Program 1000*
(END PLATE - PROGRAM NO. 1000)	*Information note*
(DATE, 08/12/99)	*Information note*
(TOOL - 01 2.00 DIA. - FACE MILL)	*Information note*
(TOOL - 02 1.500 DIA. - SPOTDRILL)	*Information note*
(TOOL - 03 .750 DIA. - HSS DRILL)	*Information note*
(TOOL - 04 1.125 DIA. - HSS DRILL)	*Information note*
N100 G00 G17 G40 G49 G80 G90	*Default Block reset codes*
N110 G91 G28 Z0.	*Incremental mode, Machine home, Z-axis*
N120 G28 X0. Y0.	*Machine home, X- and Y-axes*
(TOOL - 01: 2.00 DIA. - FACE MILL)	*Information note*

CNC Program Codes	Code Description
N130 T1 M06	*Tool 1, Auto-tool-change*
N140 M01	*Optional stop*
N150 G90 G0 X-1.5 Y-.6 S1200 M3	*Absolute mode, Rapid position, X-Y axes, 1200 RPM, Spindle on CW*
N160 G43 H1 Z.3 M08	*Length offset on, offset 1, Rapid Z-axis, Coolant on*
N170 G1 Z0. F50.	*Feed (to position), Z-axis, 50 IPM*
N180 X5.75 F24.	*Feed (milling cut), X-axis, 24 IPM*
N190 Y-2.4	*Feed (milling cut), Y-axis, 24 IPM*
N200 X-1.5	*Feed (milling cut), X-axis, 24 IPM*
N210 G0 Z1.	*Rapid, Z-axis*
N220 M05	*Spindle Stop*
N230 G91 G49 G28 Z0. M09	*Incremental mode, Length offset off, machine home, Z-axis, Coolant off*
(TOOL - 02: 1.50 DIA. - SPOTDRILL)	*Information note*
N240 G00 G17 G40 G49 G80 G90	*Default Block reset codes*
N250 T2 M06	*Tool 2, Auto-tool-change*
N260 M01	*Optional stop*
N270 G90 G0 X1.25 Y-.625 S0500 M3	*Absolute mode, Rapid position, X-Y axes, 500 RPM, Spindle on CW*
N280 G43 H2 Z1. M08	*Length offset on, offset 2, Rapid Z-axis, Coolant on*
N290 G99 G81 Z-.41 R.1 F5.	*R-return mode, Drilling cycle—1st hole, Drill depth, Retract height, 5 IPM Feed*
N300 X3.25	*Drilling cycle—2nd hole*
N310 Y-2.375	*Drilling cycle—3rd hole*
N320 X1.25	*Drilling cycle—4th hole*
N330 G99 G81 X2.25 Y-1.5 Z-.59 R.1 F5.	*R-return mode, Drilling cycle—5th hole, Drill depth, Retract height, 5 IPM Feed*
N340 G80	*Canned cycle cancel*
N350 M05	*Spindle stop*
N360 G91 G49 G28 Z0. M09	*Incremental mode, Length offset off, Machine home, Z-axis, Coolant off*
(TOOL - 03: .750 DIA. - DRILL)	*Information note*

CNC Program Codes	Code Description
N370 G00 G17 G40 G49 G80 G90	*Default Block reset codes*
N380 T3 M06	*Tool 3, Auto-tool-change*
N390 M01	*Optional stop*
N400 G90 G0 X1.25 Y-.625 S0535 M3	*Absolute mode, Rapid position, X-Y axes, 535 RPM, Spindle on CW*
N410 G43 H3 Z1. M08	*Length offset on, offset 3, Rapid Z-axis, Coolant on*
N420 G99 G81 Z-.905 R.1 F7.	*R-return mode, Drilling cycle—1st hole, Drill depth, Retract height, 7 IPM Feed*
N430 X3.25	*Drilling cycle—2nd hole*
N440 Y-2.375	*Drilling cycle—3rd hole*
N450 X1.25	*Drilling cycle—4th hole*
N460 G80	*Canned cycle cancel*
N470 M05	*Spindle stop*
N480 G91 G49 G28 Z0. M09	*Incremental mode, Length offset off, Machine home, Z-axis, Coolant off*
(TOOL - 04 1.125 DIA. - DRILL)	*Information note*
N490 G00 G17 G40 G49 G80 G90	*Default Block reset codes*
N500 T4 M06	*Tool 4, Auto-tool-change*
N510 M01	*Optional stop*
N520 G0 G90 X2.25 Y-1.5 S0355 M3	*Rapid position, X-Y axes, 355 RPM, Spindle on CW*
N530 G43 H4 Z1. M08	*Length offset on, offset 4, Rapid Z, axis, Coolant on*
N540 G99 G81 Z-1.013 R.1 F3.2	*R-return mode, Drilling cycle—one hole, Drill depth, Retract height, 3.2 IPM Feed*
N550 G80	*Canned cycle cancel*
N560 M05	*Spindle stop*
N570 G91 G49 G28 Z0. M09	*Incremental mode, Length offset off, Machine home, Z-axis, Coolant off*
N570 G90	*Absolute mode*
N580 M30	*End of program, Reset to start*
%	*Program file end code*

CNC LATHE TOOL LIBRARY EXAMPLE

Note: The following CNC tool library example includes the typical types of tools and related features that are used for CNC lathes. This list is not all inclusive or complete it is only intended to be an example.

Drills

Tool No.	Tool Dia.	Flute Length	Gage Length	Description
1000	0.125	2.0	5.500	Solid carbide
1001	0.187	2.0	5.500	Solid carbide
1002	0.125	2.0	5.500	HSS tin coated
1003	0.187	2.0	5.500	HSS tin coated
1004	0.250	3.0	6.500	HSS tin coated
1005	0.312	4.0	6.500	HSS tin coated
1006	0.375	4.0	7.250	HSS tin coated
1007	0.437	4.5	7.250	HSS tin coated
1008	0.500	3.0	5.500	HSS tin coated
1009	0.500	4.75	7.500	HSS tin coated
1010	0.531	4.75	7.500	HSS tin coated
1011	0.562	4.75	7.625	HSS tin coated
1012	0.625	5.0	7.750	HSS tin coated
1013	0.625	6.5	9.250	HSS tin coated
1014	0.687	5.0	7.750	HSS tin coated
1015	0.750	5.5	8.00	HSS tin coated
1016	0.875	5.75	8.250	HSS tin coated
1017	0.937	6.0	9.500	HSS tin coated
1018	1.000	6.5	9.500	HSS tin coated
1019	1.250	6.5	9.750	HSS tin coated
1020	1.375	7.0	10.625	HSS tin coated
1021	1.500	7.0	10.750	HSS tin coated

Insert Drills

Tool No.	Tool Dia.	Flute Length	Gage Length	Description
2011	0.750	5.5	8.00	Tin-coated carbide insert
2012	0.875	5.75	8.250	Tin-coated carbide insert
2013	0.937	6.0	9.500	Tin-coated carbide insert
2014	1.000	6.5	9.500	Tin-coated carbide insert
2015	1.250	6.5	9.750	Tin-coated carbide insert
2016	1.375	7.0	10.625	Tin-coated carbide insert
2017	1.500	7.0	10.625	Tin-coated carbide insert
2018	1.625	7.0	10.625	Tin-coated carbide insert
2019	1.750	7.0	10.750	Tin-coated carbide insert
2020	1.875	7.0	10.750	Tin-coated carbide insert
2021	2.00	7.0	10.750	Tin-coated carbide insert

Spot Drills

Tool No.	Tool Dia.	Flute Length	Gage Length	Description
3000	0.250	1.0	2.500	Solid carbide
3001	0.250	1.0	2.500	HSS tin coated
3002	0.500	1.0	3.500	Solid carbide
3003	0.500	1.0	3.500	HSS tin coated
3004	0.750	1.0	4.500	HSS tin coated
3005	1.00	1.0	4.500	HSS tin coated
3006	1.50	1.0	4.500	HSS tin coated
3007	1.50	1.0	4.500	Solid carbide

Reamers

Tool No.	Tool Dia.	Flute Length	Gage Length	Description
4000	0.125	2.0	5.500	Solid carbide
4001	0.125	2.0	5.500	HSS tin coated
4002	0.250	2.0	5.500	Solid carbide
4003	0.250	2.0	5.500	HSS tin coated
4004	0.375	2.5	6.000	HSS tin coated
4005	0.500	4.0	6.500	HSS tin coated
4006	0.625	4.5	7.000	HSS tin coated
4007	0.750	5.0	7.500	HSS tin coated
4008	0.875	5.0	7.500	HSS tin coated
4009	1.000	5.5	8.000	HSS tin coated

Taps

Tool No.	Dia.-Thrd.	Flute Length	Gage Length	Description
5000	10–32	0.500	4.750	HSS tin coated
5001	1/4–20	0.750	6.125	HSS tin coated
5002	1/4–28	0.750	6.125	HSS tin coated
5003	5/16–18	0.750	6.125	HSS tin coated
5004	3/8–16	1.125	6.750	HSS tin coated
5005	3/8–24	1.125	6.750	HSS tin coated
5006	7/16–14	1.500	7.000	HSS tin coated
5007	1/2–13	1.500	7.000	HSS tin coated
5008	1/2–20	1.500	7.000	HSS tin coated
5009	9/16–12	1.625	7.000	HSS tin coated
5010	5/8–11	1.750	7.125	HSS tin coated
5011	5/8–18	1.750	7.125	HSS tin coated
5012	3/4–10	1.750	7.125	HSS tin coated
5013	3/4–16	1.750	7.125	HSS tin coated
5014	7/8–9	1.750	7.250	HSS tin coated
5015	1.0–8	2.000	7.250	HSS tin coated
5016	1.0–12	2.000	7.250	HSS tin coated

Boring Bars

Tool No.	Tool Dia.	Bore Length	Gage Length	Description
6000	0.35–0.50	2.0	5.500	Carbide cartridge
6001	0.50–0.68	2.0	5.500	Carbide cartridge
6002	0.68–0.85	2.0	5.500	Tin-coated carbide insert
6003	0.85–0.98	2.0	5.500	Tin-coated carbide insert
6004	0.98–1.3	2.5	5.500	Tin-coated carbide insert
6005	1.1–1.6	3.0	6.500	Tin-coated carbide insert
6006	1.5–1.9	3.0	6.500	Tin-coated carbide insert
6007	1.8–2.3	3.0	6.500	Tin-coated carbide insert
6008	2.2–2.6	4.0	7.500	Tin-coated carbide insert
6009	2.5–3.0	4.0	7.500	Tin-coated carbide insert
6010	2.9–3.5	4.0	7.500	Tin-coated carbide insert
6011	3.4–4.0	4.0	7.500	Tin-coated carbide insert
6012	3.9–4.5	4.0	7.500	Tin-coated carbide insert

Turn Tools

Tool No.	Catalog No.	Insert No.	Description
9000	DCLNR-164D	CNMG-432	80° turn
9001	DCLNR-164D	CNMG-332	80° turn
9002	DDJNR-164D	DNMG-432	55° turn
9003	DDJNR-164D	DNMG-332	55° turn
9004	DVJNR-164D	VNMG-432	35° turn
9005	DVJNR-164D	VNMG-332	35° turn
9006	MTRNR-164D	TNMG-432	Triangle turn
9007	MTRNR-164D	TNMG-332	Triangle turn
9008	PROON-164D	RNMG-432	Round turn
9009	PROON-204D	RNMG-332	Round turn

Groove Tools

Tool No.	Catalog No.	Insert No.	Description
9200	NSR-164D	NG-4R	Grooving tool
9201	NLR-164D	NG-4L	Grooving tool

Thread Tools

Tool No.	Catalog No.	Insert No.	Description
9400	NSR-164D	NT-4R	Threading tool
9401	NLR-164D	NT-4L	Threading tool

Parting Tools

Tool No.	Catalog No.	Insert No.	Description
9500	GRPR-125	GS-125N	Cutoff tool
9501	GRPR-094	GS-094N	Cutoff tool

FORMULAS FOR TURNING

Note: The following formulas are typically used for CNC turning machine calculations. This only includes spindle speed and feedrate related formulas.

SPINDLE SPEED IN RPM

Required information:

CS is the cutting speed expressed in SFM—this value is selected from the appropriate chart.

D is the diameter of the workpiece (where the cutting tool is positioned to cut).

$$\mathbf{RPM} = \frac{\mathbf{4} \times \mathbf{CS}}{\mathbf{D}}$$

CUTTING SPEED IN SURFACE FEET PER MINUTE (SFM)

Required information:

RPM is the spindle speed expressed in revolutions per minute (RPM).

D is the diameter of the workpiece.

$$CS = \frac{RPM \times D}{4}$$

■ ■ ■ ■
FEEDRATES

Feedrates for turning operations, which also include drilling, tapping, boring, and reaming, are typically expressed in inches per revolution (IPR).

IPR can typically range from 0.001 inch/rev. to 0.035 inch/rev. and is determined by selecting an appropriate value from the feed charts.

In situations where the IPM feedrate is required the formula to convert IPR to IPM is as follows:

$$IPM = IPR \times RPM$$

Threading feedrate

Required information:

T is the number of threads per inch.

$$IPR \text{ (threading)} = \frac{1}{T}$$

LATHE CUTTING SPEEDS (SFM) AND FEEDRATES (IPR) FOR TURNING AND BORING OPERATIONS

The following values are for indexable carbide inserts. Value selection is based on the rigidity of the setup, the workpiece, and the toolholder. Use lower values for less rigid situations. Increase speed for interrupted cuts. Use higher SFM values for coated carbide grades. Use lower feedrates for better surface finish.

Workpiece Material	Type of Cut	Depth of Cut (inches)	Cutting Speed (SFM)	Feedrate (IPR)
Aluminum, brass, bronze, copper	Rough	0.150–0.350	650–2000	0.008–0.025
	Finish	0.010–0.120	1000–3000	0.003–0.016
Free-machining, and low carbon steels: AISI 1020 1117, 10L45	Rough	0.100–0.300	300–700	0.010–0.030
	Finish	0.010–0.100	600–800	0.006–0.012
Medium and high carbon steels: AISI 1045, 1080, 1525, 1541, 1572	Rough	0.100–0.300	200–800	0.010–0.030
	Finish	0.010–0.100	250–900	0.006–0.012
Alloy and tool steels: AISI 4140, 4150, 5120, 8640 T2, M2, A2, D2	Rough	0.100–0.300	175–750	0.010–0.030
	Finish	0.010–0.100	250–850	0.006–0.012
Gray cast iron: Class 20, 30, 35, 45, 55, and 60	Rough	0.100–0.500	200–800	0.010–0.030
	Finish	0.015–0.100	400–1000	0.004–0.015

60° V-THREAD CUTTING SPEEDS (SFM) AND FORMULAS

The cutting speeds listed are recommended starting values. Selection is based on the rigidity of the setup, the machine, the workpiece, and the tool.

CUTTING SPEEDS (SFM) for V-THREADS
(Turning with carbide insert)

Workpiece Material	Uncoated Carbide	Coated Carbide
Aluminum Low silicon	400–800	
High silicon		400–1000
Brass and bronze	250–600	
Cast iron gray	200–300	
Cast iron ductile/alloy	150–250	300–500
Mild Steel free-machining		300–700
Alloy steel 330–450 BHN		200–400
Alloy steel titanium	110–180	
Stainless steel 300 series	200–350	
Stainless steel 400 series		300–500
Nonmetallics	400–1500	

To calculate the depth of thread per side:

(OD Threads)

$0.61343 \div$ threads per inch

(ID Threads)

$0.54127 \div$ threads per inch

To calculate the number of passes:

$72 \times$ lead $+ 4$ (inch threads)
$2.8 \times$ lead $+ 4$ (metric threads)

Depth of first pass:

Total depth of cut$/\sqrt{\text{Number of passes}}$

Depth of following passes:

1st pass $\times \sqrt{\text{Next pass number,}}$
i.e., 2, 3, 4, . . .

To determine the Z-start position:

$4 \times$ thread lead or 0.250 inches,

whichever is greater.

To determine the RPM:

$$\frac{4 \times \text{Cutting Speed}}{\text{Thread OD}}$$

LATHE PROGRAM CODE DESCRIPTIONS

Note: The following is a typical CNC program used for CNC turning applications. This is only intended as a guide for interpreting CNC programs.

CNC Program Codes	Code Description
%	*Program file start code*
O3305	*Program 3305*
(BUSHING SHAFT, PROGRAM NO. 3305)	*Information note*
(DATE, 10/12/02)	*Information note*
(TOOL - 01 FIN & RGH - 80 DEG. TURN)	*Information note*
(TOOL - 02 FIN - 55 DEG. TURN)	*Information note*
(TOOL - 03 DRILL 1.00 DIA. CARBIDE)	*Information note*
(TOOL - 04 BORING BAR - .75 DIA.)	*Information note*
N10G20G40G97	*Program file start code*
(TOOL - 01 - 80 DEG. TURN)	*Information note*
(FINISH FACE)	*Information note*
N15G0T0100	*Index tool 1, cancel offset*
N20G97S1142M03	*RPM mode, 1142 RPM, Spindle On CW*

CNC Program Codes	Code Description
N25G0X3.4Z.1 T0101	*Rapid X-Z position, activate offset 1*
N30G50S3600	*Maximum 3600 RPM*
N35G96S700	*SFM mode, 700 SFM*
N40G1Z0.F.008M08	*Feed Z-position, 0.008 IPR, Coolant On*
N45X-0.064	*Feed X-position, 0.008 IPR*
N50X.136Z.1	*Feed X-Z position, 0.008 IPR*
N55G0X3.0625	*Rapid X-position*
(ROUGH OD)	*Information note*
N60G1Z-2.03F.012	*Feed Z-position, 0.012 IPR*
N65X3.162	*Feed X-position, 0.012 IPR*
N70X3.262Z-2.07	*Feed X-Z position, 0.012 IPR*
N75G0Z.1	*Rapid Z-position*
N80Z1.0M09	*Rapid Z-position, Coolant Off*
N85G28U0.W0.M05	*Machine Home X-Z, Spindle Off*
N90T0100	*Cancel offset 1*
N95M01	*Optional Stop*
(TOOL - 2 OFFSET - 2)	*Information note*
(FINISH OD TURN)	*Information note*
N100G0G20G40G97T0200	*Default codes, Index tool 2, cancel offset*
N105G97S2072M03	*RPM mode, 2072 RPM, Spindle On CW*
N110G0X2.7538Z.10T0202	*Rapid X-Z position, activate offset 2*
N150G50S3600	*Maximum 3600 RPM*
N120G96S700	*SFM mode, 700 SFM*
N125G1Z0.0F.008M08	*Feed Z-position, 0.008 IPR, Coolant On*
N130X3.0Z-.1231	*Feed X-Z position, 0.008 IPR*
N135Z-2.015	*Feed Z-position, 0.008 IPR*
N140X3.05	*Feed X-position, 0.008 IPR*
N145X3.15Z-1.965	*Feed X-Z position, 0.008 IPR*
N150G0Z.1M09	*Rapid Z-position, Coolant Off*
N155G28U0.W0.M05	*Machine Home X-Z*

CNC Program Codes	Code Description
N160T0200	*Cancel offset 2*
N165M01	*Optional Stop*
(TOOL - 3 OFFSET - 3)	*Information note*
(DRILL 1.0 DIA.)	*Information note*
N170G0G20G40G97T0300	*Default codes, Index tool 3, cancel offset*
N175G97S1500M03	*RPM, 1500 RPM, Spindle On CW*
N180G0X0.Z.25 T0303	*Rapid X-Z position, activate offset 3*
N185Z.1M08	*Rapid Z-position, Coolant On*
N190G1Z-3.15F.011	*Feed Z-Drilling Depth, 0.011 IPR*
N195G0Z.25M09	*Rapid Z-position, Coolant Off*
N200G28U0.W0.M05	*Machine Home X-Z*
N205T0300	*Cancel offset 3*
N210M01	*Optional Stop*
(TOOL - 4 OFFSET - 4)	*Information note*
(FINISH BORING BAR .750 DIA.)	*Information note*
N215G0G20G40G97T0400	*Default codes, Index tool 4, cancel offset*
N220G97S1219M03	*RPM, 1219 RPM, Spindle On CW*
N225G0X1.1632Z.10T0404	*Rapid X-Z position, activate offset 4*
N230G50S3600 M08	*Maximum 3600 RPM, Coolant On*
N235G96S700	*SFM mode, 700 SFM*
N240 G1Z0.0F.007	*Feed Z-position, 0.007 IPR*
N245G2X1.062Z-.0506R.0506	*Feed Radius, 0.007 IPR*
N250G1Z-2.9	*Feed Z-position, 0.007 IPR*
N255X1.002Z-2.87	*Feed X-Z position, 0.007 IPR*
N260G0Z.2M09	*Rapid Z-position, Coolant Off*
N265G28U0.W0.M05	*Machine Home X-Z*
N270T0400	*Cancel offset 4*
N275M30	*Program Stop*
%	

RIGHT TRIANGLE FORMULAS

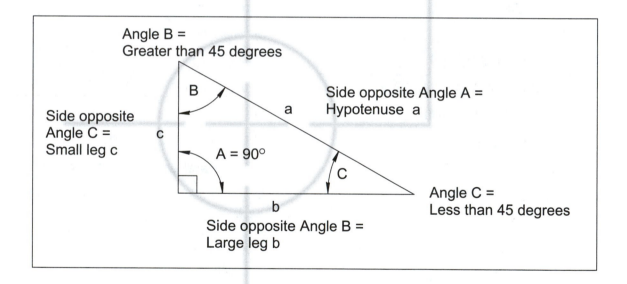

Angle B =
Greater than 45 degrees

B

Side opposite Angle A =
Hypotenuse a

a

Side opposite
Angle C =
Small leg c

c

A = 90°

C

Angle C =
Less than 45 degrees

b

Side opposite Angle B =
Large leg b

To Find Angles	Formulas	To Find Sides	Formulas
C	$\dfrac{c}{a} = \text{Sine C}$	b	$a \times \text{Sine B}$
C	$\dfrac{b}{a} = \text{Cosine C}$	b	$a \times \text{Cosine C}$
C	$\dfrac{c}{b} = \text{Tan. C}$	b	$c \times \text{Tan. B}$
B	$\dfrac{b}{a} = \text{Sine B}$	c	$a \times \text{Sine C}$
B	$\dfrac{c}{a} = \text{Cosine B}$	c	$a \times \text{Cosine B}$
B	$\dfrac{b}{c} = \text{Tan. B}$	c	$b \times \text{Tan. C}$

$$(90° - \text{Angle C}) = \text{Angle B}$$
$$(90° - \text{Angle B}) = \text{Angle C}$$

$$a = \sqrt{b^2 + c^2}$$
$$b = \sqrt{a^2 - c^2}$$
$$c = \sqrt{a^2 - b^2}$$

FRACTION AND DECIMAL CONVERSION CHART

Fraction Inch	Decimal	Fraction Inch	Decimal
1/64	0.015625	33/64	0.515625
1/32	0.03125	17/32	0.53125
3/64	0.046875	35/64	0.546875
1/16	0.0625	9/16	0.5625
5/64	0.078125	37/64	0.578125
3/32	0.09375	19/32	0.59375
7/64	0.109375	39/64	0.609375
1/8	0.1250	5/8	0.6250
9/64	0.140625	41/64	0.640625
5/32	0.15625	21/32	0.65625
11/64	0.171875	43/64	0.671875
3/16	0.1875	11/16	0.6875
13/64	0.203125	45/64	0.703125
7/32	0.21875	23/32	0.71875
15/64	0.234375	47/64	0.734375

Fraction Inch	Decimal	Fraction Inch	Decimal
1/4	0.2500	3/4	0.7500
17/64	0.265625	49/64	0.765625
9/32	0.28125	25/32	0.78125
19/64	0.296875	51/64	0.796875
5/16	0.3125	13/16	0.8125
21/64	0.328125	53/64	0.828125
11/32	0.34375	27/32	0.84375
23/64	0.359375	55/64	0.859375
3/8	0.3750	7/8	0.8750
25/64	0.390625	57/64	0.890625
13/32	0.40625	29/32	0.90625
27/64	0.421875	59/64	0.921875
7/16	0.4375	15/16	0.9375
29/64	0.453125	61/64	0.953125
15/32	0.46875	31/32	0.96875
31/64	0.484375	63/64	0.984375
1/2	0.5000	1	1.0000

LETTER AND NUMBER DRILL SIZE CONVERSION CHART

Letter Drill Size Chart

Letter	Size	Letter	Size	Letter	Size	Letter	Size
A	0.234	H	0.266	O	0.316	V	0.377
B	0.238	I	0.272	P	0.323	W	0.386
C	0.242	J	0.277	Q	0.332	X	0.397
D	0.246	K	0.281	R	0.339	Y	0.404
E	0.250	L	0.290	S	0.348	Z	0.413
F	0.257	M	0.295	T	0.358		
G	0.261	N	0.302	U	0.368		

Number Drill Size Chart

No.	Size	No.	Size	No.	Size	No.	Size
1	0.2280	21	0.1590	41	0.0960	61	0.0390
2	0.2210	22	0.1570	42	0.0935	62	0.0380
3	0.2130	23	0.1540	43	0.0890	63	0.0370
4	0.2090	24	0.1520	44	0.0860	64	0.0360
5	0.2055	25	0.1495	45	0.0820	65	0.0350
6	0.2040	26	0.1470	46	0.0810	66	0.0330
7	0.2010	27	0.1440	47	0.0785	67	0.0320
8	0.1990	28	0.1405	48	0.0760	68	0.0310
9	0.1960	29	0.1360	49	0.0730	69	0.0292
10	0.1935	30	0.1285	50	0.0700	70	0.0280
11	0.1910	31	0.1200	51	0.0670	71	0.0260
12	0.1890	32	0.1160	52	0.0635	72	0.0250
13	0.1850	33	0.1130	53	0.0595	73	0.0240
14	0.1820	34	0.1110	54	0.0550	74	0.0225
15	0.1800	35	0.1100	55	0.0520	75	0.0210
16	0.1770	36	0.1065	56	0.0465	76	0.0200
17	0.1730	37	0.1040	57	0.0430	77	0.0180
18	0.1695	38	0.1015	58	0.0420	78	0.0160
19	0.1660	39	0.0995	59	0.0410	79	0.0145
20	0.1610	40	0.0980	60	0.0400	80	0.0135

METRIC CONVERSION

UNITS OF LENGTH AND THEIR ABBREVIATIONS

1000 Meters	= Kilometer	(km)
100 Meters	= Hectometer	(hm)
10 Meters	= Dekameter	(dam)
1 Meter	= Meter	(m)
1/10 Meter	= Decimeter	(dm)
1/100 Meter	= Centimeter	(cm)
1/1000 Meter	= Millimeter	(mm)

METRIC AND ENGLISH CONVERSION FORMULAS AND TABLES

To Convert millimeters to inches:

Divide millimeter by 25.4

$$\frac{\mathbf{mm}}{\mathbf{25.4}}$$

To Convert inches to millimeters:

multiple inches by 25.4

(Inches × 25.4)

ENGLISH TO METRIC

1 Yard = 0.9144 meters
1 Foot = 0.3048 meters or 304.8 millimeters
1 Inch = 25.40 millimeters or 2.54 centimeters

METRIC TO ENGLISH

1 Meter = 39.37 inches
1 Centimeter = 0.3937 inches
1 Millimeter = 0.03937 inches

ENGLISH TO METRIC FOR TURNING SFM

For G96 SFM: Convert feet per minute to metric

Example: 500 SFM = (.3048 × 500)

= 152 MPM

GLOSSARY

A-Axis The axis of a circular motion of a CNC machine tool member or slide about the X-axis.

Absolute Dimensioning The method of dimensioning that defines dimensions from the zero origin with the sign dependent on which quadrant the dimensions are located in.

Absolute Programming The method of programming that defines coordinates from the part origin with the sign dependent on which quadrant the tool is moving to.

Accuracy The trueness of a measured feature or location in relation to the known dimension.

Adaptive Control A technique for achieving optimum cutting conditions. The speed and feed of a tool are continuously adjusted by a computer acting on sensor feedback data, which measures tool stress, heat, and vibration.

Address A letter, group of letters, or numbers that are used to identify to the computer the type of information that follows.

Allowance The intentional minimum clearance or maximum interference assigned between mating parts.

Alloy A mixture of metals that have been combined to form the alloy in order to gain desired characteristics.

Alloy Steel A steel made up of several materials in addition to carbon and iron.

Alphanumeric Code A system of entering information using alphabetic characters (A–Z), numeric characters (0–9), and special characters (+, −).

Analog Refers to a system in which data is gathered continuously from a sensor, which monitors some physical activity. A tachometer is an analog device that produces an output voltage signal in direct proportion to sensed speed.

Angle-Plate Fixture A modified form of plate fixture used to machine a part at an angle to its location.

Angularity The geometric characteristic that specifies a specific angular relationship between two surfaces of a part.

ANSI The abbreviation for American National Standards Institute, which specifies drafting standards for parts prints.

APT The abbreviation for Automatically Programmed Tools. APT was the first and remains today the most powerful computer-aided part programming language.

Arc A circular motion of the tool generated by two axes in a CW (G02) or CCW (G03) direction.

Arc Center The incremental distance from the arc start point to the center point of the arc that is represented by I, J, and K CNC codes.

ASCII The abbreviation for American Standard Code for Information Interchange. With this system the computer represents each character, internally, as a unique series of seven bits.

Assembly Drawing A mechanical drawing that shows all the parts and elements of a tool in their assembled form.

Automap The abbreviation for automatic machining programming. This computer-aided programming language is a subset of APT.

Automatic Tool Changer The CNC machine component that is used for automatically changing tools usually with the tool change arm from the magazine.

Auxiliary Functions Additional programmable functions of the machine other than coordinate movement of the tool. These are referred to as M codes in word address format.

Axis A line of reference for coordinate system. In CNC programming, the X-, Y-, and Z-axes extend out at right angles to each other from a common origin. They are used to specify a direction for moving a tool.

Axis Inhibit A feature of a CNC unit that enables the operator to withhold command information from the machine tool slide.

Basic Dimension A theoretically perfect dimension used to locate part features or datums. This type of dimension is shown enclosed in a box on the part drawing.

B-Axis The axis of a circular motion of a CNC machine tool member or slide about the Y-axis such as for the HMC table.

BHN The abbreviation for the Brinell Hardness Number. This decimal number placed to the right of BHN specifies the hardness the material is to exhibit when indented by the test's hardened steel ball.

Bilateral Tolerance Tolerance value that may go in both a plus and minus direction from the basic size.

Block A group of words representing a complete operational instruction. A block is ended by an end-of-block (EOB) character.

Block Delete A slash (/) entered in the front of a block directs the CNC system to ignore the block in a program.

CAD The abbreviation for computer-aided design or drafting.

CAD/CAM The abbreviation for a computer system that is capable of design, drafting, and manufacturing.

CAM The abbreviation for computer-aided manufacturing.

Cancel Code A command that will discontinue any modal code, canned cycle, or offset control.

Canned Cycle A preset sequence of events that is executed by issuing a single command on a CNC machine.

Cap Screw A type of mechanical fastener used to assemble the parts of a jig or fixture. Typically, these fasteners have a cylindrical head with an internal hexagon drive.

Carbon Steel A specific type of steel composed mainly of iron and carbon. The three standard types or designations of carbon steel are low, medium, and high carbon.

Cartesian Coordinate System A system for defining a point in space relative to (X, Y, Z) zero origin position.

Cast Iron A cast ferrous material with a higher carbon content than carbon steel; it is used for both manufactured parts and tooling. Composed mainly of iron, carbon, and other elements.

C-axis The axis of a circular motion of a CNC machine tool member or slide about the Z-axis.

Cemented Carbide Insert An indexable cutting tool tip material composed mainly of tungsten carbide or titanium carbide and cobalt as a bonding agent.

Ceramics A cutting tool tip material composed mainly of cemented oxide tool materials that can cut at higher temperatures and cutting speeds.

Chamfer A beveled edge machined to break a sharp machined corner.

Character A number-, letter-, or symbol-coded representation that is programmed into the computer.

Chuck A commercially available workholder, used mainly on a lathe or grinder, which holds round parts with a series of jaws.

Circular Interpolation A code and coordinate that directs the system to cut an arc or circle on a CNC machine.

Circularity The geometric characteristic that specifies a condition of roundness in a part feature.

Circular Runout The geometric characteristic that specifies a specific relationship between two or more concentric diameters of a part measured in single line elements around the part.

Clamp A mechanical device used to hold a part securely in a jig or fixture.

Clamping Force The force exerted by the clamping device while holding the part in the workholder.

CMM The abbreviation for a coordinate measuring machine.

CNC The abbreviation for computer numerical control.

Collet A mechanical device used to hold parts, normally of a specific diameter or size, which are to be turned or rotated around a central axis.

Command A signal or group of signals for initiating a step in the execution of a CNC program.

Command Readout The display of the table slide position resulting from the control system.

Compact II A computer-aided programming language having commands that are more English-like than APT.

Concentricity The geometric characteristic that specifies a concentric relationship between two or more diameters of a part.

Continuous Path A programming method used for numerically controlled machine tools whereby the position of the tool is constantly controlled throughout the complete machining cycle.

Coordinate Measuring Machine A machine with a computer system that measures part features then performs calculations and outputs the results.

CPU The abbreviation for the central processing unit of a computer.

Critical Dimension A part dimension that controls the size or location of a part feature that is very important to the overall function of the part.

CRT The abbreviation for cathode-ray tube. This device is used for displaying all programmed input and corresponding output of text and graphics from the MCU.

Cubic Boron Nitride (CBN) A cutting tool material that can be used at feeds and speeds three times faster than carbide.

Cutter Diameter Compensation A feature of a control system in which the cutter radius data is input to automatically place the tool on the part boundary. This is also a method of roughing and finishing a part using the same program data.

Cutter Offset The distance from the part profile to the center of the cutter.

Cutting Forces The forces exerted against the part by the cutter during the machining cycle.

Cutting Speed The rate of speed usually in surface feet per minute (SFM) that material is removed by the cutter during the machining cycle.

Cycle A sequence of operations that is repeated regularly.

Datum   A point, line, surface, or feature considered to be theoretically exact and which is used to locate the part or the geometric characteristic features of the part.

Datum Dimensioning   A system of dimensioning based on a common starting point.

Datum Feature Symbol   The datum reference letter contained in a rectangular box and that is used to note the datum surfaces on the part.

Debug   The process of checking, locating, and correcting errors in a program.

Depth of Cut (DOC)   Refers to the machining amount in depth of the cutting tool as it is engaged in the part.

Diameter (Dia.)   A round feature typically used to dimension a hole or shaft.

Diamond Pin   A specific type of relieved locating pin used for jigs and fixtures. The pin, when viewed from the end, has a diamond shape.

DNC   The abbreviation for direct numerical control, which consists of several CNC machines that are connected to and receive programs from a main or host computer.

Dowel Pin   A hardened steel pin used to locate assembled parts or as a locating device to accurately position the parts in a jig or fixture.

Drill Bushing   A hardened steel or carbide bushing used to guide and support a cutting tool, such as a drill, throughout its machining cycle.

Dry Run   The process of checking, locating, and correcting errors in a CNC program.

Dwell   A time delay created by a command in a CNC program.

Edit   The process of modifying a CNC program stored in memory.

EIA Code   The abbreviation for Electronics Industries Association code set. An older code system used for tape-punched programming systems.

End-of-Block   A character placed in a CNC program that signals the end of a command line or block.

End-of-Program   A miscellaneous function code (M30) placed at the end of each program. This code signals the control to reset to the start of the CNC program.

Executive Program   A set of instructions that enables the MCU to respond to CNC commands specific to the type of CNC machine that it controls.

F Code   The address code used to specify the feedrate in a CNC program.

Feature   Refers to any surface, angle, line, hole, etc., which is to be controlled for production accuracy.

Feature Control Symbol   The symbol used in geometric dimensioning and tolerancing, which contains the geometric characteristic symbol, tolerance value, and datum reference.

Feed   Positioning the slides at a specified rate of movement in either inches per minute (IPM) or revolution (IPR) to cut the workpiece.

Feedrate   The rate of movement of the tool into the work as related to a particular machining operation in IPM for milling, IPR for drilling.

Feedrate Override   A manual switch that enables the operator to alter the programmed feedrate during a cutting operation.

Fixed Renewable Bushings   Easily replaceable drill bushings designed to be used in a linear bushing and which are held in place with a lock screw or clamp.

Fixed Stop Locator   A block or pin that is set up in a group to locate the workpiece in a manner that is not adjustable. Each of the locators is fixed in order to maintain repeatability of the production run.

Fixture   A workholding device, which holds, supports, and locates the workpiece while providing a referencing surface or device for the cutting tool.

Fixture Key   Square or rectangular block attached to the base of a jig or fixture, which locates and aligns the workholder in the "T"-slots of the machine tool on which it is used.

Flatness   The geometric characteristic that specifies an amount of permitted variation from a perfectly flat surface.

Form Character   The geometric characteristics that control the form of a part. These include straightness, flatness, circularity, and cylindricity.

Functional Gauging   Refers to a complete gauge fixture that is especially designed to inspect a part for compliance with a GDT tolerance specification.

Gauge   A mechanical device of a known and precise size, used to check, or compare, part sizes (also *gage*).

Gauge Height   A preset height above the work to which the tool retracts after an operation.

G Codes   Preparatory codes designated by the address "G," which are used to treat programmed information in a specific mode.

Geometric Dimensioning and Tolerancing (GD&T)   A dimensioning method used to show the exact specifications of part size and form through the use of geometric symbols.

Geometric Symbols   The symbols used to show the specific type of feature control required for the part.

Geometry Offset   The offset storage area used to compensate for the diameter and length differences in tools of a CNC turning machine.

Hardness   The property of a material that permits it to resist penetration or indentation.

Hardware   All physical units that make up a computer or control system.

H Code   The address code used to specify the tool height offset of a tool in a CNC program.

Heat Treatment   The process of using heat to modify the properties of a metal.

Helical Interpolation   The software feature directing the system to cut a circular interpolation while simultaneously interpolating a linear move of a third axis.

High Speed Steel (HSS)   An alloy steel, which is made of tungsten and other materials, that is used to manufacture drills, end mills, and taps.

Home Position   The fixed X, Y, Z starting point on a machine from which all subsequent tool motion is measured.

Horizontal Machining Center (HMC)   A CNC machine that is designed with the spindle in the horizontal position.

ID   The abbreviation for inside diameter.

Inch   The customary system of measurement used for linear measurements in the United States.

Inches Per Minute (IPM)   Used to specify a feedrate per time, usually for milling applications.

Inches Per Revolution (IPR)   Used to specify a feedrate per spindle revolution, usually for single-point tools such as drills, taps, and boring bars.

Incremental Dimensioning   A system of specifying part dimensions whereby each new dimension is taken relative to the last dimension given.

Incremental Programming   The method of programming that defines an incremental coordinate value relative to the point that the tool is at. The sign is determined by the movement direction of the tool.

Indexer   A device that accurately spaces holes or other features around a central axis.

Input   All external information entered into the MCU control unit.

Inspection   The process of checking a workpiece to verify its conformity to the specifications in the part print.

Interpolation   The process of determining an intermediate point located between two given data points.

Jaws   The mechanical devices or elements of a workholder, such as a vise or chuck, that are used to grip the part being machined.

Jig   A workholding device that holds, supports, and locates the workpiece while providing a guiding device for the cutting tool.

Jog   A manually operated device that is used to move the tool in the X-, Y-, and Z-direction of a machine.

Leading Zero   A programming format that does not use a decimal point and eliminates all the leading zeros to the left of the first significant digit in a coordinate word (Y-0.505 is input Y-5050).

Linear Interpolation   A feature of control system to accept two positioning points on a part profile and fit a straight line between the points.

LMC   The abbreviation for least material condition. Refers to the condition at which an external feature (such as a shaft) is at the smallest size permitted and an internal feature (such as a hole) is at the largest size permitted.

Locator   A device used to establish and maintain the position of a part in a setup or fixture; may be fixed or adjustable.

Loop   An instruction or series of instructions to be executed repeatedly to produce a desired operation.

Machinability   The measure of a material's ability to be machined. Some materials are soft and have good machinability characteristics, while others are harder and have poor machinability.

Machine Vise   A vise designed and intended to be used on a machine, rather than on a bench. The most common types of machine vises are the milling machine vise and the drill press vise.

Machining Center   A CNC machine that in one setup is capable of executing various operations such as milling, drilling, tapping, reaming, on one or more surfaces of the part.

Macro   A complete set of instructions for executing a particular operation.

Magnetic Tape   A plastic tape that is coated with magnetic material for storing information.

Major Diameter   The largest diameter on external or internal threads.

Manual Data Input (MDI)   A feature of a control system that allows the operator to manually key in a program or a command.

Manual Part Programming   The preparation of a manuscript in machine control language and format to define a sequence of commands for use on a CNC machine.

Manuscript   A form, usually paper, that details the CNC part program.

M Codes   The miscellaneous function codes designated by the address "M" that are used to treat programmed information in a specific mode of the machine such as the Spindle On/Off, Coolant On/Off, and so on.

Memory   A unit that is used by a computer to store information.

Metric   The customary system of measurement used for linear measurements in most European and South American countries.

Microinch   Is one millionth of an inch or 0.000001 inch.

Minor Diameter   Refers to the smallest diameter on external or internal threads.

MMC   The abbreviation for maximum material condition. Refers to the condition of a part feature in which it has the maximum amount of material permitted by the tolerance. It is the smaller size of a hole or the largest size of an external part feature.

Modal Code   Refers to the codes that stay active until replaced by a cancellation code or newer modal code.

Modular Tooling   A system of individual parts assembled for a variety of different fixtures.

N code   The address code used to specify the block numbers in a CNC program.

Nonmodal Code   Refers to the codes that do not stay active when read and executed by the control.

Numerical Control (NC)   An automated manufacturing system that uses a numerical data system to operate a variety of machine tools.

Nut   A mechanical fastening device used to secure bolts in assembled units.

O Code   The address code used to specify the CNC program number.

OD   The abbreviation for outside diameter.

Offset   The compensation in axis movement for the diameter and length differences in tools.

Open-Loop System   A control system that has no means of comparing the output with the input for control purposes.

Optional Program Stop   A miscellaneous (M01) command in a CNC program that is ignored unless the operator has previously pushed a button to initiate the command.

Orientation   The geometric characteristics that control the orientation of part features such as angularity, parallelism, and perpendicularity. Also the position of the machine spindle that locates the spindle keys for automatic tool changing.

Origin   The zero reference point from which all coordinates in X, Y, and Z are referenced.

OSHA   The abbreviation for Occupational Safety and Health Administration, a government agency that manages and inspects health and safety concerns in industry.

Parallelism   The geometric characteristic that specifies a relationship between two parallel surfaces of a part.

Part   The workpiece or object to be manufactured or machined.

Part Drawing   The drawing, engineering drawing, or print of an object to be manufactured or machined.

Part Program   A complete set of instructions written by a CNC programmer in numerical programming language.

Peck Drilling   A method of drilling deep holes that is activated by the CNC program when it reads a G83 peck drilling canned cycle code.

Perpendicularity   The geometric characteristic that specifies a perpendicular relationship between two surfaces of a part.

Pin   A cylindrical fastener that is normally pressed into a premachined hole for the purpose of alignment or holding parts of an assembled workholder.

Pitch   The distance between the adjacent crests or roots of thread teeth.

Plate Fixture   A fixture that has a plate as its main structural component.

Polar Coordinates   A method of locating a point by specifying the length of a line from an origin to the point and the angle the line makes with the positive X-axis.

Polycrystalline Diamond (PCD)   A cutting tool material made from synthetic diamonds that is used mainly for finishing operations of nonmetallic materials. PCD tools cut faster and last longer than carbide.

Position   The geometric characteristic that specifies a relationship between a part feature and a central axis, or center. Also the axis location on a CNC machine.

Positioning (Contouring)   A control system in which the tool can remain in continuous contact with the part as it is moved from point A to point B.

Positioning (Point-to-Point)   A control system in which the tool is not a continuous contact with the part as it is moved from point A to point B.

Power Clamping   A clamping system, either hydraulic, pneumatic, or a combination of both, used to activate and hold the clamping pressure.

Preparatory Function   A command for changing the mode of operations of the control. Expressed by a "G" and two digits to specify a code.

Primary Locators   The locators used to locate, or reference, the primary locating surface of the part.

Process Planning   The act, or process, of planning the step-by-step procedure to be used to manufacture or machine a workpiece.

Production Plan   The document specifying the exact means and methods to be used to manufacture a part or product.

Production Run   A specific group, or number, of parts to be manufactured at one time.

Profile   The geometric characteristics that control the form, or shape, of a part profile. These include profile of a line and profile of surface.

Program Stop   A miscellaneous (M00) function entered in a program to stop the automatic cycle of a CNC machine.

Projected Tolerance Zone   The imaginary zone projected above, a part feature, such as a hole, to ensure assembly with a mechanical fastener passing through both the tolerance feature and the projected thickness of the mating part.

Proveout   The process of executing a new CNC program in "single block" to locate and correct program or setup errors.

Quadrant   One of the four sections in the Cartesian coordinate system.

RAM   The abbreviation for random-access memory, which can be used to store, access and edit files.

Rapid Traverse   Positioning the slides at a high rate of speed in inches per minute (IPM) before a cut is started.

Reference Dimension   A dimension shown on a part print for informational purposes and used for reference, rather than manufacture. Such a dimension is shown inside parentheses ( ), or with the abbreviation REF.

Repeatability   The ability of a workholder that permits the parts being machined to be duplicated within the stated limits of size, part after part.

Reset   Returning a program or machine function to a specified initial condition or value.

Rockwell Hardness   A specific measure and rating of a material's resistance to penetration or indentation (Rc).

ROM   The abbreviation for read-only memory which can be read but cannot be changed.

Root   Refers to the bottom of the thread teeth for external threads and the top of the thread teeth for internal threads.

Roughness   Refers to the finer irregularities created on a part's surface by the production process.

Roughness Average (Ra)   Refers to the average roughness value over a control distance on a part's surface.

RPM   The abbreviation for revolutions per minute.

Runout   The geometric characteristics that control the runout of a part. These include circular runout and total runout.

S Code   The address code used to specify the spindle speed in a CNC program.

Screw Clamp   A clamping device that uses a screw thread to clamp the part.

Secondary Locator   Used to reference the secondary locating surface of the part.

Sequence Number   Consists of the letter N followed by a one- to four-digit number in a CNC program.

Set Block   A gauging device included on many fixtures to aid in setting the cutters to suit the workpiece.

Set Screw   A mechanical fastening device that is generally made without a head and that is used for adjustable locators or as a locking device for other applications.

Setup   The preparation of the CNC machine, or other machine, for a production run. This may include assembling and loading the tools, aligning and fastening the workholding device, loading the CNC program, setting the origins, entering the offsets, and completing the proveout.

Setup Gauge   A special purpose gauge used to establish an origin or an actual cutter position.

Slot Nuts   The name used to describe the type of nut that is held in the machine table slots that are shaped like a "T". Also called T-nuts.

Socket Head Cap Screw (SHCS)   A type of mechanical fastener used to assemble the parts of a setup or fixture. Typically these fasteners have a cylindrical head with an internal hexagon drive.

Software   Part programs or instructions that are used to drive the computer's circuits in executing a particular operation.

Solid Locators   Locators used on fixtures that are solid and nonadjustable.

Spindle   The cylindrical machine component that enables the cutting tool or the workpiece to rotate at the programmed speed during a cutting operation.

Spindle Speed Override   A manual switch that enables the operator to alter the programmed spindle speed during a cutting operation.

Spotface   An enlargement machined at the end of a hole to a shallow depth intended for seating a washer or the head of a bolt.

Storage   The process of holding information in memory until it is needed by the MCU.

Straightness   The geometric characteristic that specifies a relationship between a part feature and a true straight plane or line.

Strap Clamp   A specific type of clamping device that uses a metal bar, similar to a strap, to hold the part securely to the workholder. The elements are the strap, hold-down bolt, and the heel block.

Subplate   A plate containing various drilled and tapped holes to which a variety of different components and clamps are to be attached to form a crude type of modular fixture.

Subprogram   A CNC program that is executed automatically during the execution of another CNC program.

Subroutine   A set of instructions that executes a complete operation and is assigned a name.

Supports   Locators placed under a workpiece to eliminate part deflection.

Surface Feet Per Minute (SFM)   The linear measurement of a moving tool per time in minutes.

Surface Finish   The texture of a workpiece that typically can range from smooth to rough depending on the type of manufacturing process.

Swing Clamp   A variation of a screw clamp consisting of an arm that swings about a stud at one end and contains the screw clamp at the other end.

Tap   Refers to a tool with fluted cutting edges used to cut internal threads.

Tap Drill   A drill used to make a hole in a part to accommodate a thread cutting tap.

Tape   A perforated paper or magnetized medium used for storing information in the form of holes or magnetized spots.

T-Bolt   A mechanical fastener that has a head designed to fit into the "T"-slots of machine tools.

T Code   The address code used to specify the tool number in a CNC program.

Tertiary Locator   Used to reference the tertiary (third) locating surface of the part.

Thread Depth   The perpendicular distance between the crest and root of thread teeth.

Thru Hole   A hole drilled through the entire material of the part.

TIR   The abbreviation for total indicator reading.

T-Nuts   Nuts made specifically for use in the "T"-slots of machine tools.

Toggle Clamp   A commercially available style of clamping device that uses a lever mechanism to operate and hold the position of the clamp.

Tolerance   The amount of variation, or deviation, permitted in the size of a part.

Tombstone   A commercially available tooling block that is mounted on machining centers and similar machine tools to provide multiple mounting surfaces for workholders.

Tool Cutter   The implement that cuts and shapes the workpiece on a machine. Tool cutters generally are designed with a sharpened edge or edges depending on the type of cutting operation (drill, mill, turn).

Tool Drawings   The mechanical drawings made and used to build or construct tools.

Tool Length Offset   The distance that represents the amount to be compensated by the control in the Z-axis. This amount is stored in the tool geometry offset memory under the corresponding offset number.

Tool Setter   A type of sensing or measuring device that automatically or manually references the dimensions of each tool in a setup for offset storage.

Tool Steel   A type of alloy steel made to very close specifications and primarily used for tools or similar types of parts.

Total Runout   The geometric characteristic that specifies a relationship between two or more concentric diameters of a part measured totally around and across the part feature.

Touch-Off   The method of setting the tool lengths by touching the tool tip to the specified workpiece Z-zero surface.

Trailing Zero Format   A programming format that does not use a decimal point and eliminates the trailing zeros to the right of the first significant digit in a coordinate word (Y-5.000 is input Y-05).

Turret   The machine component that enables the cutting tools to be mounted and fastened and later perform a cutting operation.

Unilateral Tolerances   Tolerance values that may go only in a single direction, either plus or minus, from the basic size.

V-Block   A workholding device that has a "V-shape" form for aligning and holding cylindrically shaped parts.

Vector   This is the term used to describe the coordinates of a point.

Vertical Machining Center (VMC)   A CNC machine that is designed with the spindle in the vertical position.

Virtual Condition   Referred to as the mating condition of the part.

Vise-Jaw Tooling   Refers to special purpose fixtures made from blanks that are the same dimensions as the replaceable jaws of a machine vise.

V-Locators   Locating devices that have a "V-shape" form for locating cylindrically shaped parts.

Width of Cut (WOC)   Refers to the machining amount in width of the cutting tool as it is engaged in the part.

Word   A combination of a letter code and a numerical value such as S1200 and any other combination.

Word Address Format   A system of coding instructions whereby each word in a block is addressed by using one or more alphabetic characters identifying the meaning of the word.

X-Axis   The CNC machine axis that moves the table on CNC machining centers and moves the turret on CNC turning machines.

Y-Axis   The CNC machine axis that moves the table on CNC machining centers.

Z-Axis   The CNC machine axis that moves the spindle for machining centers and the turret on CNC turning machines.

Zero Offset   A feature of an MCU that allows the programmer to shift the zero or starting point for movements to a new position over a specified range.

INDEX